OUTLOOKING BEYOND NANOELECTRONICS AND NANOSYSTEMS

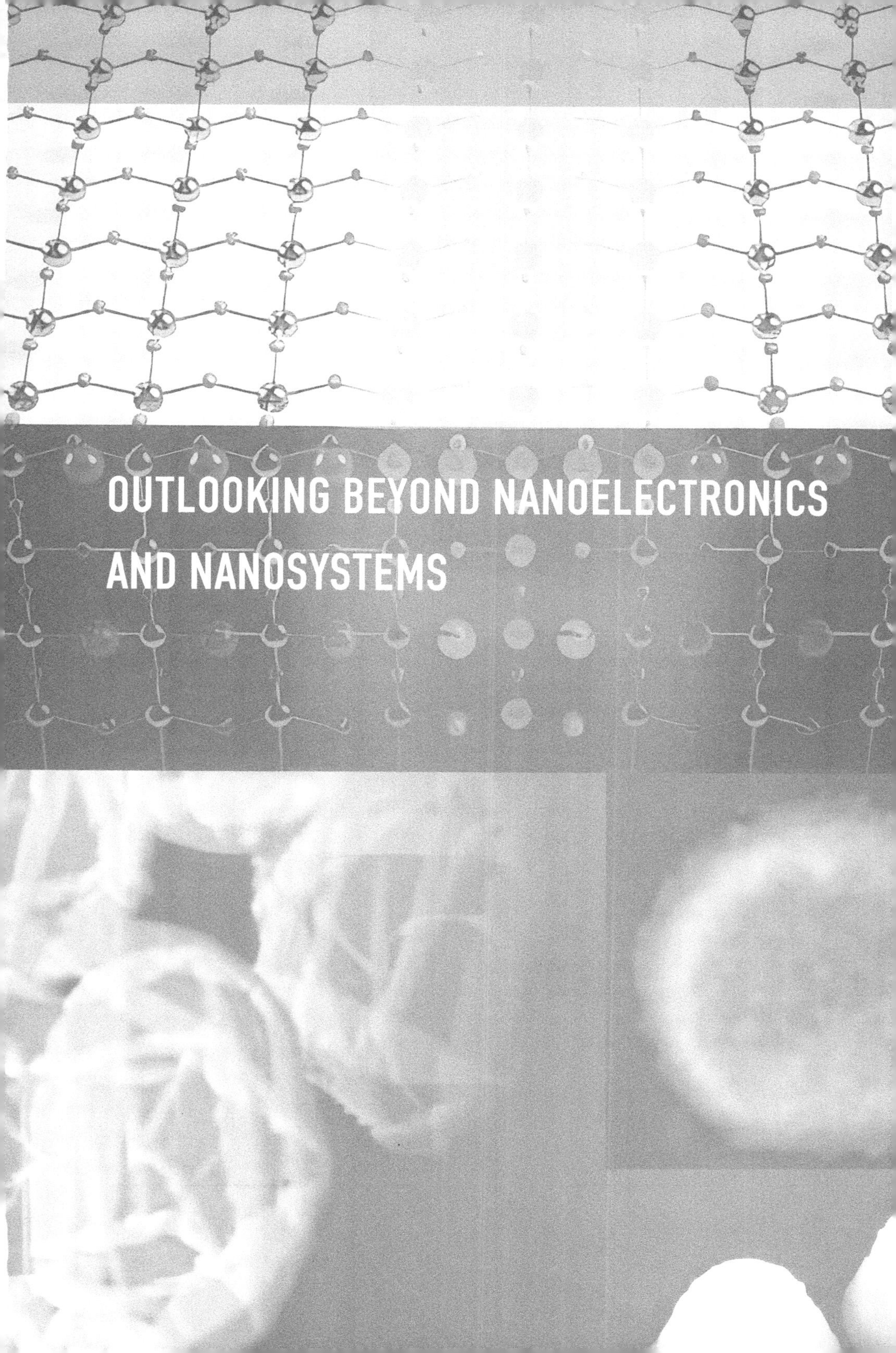

Jenny Stanford Series on Intelligent Nanosystems

Series Editor

Simon Deleonibus

Titles in the Series

Vol. 1

Intelligent Integrated Systems: Devices, Technologies, and Architectures

Simon Deleonibus, ed.

2013

978-981-4411-42-4 (Hardcover)

978-981-4411-43-1 (eBook)

Vol. 2

Integrated Nanodevice and Nanosystem Fabrication: Materials, Techniques, and New Opportunities

Simon Deleonibus, ed.

2017

978-981-4774-22-2 (Hardcover)

978-1-315-18125-7 (eBook)

Vol. 3

Emerging Devices for Low-Power and High-Performance Nanosystems: Physics, Novel Functions and Data Processing

Simon Deleonibus, ed.

2019

978-981-4800-11-2 (Hardcover)

978-0-429-45873-6 (eBook)

Vol. 4

Convergence of More Moore, More than Moore and Beyond Moore

Simon Deleonibus, ed.

2021

978-981-4877-12-1 (Hardcover)

978-1-003-04303-4 (eBook)

Vol. 5

Outlooking beyond Nanoelectronics and Nanosystems: Ultra Scaling, Pervasiveness, Sustainable Integration, and Biotic Cross-Inspiration

Simon Deleonibus, ed.

2025

978-981-5129-32-8 (Hardcover)

978-1-003-50990-5 (eBook)

Jenny Stanford Series on Intelligent Nanosystems Volume 5

OUTLOOKING BEYOND NANOELECTRONICS AND NANOSYSTEMS

Ultra Scaling, Pervasiveness, Sustainable Integration, and Biotic Cross-Inspiration

edited by
Simon Deleonibus

JENNY STANFORD
PUBLISHING

Published by

Jenny Stanford Publishing Pte. Ltd.
101 Thomson Road
#06-01, United Square
Singapore 307591

Email: editorial@jennystanford.com
Web: www.jennystanford.com

British Library Cataloguing-in-Publication Data
A catalogue record for this book is available from the British Library.

Outlooking beyond Nanoelectronics and Nanosystems: Ultra Scaling, Pervasiveness, Sustainable Integration, and Biotic Cross-Inspiration

ISBN 978-981-5129-32-8 (Hardcover)
ISBN 978-1-003-50990-5 (eBook)

Contents

Preface

The fifth volume of the Jenny Stanford Series on Intelligent Nanosystems, entitled *Outlooking Beyond Nanoelectronics and Nanosystems,* collects global reviews on (1) the evolution of nanoelectronic and nanosystem components and the necessity for sustainable development in the future; (2) new materials, techniques, and pervasive applications out of mainstream; (3) memristor foundation and biological engineering at the crossroads with nanotechnologies. The volume is thus sub-titled *Ultra Scaling, Pervasiveness, Sustainable Integration, and Biotic Cross-Inspiration.*

In the **introduction chapter**, ***Technology as a Science Intensive Enabling Social Tool,*** I assess that technology has been one of the main foundations of society since the Paleolithic era, when the first hominoid groups appeared. Nowadays, it is expressed as a science intensive enabling social tool due to its wide adoption worldwide and its impact on daily life. Main challenges show up because of the pervasive character and social and environmental impact of technology, appealing to sustainability and ethical questions.

The book is divided into three parts:

Part I, ***Nanoelectronics, Nanosystems, and Sustainability,*** contains two chapters on the *Past and Future of Integrated Circuit Technologies* followed by a proposal for *Life Cycle Assessment and Ecodesign* in the frame of the Internet of Things Era. Nanoelectronics and nanosystems are deeply shaping our way of life and have a great impact on the economy and the environment. Their massive usage, such as artificial intelligence or intensely connected internet of things, is gaining popularity. They depend not only on software or system architectures but basically on hardware and thus available components. However, CMOS devices cannot be scaled linearly anymore; alternative paradigms, either evolutionary or groundbreaking, are being developed. Materials' scarcity, energy availability, and detrimental carbon footprint have to be considered as well in our future technology research and developments.

Part II, *Off the Beaten Tracks for Future Pervasive Nanosystems Technology,* reports, in three chapters, prospective fabrication techniques and devices capable of manipulating disorder and defects to offer new options for integrated nanosystems. The chapters in this part report on three emblematic examples. *Helium ion implantation* is reviewed as a promising technique for characterization and fabrication to control graphene or superconductors electrical properties. *Quantum diamond sensors* based on vacancies spin qubits can be used for biomedical or automotive applications. Multifunctional *thin film batteries* can store information as well as energy on the same chip. Pervasiveness, multifunctionality, and high-resolution sensing will be among the guides to introduce new technologies and device architectures. Going off the beaten tracks will appear to be more and more accessible and foster new approaches in fabrication technologies and new functionalities.

In Part III, *Inspirations from Nature and Unifying Concepts for Future (Beyond) Technologies*, the authors discuss the benchmarking the non-linear behavior of electron devices with existing natural systems and the way to build nanostructured systems thanks to biological engineering. The first question deals with our understanding of non-linear electron devices' behavior and their analogies with living systems. This question has been an enigma since Galvani's experiment on the accidental electrical excitation of a frog's leg in 1791. How much does the electrical behavior of a *memristor* resemble the action of human or squid *neuron* axon ion channels? Non-linear behaviors observed in nature can be modeled by devices reaching the kernel of the *edge of chaos,* which is the expression of the so-called local activity. Memristors are a strong leverage to neuromorphic-computing based system development. The last chapter reviews *biological engineering* and the utilization of biological components in the assembly of *smart nanostructured systems.* Biological proteins, molecules and lipids combined with synthetic materials are the basic necessary components to build them. The resulting systems are targeted for diagnostic, therapeutic, biofuel cell, or tissue enhancement/replacement in the body. Socio-ethical and safety questions are reviewed, as they cannot be avoided in the field of biological engineering.

Simon Deleonibus
Jenny Stanford Publishing Invited Editor

Acknowledgements

I wish to congratulate my co-authors of the seven chapters of *Outlooking Beyond Nanoelectronics and Nanosystems: Ultra scaling, Pervasiveness, Sustainable Integration and Biotic Cross-Inspiration*, fifth volume published in the frame of the Jenny Stanford Series on Intelligent Nanosystems, for the results of the invaluable quality of their research, as well as for their excellent contributions that include a review of the state of the art in their respective fields. They are globally recognized researchers and professionals, belonging to top-level renowned universities, institutes, or companies. Their determination and energy are of utmost importance to the success of this ambitious ongoing project.

It is a great honor to receive the support for the promotion of the book from Professor Mansun Chan, Hong Kong University of Science and Technology, Hong Kong, Professor J. Joshua Yang, University of Southern California, USA, and Professor Cor Claeys, KU Leuven, Belgium. Their influence in the world of Nanoelectronics, Nanotechnology, and Nanoscience underlines the high scientific level of various contributions.

I have special thanks to address to Ms. Jenny Rompas, Mr. Stanford Chong, Mr. Arvind Kanswal, and staff members of Jenny Stanford Publishing for their responsiveness throughout the whole process from the definition of the book content to the final edition and the promotion of this book. I wish to deeply thank them for their trust and enthusiasm, after many years of collaboration and along the whole long process of the Series publication.

I wish to highlight the constant support and patience of my wife, Geneviève, and my son, Tristan, and the encouragement from my friends and colleagues while I work on the Series. Their positive attention has been essential to motivate the efforts to overcome the encountered difficulties. I wish to dedicate this work to them.

Simon Deleonibus
Past Chief Scientist CEA-LETI, Life Fellow IEEE,
Emeritus Fellow Electrochemical Society, CEA Research Director,
Jenny Stanford Publishing Invited Editor

Introduction

Technology as a Science Intensive Enabling Social Tool

Simon Deleonibus

Past Chief Scientist, Université Grenoble Alpes, CEA Leti, Grenoble, France

simonedit2016@gmail.com

The development of information and communication technologies is a unique example of the main technology revolutions in human history, such as the invention of iron tools or vapor engines. The 5th volume of the *Series on Intelligent Nanosystems* discusses device architecture and downsizing, ultra-thin film processing, quantum sensing, non-linearities and chaos, biotechnologies, energy efficiency, technology pervasiveness, and sustainable integration, which will shape future groundbreaking inventions.

I.1 From Stone Tools to High Abstraction Levels

Technology is one of the pillars of human society foundations.

Tools made from bones, probably made by hominid groups, dating more than 3 million years ago, have been found in Africa.

Outlooking beyond Nanoelectronics and Nanosystems: Ultra Scaling, Pervasiveness, Sustainable Integration, and Biotic Cross-Inspiration
Edited by Simon Deleonibus

ISBN 978-981-5129-32-8 (Hardcover), 978-1-003-50990-5 (eBook)
www.jennystanford.com

In the further evolution through the ages, transforming matter and natural resources, including from animals, allowed humans to produce food and clothing, build shelters and homes, and trade and communicate including by artistic expression. The meaning of the early drawings and engravings in the Chauvet or Altamira caves (successively in France and Spain) realized in the Paleolithic age (as far as 33,000 years before our age) is still mysterious or not definitely settled among experts. Their crafting was the expression of an abstraction necessity, which was possible thanks to the technologies of charcoal and pigmentation taken from the surrounding nature. The new skill requested, as well, the interaction of the human body with the environment, thanks to the "invented" tools. Prehistoric carvings and sculptures later needed a more sophisticated technology based on the invention of new tools such as a stylus, probably needed for hunting, handcrafting, agriculture, or construction. Some of these objects have been found in the surroundings of the artworks. Technology and abstraction, the base of knowledge and science, are quite strongly linked to and challenging each other by permanent feedback through time. In turn, they are questioned on their viability or the notion of scientific truth, by facing experimental evidence. Capitalizing on their knowledge, humans have conducted huge efforts to serve different fields or improve the interface with the surrounding world. Making functional devices or systems to mimic or interface living species, whether animal or vegetal, has always been fascinating. On another hand, human capabilities were overcome or assisted by instruments and machines that supply many services or introduce new usages, including material or intellectual activities.

Such a behavior has been perpetrated for ages, until nowadays in many fields of interest such as information and communication technologies (ICTs), medicine and healthcare, mobility and transport, agriculture, manufacturing, education and defense. These activities presently assist humans' daily lives. Thanks to electron devices scaling into large-scale integration, silicon microelectronics strongly emerged in the 20th and 21st centuries as an enabling technique included and pervaded in all modern functional systems, and co-integrating hardware and software. Nowadays, the huge calculating power computers are reaching the complexity of the human brain. However, they are very far from competing with its figures of merit, such as energy efficiency. In the late 2010s, exascale range (10^{18} floating point operations per second (FLOPS)) computing power has been overcome. The level of abstraction in system design has kept increasing dramatically, as never before. Increasing the level of abstraction has been a constant quest throughout history for all technologies that humans have invented.

Contemporaneously, ICTs represent major energy consumers (5–10% of world electric consumption) of the economy and thus contribute to CO_2 emissions. As well, ICTs and energy production machines are employing an increasing number of chemical elements, they pose a sustainability question. Their rarefaction and presently known level of reserves represent a severe geopolitical issue due to their unequal distribution around the world.

A balance will be needed in the future to promote sustainability at the very early stages of new research and development. Such an effort is of course challenging to avoid curbing creativity. Such complexity also appeals to the reconsideration, with their opportunities and threats of many human activities, such as robotization by hardware or software (i.e., artificial intelligence (AI)).

I.2 The Silicon Age, Its Marvels and Benefits Thanks to MOSFET and Integrated Circuits Technology

The history of microelectronics and silicon-based devices is an emblematic period of history corresponding to the age of information, following the different ages of mankind's evolution through the stone, bronze, and iron ages, great discoveries, and industrial eras. Silicon is the second most abundant chemical element in the earth's crust behind oxygen. Silicon is widely available in nature in its various silicate forms, and that makes it difficult to isolate as a simple chemical element. It was certainly not by chance that silicon was known by the Egyptians in antiquity through its compounds entering the composition of jewels or other art crafts. Following unfortunate tentative by many famous chemists such as A. Lavoisier, J.-L. Gay-Lussac, and H. Davy, silicon was isolated and recognized as a simple chemical element in 1823 by J.J. Berzelius.

Silicon was definitely adopted as a major contributor to electron devices during the period of 1940s–1960s, contemporaneously with the successive inventions of bipolar and metal oxide semiconductor field effect transistors (MOSFETs). Germanium was directly competing against silicon thanks to the faith of scientists in higher low-field electron and hole mobilities. However, the high affinity of the silicon surface with oxygen made it a very good candidate for fabricating insulated gate devices. MOSFETs, featuring a high-quality silicon dioxide-to-silicon interface, were thus viable for high-volume manufacturing. MOSFET-based integrated circuit technology has gone into many breakthroughs to respond to its linear scaling and to continue Moore's law. For example, in the most advanced circuits, Copper has replaced aluminum for interconnects; SiO_2 has been replaced by a high dielectric constant (HiK) insulator HfO_2 (see Chapter 1).

Because linear scaling of bulk transistors has reached its limits, new devices and circuit architectures have been introduced. Planar channels are replaced by gate-wrapped around channels (FinFEts, silicon on insulator, or nanowire-based) to support advanced scaling. 3D integration at the packaging level or inside a chip is giving a new breath to Moore's law, ... Other options using MOSFETs with new actuating modes (for example, tunnel FET) or computing architectures (adiabatic, quantum, neuromorphic, ...) are developed (see vols. 1 and 3 in [1]). Each newly introduced innovation has requested increasing R&D efforts and development time [1]. Roadmaps were needed to discuss and forecast the next generations by international committees involving industry and academia [2]. The microelectronics community has been doing so, since 1998. The continuation of Moore's law gives the possibility to innovate at the system and application levels. Thanks to such historical development, high-end application complexity has been increasing since the invention of integrated circuits. The number of transferred instructions at higher speed has been doubling every generation, reaching ExaFLOPS levels. The integrated circuits can process digital and analog signals and their technology has pervaded in many sectors ranging from professional, military, biomedical, transport, to telecommunication, ... to all activities in our daily lives. Finally, making integrated circuits to serve and assist human beings in their activities has become a reality.

The mobile phone is a typical example of the success of integrated circuits technology and its pervasion into different diversified technologies such as microsystems. It has become a companion for many duties in our daily life. Other emblematic applications assist human beings such as robots, the internet of things, and AI. New paradigms will be necessary to face the challenges that the increasing number of new applications will pose, accompanied by the nomadism of the users and their demand for enhanced communication. In such a context, the energy challenge and its harvesting in the environment have become a significant topic to deal with (see vols. 3 and 4 in [1]). The energy supply of autonomous mobile systems, from worn and handled to automotive devices, will become among the top concerns in future smart city organizations. From the nano to the macro scales, energy and battery engineering have become a crucial topic (see Chapter 5).

I.3 New Way of Life, Sustainability, and Ethics

It has been a long way to reach today's so-called information society. Finding the Graal-type material or device to compute, sense, actuate, and communicate has been paving the history of electrical engineering since the 18th century. The use of heterostructures and semiconductor-doped homojunctions allowed to exploitation of non-linearities in device characteristics. Many trials and errors were certainly necessary to finally end in the generalized use of integrated circuits. Objects containing nanoelectronics devices and nanosystems have massively penetrated into society worldwide. As an example, the number of smartphones circulating is equivalent to half of the earth's population or so. They can be used for many applications, not only to place a phone call (!) but also access to the internet, video conferences, games, databases for many services, domotics, information and knowledge, music and video streaming, ... In 2023, more than 15 billion connected devices were settled worldwide. The number might be doubled by 2030. The business generated 300 billion US dollars revenue, reaching a yearly growth higher than 10% [3] since 2020.

The massive penetration of new devices and systems, besides bringing many virtuous advantages in our daily lives, possibly generates detrimental moral or environmental issues if no rule is established. In the first case, education, legislation, and regulations are necessary to promote ethical behaviors. Second, the carbon footprint of any technical and commercial activity has been pinpointed because of its huge and sensitive impact on global warming. The contribution of ICTs to newly generated CO_2 is estimated today to be in the range of 4%, comparable to the contribution of aviation, but still much less than carbonated energy production, in a range of 30–40%. Among the applications of ICTs, videos account for 80% of worldwide flows and 80% of their volume growth [4]. That is explained by the multiplication of video usages (business, education, entertainment, family uses, etc. ...) but also by its nature. Video is a dense support, for example, 10 hours of high-definition movies represent the data volume of the whole English written papers in the encyclopedia Wikipedia. However, video is an excellent support to investigate the required solutions for installing practically digital sobriety – video helps save approximately 30% of travel expenditures and did much more during the crucial phases of the COVID-19 pandemic (2020–2022).

Such a massive use of electron devices and systems, unfortunately, generates waste due to an encouraged reduction of product lifetime. It has become a global issue. Regulation was felt to be necessary, for example, the European Union (EU) has decided to avoid endless mounds of mobile phones or other electronic trash. Obligations are given to these goods manufacturers to fight against waste of scarce or critical materials at the stage of their fabrication and during their life cycle. In this frame, the management of resources, such as scarce and critical metals, water, and energy, has become a priority concern [5]. In 2022, a list of 30 critical materials (increasing to almost 60 in 2030) that might be of strategic importance to the industry especially in the fields of electronics and energy has been updated. Essential very popular materials have been listed such as titanium, tungsten, lithium, or rare earth metals [6]. In some specific cases, recycling by circular economy has been proposed. However, materials are dispensed in small quantities in each system, conclusively, they might be difficult to recycle.

On the optimistic side, techniques such as atomic layer deposition (ALD) have emerged and allow to save reacting precursors; for example, in hafnium oxide deposition (HfO_2) to replace SiO_2. Catalysis is also an appealing technique in the area of thin film deposition thanks to its high mass efficiency and positive impact on waste management (in [1], vol. 3). Taking into account the ecological aspects, sustainability will become a necessity. So-called "ecodesign" is developed to support life cycle assessment at the very beginning of any product design (see Chapter 2).

The EU has been at the forefront of questioning and regulating the ethical side and consumers' personal data protection, in particular, to prevent any abusive intrusion into privacy. Many countries and continents are inspired and follow the EU example of the General Data Protection Regulation (GRDP) introduced in 2016 [7]. Recently, the Digital Markets (May 2023) and the Digital Services Acts (Aug 2023) have been approved to foster higher consumer safety, protection, and information fairness [8].

I.4 Getting Off the Beaten Tracks Toward Technology and Nanosystem Innovation

Innovation in device technology has been constant since the invention of integrated circuits (see [1] vol. 4). At each generation jump, the so-called Moore's law was possible thanks to the introduction of new process modules in lithography, material engineering, and device process integration, design, or system architecture. Following, Moore's law requested huge investments and research efforts to increase the number of devices on a chip and to push back the limits of technology. Exploratory studies have been led by the microelectronics community at least 10–15 years in advance of the introduction in manufacturing. That was the case for example for the successful replacement of SiO_2 by high dielectric constant dielectrics (HiK), which brought new constraints and changes in the transistor process flow and structure. It was necessary to replace polysilicon with a metal as gate material, to properly control work function engineering. Identically, the introduction of strained channels or copper interconnect required a huge number of studies and tries worldwide on material growth and process integration as well as their modeling at the molecular and atomic levels (see vols. 1 and 2 in [1]).

Semiconductor process integration is at the crossroads of many know-how coming from different disciplines in chemistry, physics, electronics, and mechanical engineering. Nowadays, biology is coming into play as a new challenge for device and system engineering.

Many applications have been made possible thanks to cross-disciplinarity. A few examples are provided hereafter. The investigation of new pathways in materials science and interactions between matter and radiation has been crucial to combining know-how. The resulting techniques can either be set to extract information or as fabrication tools. For example, electron beams are essential for imaging by scanning or transmission microscopy. As well, they allow us to enjoy mask fabrication or direct transfer of patterns on wafers. Techniques using ion beams, including a single particle, plasmas, and their combinations are other examples used either to dope a material or transform its local topography, depending on the energy of the incident ions. Plasma etching, reactive ion etching, sputter deposition, and ion implantation use plasma source generation. The energy – ranging from a few eVs to a few MeVs – that ions will reach on the substrate or the target during the process will be one of their signatures. Nowadays, single ion implantation and scanning probes microscopy (SPM) are possible and offer the same type of versatility. For these two mentioned techniques, see vols. 1 and 2 in [1].

Taking advantage of existing defects to sense or activate is another technique, which has led to many innovations. That is the case in the use of vacancies to produce radiative effects due to trapping or generation-recombination-related luminescence. In wide bandgap materials (SiC, GaN, or diamond), it might be difficult to dope in conventional ways. Defects appearing during the material growth are interesting to exploit. Opportunities are shown in this book (see Chapter 4) for applications in quantum sensing thanks to NV centers in diamond-based devices for biomedical or automotive applications. The conventional doping issue in single atomic sheets can be overcome by ion irradiation to implement artificial bandgaps, for example, nanowires, nanostructures arrays by etching, or local irradiation generated defects. Such a technique, applied to superconductors, generates insulating materials. Examples are given with the He ion-based technique, initially studied for microscopy purposes, giving promising results in material and device fabrication, as shown in this volume (see Chapter 3).

Electron devices can easily fulfill multiple uses and be versatile. For example, MOSFETs or bipolar transistors can indeed transfer digital or analog signals to compute or memorize information. They serve multiple applications with the same basic design. Examples in nature are very inspiring and challenging in this respect. That is the case for neural cells in our nervous system as a whole, capable of information storage and processing. Mimicry of the nervous system operation and efficiency is a dream for device engineering to apply in robotics, biomedicine, and high-performance computing. In the field of electron or electrochemical devices, memristors could be used as memory elements or nano-batteries to store energy locally, in circuits operating at low voltage (see Chapters 5 and 6). Such a possibility opens new ways to manage power in integrated circuits, either to save energy or enhance computing efficiency. In many examples, 3D assembly makes sense to integrate new architectures leading to high performance, low power consumption, and cost-efficient systems (see vol. 3 in [1]).

I.5 Artificial, Inspired by Nature or Hybridized Systems and Their Relevance

Since the beginning of the 20th century, computer engineering has evolved to lower power consumption to process elementary information in Boolean logic, thanks to the invention of integrated circuits and miniaturization of electron devices. Nowadays, optical components or interconnects are co-integrated at the system or network levels and they do improve the expected figures of merit to increase the number of operations per amount of consumed energy at a higher speed.

Computer engineering enables us to improve, complement, and overcome the limited processing capabilities of humans dealing with a large amount of data. However, digital machines are far from the ability to have consciousness or global vision without learning algorithms introduced by humans! They are still far from competing with humans in such fields. A human brain contains 100 billion neurons and achieves 1000 billion operations with a power budget of 20 W. A digital machine with the same number of components would, in 2023, request a small hydroelectric power plant (ca. 50 MW) to obtain the same result [9]. Nevertheless, consciousness capabilities would be missing. Is there a virtuous approach that would lead us to replace humans or animals with robots, bio-inspired devices designed by humans? Mimicry in various ways has fascinated humans since antiquity. Archytas of Tarantum invented the flying pigeons' automats in the 5th–4th century B.C. In the Renaissance period, Leonardo da Vinci's first android robot could coordinate its arms, legs, and jawbones. More generally, the biomimicking question has become more and more timely nowadays to benchmark human activities, typically for manufacturing or social assistance [9].

It is important to understand how nature operates to process any elementary input coming from the environment. Modeling nature behavior is far from being a completely mastered question, due to the high complexity of existing diversity. We could nowadays try to understand or model some natural behavior with existing tools of our scientific representations. Electrical analogy models might help understand the basic functional mechanisms. For example, in 1952, Hodgkin and Huxley obtained the Nobel Prize for proposing their electrical equivalent circuit model of the neuron axon. They suggested to represent the potassium or sodium-ion channels spread along the axons by circuit elements. Still, the time dependence of the conductance is revised today and would rather be replaced by memristors. The brain synaptic architecture has been modeled by using electrical cellular neural non-linear networks used for image processing applications at a very low energy cost [10] (see Chapter 6).

Biomimicry, bio-assisted, digitally, nanostructured, and interfaced bio-systems become topics of interest to scientists and technologists to extend today's systems capabilities. Biocompatibility is as well a concern each time, living matter and artificial devices come close or get into contact outside or inside a living body (i.e., dental prostheses, pacemakers, insulin pumps, ...). How new hybrid systems can replace existing machines or parts in human bodies is also a difficult and exciting question – to produce energy, drugs, compute, etc. ... Bioengineering is a discipline that will certainly boom up in the future (see Chapter 7). Besides their opportunities and cost issues, the boundary limits in the use of such technologies very quickly encounter ethical, social, and legal concerns. This is, for the time being, the necessary expectation by society from scientists and technologists.

I.6 The Fifth Volume of the Jenny Stanford Series on Intelligent Nanosystems

In the fifth volume of the Jenny Stanford Series on Intelligent Nanosystems entitled ***Outlooking Beyond Nanoelectronics and Nanosystems***, a critical review of the evolution of nanosystems and their components is given through historical and prospective aspects. The volume is sub-titled *Ultra Scaling, Pervasiveness, Sustainable Integration, and Biotic Cross-Inspiration.*

The book is divided into three parts titled:

Part I *Nanoelectronics, Nanosystems, and Sustainability*

Part II *Off the Beaten Tracks for Future Pervasive Nanosystems Technology*

Part III *Inspirations from Nature and Concepts Unification for Future (Beyond) Technologies*

Part I collects Chapters 1 and 2.

In Chapter 1, ***Hiroshi Iwai,*** *Professor at National Yang Ming Chiao Tung University (Taiwan) and Tokyo Institute of Technology (Japan)*, reminds us how electron devices based on semiconductor heterostructures emerged in the middle of the 19th century. He describes how the initial device evolved in time to finally give birth to bipolar transistors and MOSFETs. They became very popular, thanks to the use of silicon and its oxide in the 1960s. The invention of the integrated circuit in 1958 drastically gave a boost to the idea of the collective fabrication on a single piece of semiconductor, the so-called “chip”. Linear scaling of transistors first led to an accelerated progress of microelectronics and nowadays approaches to an end, due to physical geometrical limitations, such as tunneling in dielectrics and between device electrodes. Because silicon micro/nanoelectronics has a huge societal impact, different pathways are explored toward continued integrated circuits progress: new materials, new device architectures, or bioinspired paradigms might be introduced.

In Chapter 2, *Thibault Pirson et al.* *of Université Catholique de Louvain (Louvain La Neuve, Belgium)* in a joint effort with co-authors affiliated to *Université Grenoble Alpes, Grenoble-INP and CEA Leti (Grenoble, France)*, provide an overview on the use of life cycle assessment (LCA) to support judicious ecodesign in the context of the IoT. By putting interested designers on concrete tracks to integrate environmental considerations, LCA helps in improving the ecological dimension of IoT systems over their entire life cycle. Several questions are addressed: (i) available tools, (ii) examples of existing works, (iii) encountered hurdles, and (iv) already applicable solutions and future directions. LCA is a quantitative methodology based on already-existing tools usable by engineers during the technology design phase. The authors propose a streamlined overview of current practices and perspectives to foster life cycle thinking in the field of IoT systems.

Part II features Chapters 3, 4, and 5.

In Chapter 3, *Shinichi Ogawa,* *Research Director at Advanced Industrial Science and Technology, Tsukuba (Japan)* and co-authors of the *Helmholtz-Zentrum Dresden-Rossendorf, Dresden (Germany),* the *Ocean University of China, Qingdao (People's Republic of China),* the *National Institute for Materials Science, Tsukuba (Japan)* and the *University of California at Riverside, Riverside – CA (United States of America)*, provide an overview of the helium ion microscopy (HIM) technique and its use as a characterization and fabrication tool. HIM presents unique features thanks to the superior physical properties of helium (He) for gas field ion source (GFIS) microscopy. He is the gas with the highest ionization potential. He is thus the best choice for a GFIS with respect to source operation. Once a trimer is formed at cryogenic temperatures and the best imaging voltage has been selected, only He atoms can reach the trimer location and are subsequently ionized. The authors present the HIM technique and the tool for different applications in characterization, nanofabrication of electronic material processing, and device engineering. For instance, they modify the electrical conduction properties in graphene films by creating nanopore arrays or point defects by He-ion-beam irradiation. They achieve Josephson junctions in superconductors by manipulating the disorder of atomic arrangement with the same technique.

In Chapter 4, *Mutsuko Hatano,* *Professor at Tokyo Institute of Technology, Tokyo, Japan,* draws our attention to the great attractivity of diamond-based quantum sensors. Among many solid-state spin systems, nitrogen-vacancy (NV) centers in diamonds have attracted considerable attention in the fields of quantum technologies, from quantum sensing to information processing and communication. Their quantum coherence (up to a few milliseconds) is preserved even at room temperature. The energy levels of NV centers are sensitive to magnetic fields, electric fields, strain, and temperature. Their spatial resolution is scalable from nanometers to millimeters. Diamond quantum sensors are multimodal and span scalable application fields. To realize an ideal quantum sensor, the technical challenges in the sensor materials and the quantum control techniques, such as coherent microwave spin manipulation, effective optical spin-state initialization, and readout, have been developed. This chapter briefly explains the sensing principle using NV centers and then introduces sensor systems for medical applications, i.e., millimeter-scale magnetocardiography on small animals, and monitoring of batteries for electric vehicles and internal power devices.

In Chapter 5, *Sami Oukassi*, *Research Director at CEA-LETI (France),* reviews the possibility of thin film batteries for energy or information storage. In the quest for the highest efficiency in energy consumption or multifunctional devices, thin film batteries are good examples to focus on. His developments on LiPON/$LiCoO_2$-based integrated batteries offer the possibility to store information as well. An innovative concept of a hybrid dual $CuTe_x|Ta_2O_5|W$ resistive random-access memory, behaving also as a battery, has been demonstrated. These electrochemical nanosystems would offer the possibility to store energy and data at different levels. They advantageously use their combination of both functions in hybrid 3D crossbar arrays. As an example, synaptic transistors can be an emblematic application in artificial neural networks for neuromorphic computing with high learning precision.

Part III compiles Chapters 6 and 7.

In Chapter 6, *Leon Chua*, *Professor at Berkeley University (California, USA),* reminds us that non-linearities can turn into instabilities; "all or none" responses appearing in nature are the consequence of an action potential. It defines the principle of local activity including an edge of chaos kernel. He suggests that the local activity principle is the missing complement of the 2nd law of thermodynamics. He gives an interpretation of a 240-year unsolved paradox questioned by Galvani on the all or none phenomenon that he observed on a frog's leg muscle response to electrical excitation. He gives a critical review of the various electrical models proposed since then. He illustrates the meaning of these phenomena by the necessity to introduce a time-invariant memristor instead of a time-varying conductance in the equivalent circuit models. In 1971, Professor Chua originally identified the memristor as the 4th electric circuit element, establishing the axiomatic relation between electric flux and charge, adding thus a time-dependent state variable in Maxwell's equations. Since then, many electrical and natural phenomena have been identified as illustrating the properties of a memristor such as ion channels in living cells, neuron synapses, and similar metal/insulator hybrid systems featured in alternative resistive non-volatile memories. Finally, he suggests that the existence of the boundary of the edge of the chaos domain defines the transition from stable to unstable cycles of many physical phenomena, such as the evolution of life.

In Chapter 7, *Donald Martin,* *Professor at Université Grenoble Alpes (France),* gives an overview of the latest developments in biological engineering. Biological engineering covers areas going from civil engineering to biomedical engineering with emblematic applications such as waste treatment remediation, the interface between living systems and sensors, as well as physical and energetic interfaces between physically engineered systems and human living systems. His approach utilizes biological components in the assembly of smart nanostructured systems, in contrast with the one that constructs synthetic material inspired by nature. He refines the discipline of "biological engineering" to be the one that utilizes biological proteins, molecules, and lipids in combination with synthetic materials to assemble smart nanostructured systems. This discipline is targeted for applications such as diagnostic, therapeutic, biofuel cell, or tissue enhancement/replacement in the body. Hybrid structures associated with genetic manipulation can be envisaged for many therapeutic applications such as insulin control in pancreatic β cells or energy supply in vivo, as well as "symbio-bot" medical devices that incorporate proteins for ion transport. Beyond economics, ethics and biocompatibility play the roles of safeguards paving the way for new application developments.

References

1. S. Deleonibus, Editor, Jenny Stanford Publishing, *Series on Intelligent Nanosystems*, vol. 1 (2014), vol. 2 (2017), vol. 3 (2018), and vol. 4 (2021), Singapore.

2. Roadmap - IEEE IRDS™ (https://irds.ieee.org/editions) and ITRS Reports (itrs2.net) (http://www.itrs2.net/itrs-reports.html).
3. IoT connected devices worldwide 2019–2030 | Statista (https://www.statista.com/statistics/1183457/iot-connected-devices-worldwide/).
4. Video Conferencing Market Size & Growth | Forecast [2030] (https://www.fortunebusinessinsights.com/industry-reports/video-conferencing-market-100293).
5. The 2020 EU Critical Raw Materials List – Policies - IEA (https://www.iea.org/policies/15274-the-2020-eu-critical-raw-materials-list).
6. G. Pitron, *The Rare Metals War: The Hidden Face of the Energy and Digital Transition*, London, Scribe, translation Bianca Jacobsohn, 2020, 288 p., ISBN 978-1-912-85426-4 (https://en.wikipedia.org/wiki/Special:BookSources/978-1-912-85426-4).
7. The general data protection regulation - Consilium (https://www.consilium.europa.eu/en/policies/data-protection/data-protection-regulation/).
8. EU Digital Markets Act and Digital Services Act Explained | News | European Parliament (https://www.europarl.europa.eu/news/en/headlines/society/20211209STO19124/eu-digital-markets-act-and-digital-services-act-explained).
9. Learning from the brain to make AI more energy-efficient (humanbrainproject.eu); Neuromorphic Computing (https://www.humanbrainproject.eu/en/follow-hbp/news/2023/09/04/learning-brain-make-ai-more-energy-efficient/).
10. L. O. Chua, The Chua lectures: From memristors and cellular nonlinear networks to the edge of chaos, in *World Scientific Series on Nonlinear Science*, Ed. L. O. Chua, Series A, vol. 98, World Scientific, Singapore (2021).

Part I

Nanoelectronics, Nanosystems, and Sustainability

Chapter 1

Past and Future of Integrated Circuit Technologies

Hiroshi Iwai

National Yang Ming Chiao Tung University, Hsinchu, Taiwan
Tokyo Institute of Technology, Yokohama, Japan

hihi20190425@gmail.com

Semiconductor integrated circuit technologies, which began in the late 1950s, have evolved into nano CMOS technology, ushering in the era of our current super-intelligent human society. In this chapter, I will delve into the historical evolution of semiconductor integrated circuit technology and explore its promising future.

1.1 Introduction

The advancement of human society's intelligence owes much to the evolution of information and communication technology (ICT). Throughout history, there have been three significant waves of ICT innovation. The first wave emerged in ancient times with the invention of language, enabling the exchange of intricate ideas and fostering profound thinking. Around 5000 years ago, the second wave arrived with the creation of written characters,

Outlooking beyond Nanoelectronics and Nanosystems: Ultra Scaling, Pervasiveness, Sustainable Integration, and Biotic Cross-Inspiration
Edited by Simon Deleonibus

ISBN 978-981-5129-32-8 (Hardcover), 978-1-003-50990-5 (eBook)
www.jennystanford.com

facilitating precise communication across vast distances and generations. Then, approximately 175 years ago, the third wave emerged with the invention of electric information. This break-through allowed instantaneous communication over great distances and the processing of massive amounts of data, transcending our natural human capacities. As our ability to process electric information expanded, the development of novel electric and electronic devices further accelerated the progress of intelligence in society. Today, we find ourselves in a super-intelligent 21st-century society, thanks to nano-CMOS VLSIs (nano-complementary metal oxide semiconductor very large-scale integration) circuits. This chapter extensively explores the history and future of LSI technology, drawing on valuable insights from historical patents, papers, and articles commemorating the transistor's 75th anniversary [1–6], with particular emphasis on reference [4].

1.2 Pre-History Transistor/LSI Development

Electrical engineering, which emerged in the early 19th century, introduced a groundbreaking technological revolution with the development of electric wires capable of instantaneously transmitting electric signals (or information) and electrical energy over long distances, as depicted in Figure 1.1 [4]. This marked a significant milestone in the field. The practical implementation of electric telegraph technology, pioneered by Samuel Morse and others in the late 1830s to the mid-1840s, revolutionized information transportation [7]. By the end of the 19th century, the world was connected through submarine cables, establishing a global electric telegraph network. Until the 1940s, the electric telegraph remained the primary telecommunications tool, even as the transmission medium transitioned from wired to wireless.

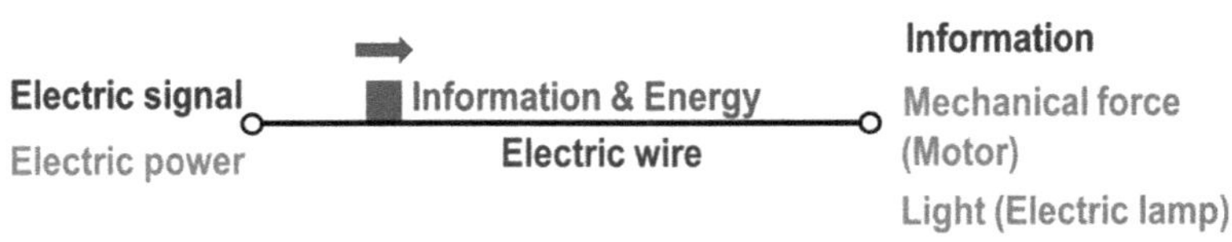

Figure 1.1 Information and energy transmission by an electric wire [4], copyright 2023 JSAP.

Electric telegraph uses the propagation of a combination of short and long electric pulses modulated by Morse code (Figure 1.2 [4, 7]), and this was the first digital data transmission. Because of the decay of the electric pulses on the long electric wire, amplification of the pulses in the middle of the wire was necessary. This amplification was accomplished by a relay, which was invented between the 1830s and the 1840s (Figure 1.2) [4, 7]. The relay is a kind of three-terminal mechanical-switching device, operating by the electromagnetic force controlled by the input current flowing in the magnetic coil. The decayed electric signal on the 1st wire is transferred to the 2nd wire with amplification. The relay was the first electric signal amplifier as well as an electric switching device and had been used for telephone exchange and machine control for many years. Although the operating frequency is very low because of the mechanical switching, relays can control huge currents because of massive metal switches, and most of the elevators used relay controllers even in the 1970s.

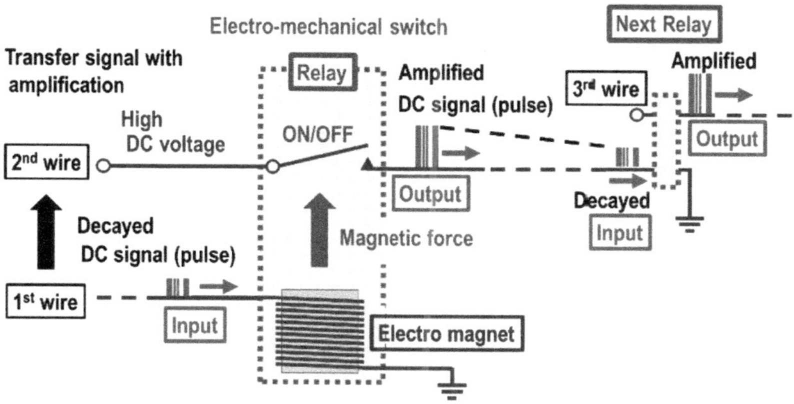

Figure 1.2 Digital signal amplification by relay [4], copyright 2023 JSAP.

Édouard Branly made a significant discovery in 1890, finding an electromagnetic wave detector [8]. His research focused on the behavior of materials exposed to electromagnetic waves. He noticed that when oxidized metallic particles or flakes were placed inside a cylindrical container, their resistance decreased significantly, dropping by several orders of magnitude upon receiving electromagnetic waves. The particles would physically connect through welding when subjected to an electromagnetic field, leading to an overall reduction in resistance. Removing the electromagnetic wave did not restore the decreased resistance, but it could be recovered by applying a mechanical shock or impact to the container, effectively separating the particles.

In 1894, Oliver Joseph Lodge created a circuit capable of automatically detecting the decrease in resistance and delivering a mechanical shock to the cylinder [9]. This invention, referred to as the 'coherer' (Figure 1.3 [4]), marked the initial utilization of a device as a receiver for wireless electric telegraphy. Following this breakthrough, Guglielmo Marconi and other pioneers further developed practical wireless electric telegraphy in 1895 [10].

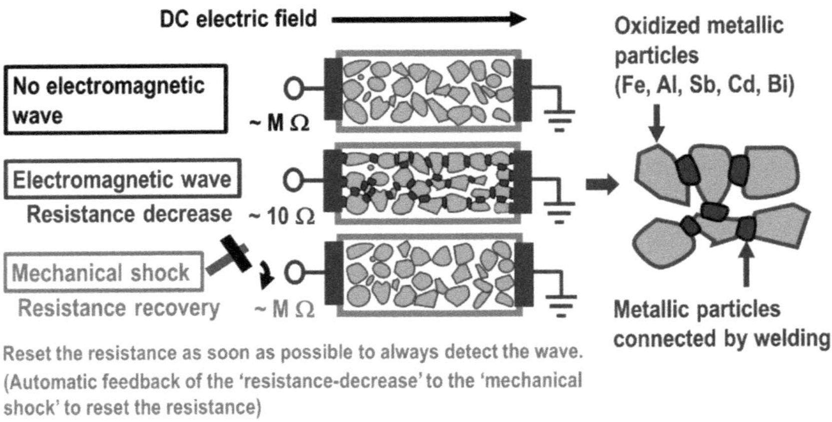

Figure 1.3 Coherer: Electro-magnetic wave detector [4], copyright 2023 JSAP.

The coherer was capable of identifying the existence of an electromagnetic wave but lacked the ability to measure its amplitude. Consequently, it found predominant use as a digital signal pulse detector in wireless telegraphy. Nevertheless, for wireless voice telecommunication, like broadcasting and wireless telephony, an analog signal detector was indispensable. This was achieved by implementing a demodulation circuit that extracted the low-frequency voice signal from the high-frequency electromagnetic wave, necessitating a crucial circuit element called a diode or rectifier.

As far back as 1874, the phenomenon of point-contact metal-to-semiconductor rectification had been independently discovered by both Karl Ferdinand Braun [11] and Arthur Schuster [12]. However, during that period, it held no practical applications. Around 1895, Jagadish Chandra Bose extensively investigated materials for the coherer and stumbled upon the rectification phenomenon occurring at the point of contact between different materials. Utilizing this discovery, he conducted experiments using it as a receiver for electromagnetic waves, as depicted in Figure 1.4 [4, 13].

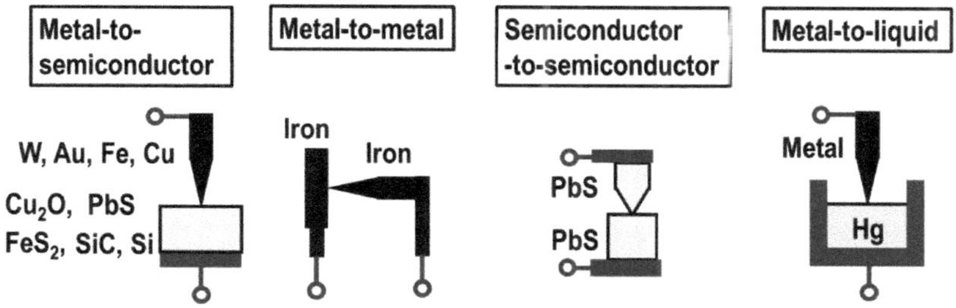

Figure 1.4 Various point contact rectifiers.

During the 1910s, the utilization of point-contact metal-to-semiconductor receivers became prevalent in economical radio devices and earned the name 'crystal' rectifiers. By the 1930s, Si crystal rectifiers emerged as the most suitable high-frequency receivers, outperforming even vacuum tubes [14], and they served as crucial components for microwave receivers until the 1940s. However, point-contact rectifiers are two-terminal devices and do not possess inherent signal amplification abilities. This presented difficulties for transmitting high-power signals and receiving weak signals over long distances in wireless communication.

Lee de Forest introduced the initial analog and digital amplification device in 1906, known as the triode vacuum tube (a 3-terminal device), as depicted in Figure 1.5 [4, 15]. The triode functioned by modulating the thermal electron current emitted from the cathode through the grid potential, enabling the applied

AC/DC signals on the grid manifest as amplified AC/DC signals in the anode current.

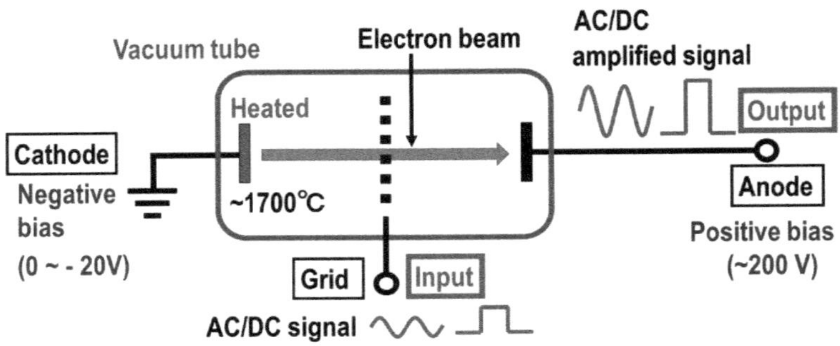

Figure 1.5 Triode vacuum tube [4], copyright 2023 JSAP.

Vacuum tubes were highly successful electron devices that found application in a wide range of electronic devices. They served as the primary components for RF signal transmitters/receivers, machine controllers, and computers from the late 1910s until the 1960s. The triode vacuum tube is widely recognized as 'the first electron device'. However, as applications advanced, vacuum tubes started to exhibit limitations in certain areas. Specifically, achieving a substantial reduction in the size of vacuum tubes proved challenging, primarily due to structural constraints. Consequently, adequately decreasing the electron transit time between the cathode and anode—a crucial requirement for microwave-range operation—posed difficulties [14]. The high operating voltage and the need for cathode heating became issues in terms of power consumption reduction and heat insulation within the overall system.

1.3 From the Idea of MOSFET to Real Device

As a solution to the problems posed by vacuum tube devices, it was only natural to consider a 3-terminal semiconductor device that could amplify signals, drawing an analogy to the triode vacuum tube. However, unlike vacuum tubes, semiconductor materials contain a multitude of carriers, rendering it impossible to control the current by inserting a grid inside the semiconductor. Therefore, devices were proposed and studied that could regulate the lateral current flow in an ultra-thin semiconductor film using a vertical electric field generated by an isolated gate electrode through a capacitor, as depicted in Figure 1.6. This type of device later became known as the field effect transistor (FET).

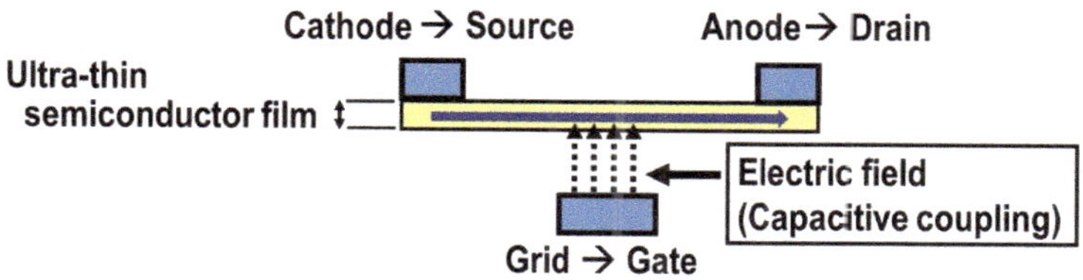

Figure 1.6 Idea of field effect transistor (FET).

In 1925, the first concept of FET was patented by Julius Edgar Lilienfeld, as shown in Figure 1.7 [4, 16]. This concept involved a type of FET with a Schottky metal gate, also known as a metal-semiconductor FET (MESFET), and it marked the world's first solid-state amplifier idea. By biasing the gate electrode in reverse mode, the flow of direct current (DC) between the gate and channel could be suppressed. Lilienfeld proposed the use of a SiO_2 glass substrate.

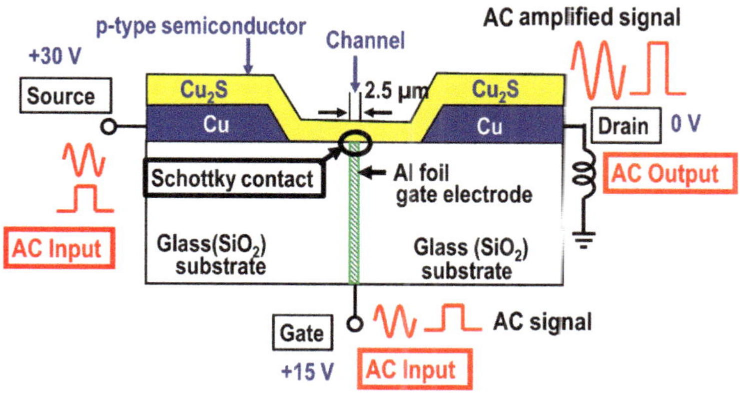

Figure 1.7 First metal-semiconductor FET (MESFET) idea by J. Lilienfeld in 1925.

The glass substrate was initially fractured into two parts, which were then reassembled to form a single substrate with an

Al foil placed between them. The glass surface was polished, and Cu electrodes were deposited on the glass to serve as the source and drain. A thin cuprous sulfide (Cu_2S) film was deposited onto the structure. In his patent, an AC input signal was applied to both the Cu source and the Al gate electrodes. The current flowing through the Cu_2S film was expected to be modulated by the voltage changes in the Al foil film. To minimize parasitic capacitance, the thickness of the Al foil film was kept as small as 2.5 μm. By biasing the Al foil film positively, any leakage between the Al and Cu_2S films became negligible.

In 1928, another type of FET, namely the metal-oxide-semiconductor field-effect transistor (MOSFET), was proposed by J. Lilienfeld, as shown in Figure 1.8 [17]. In this design, a 100 nm thick layer of Al_2O_3 was formed on a 30 μm thick Al substrate. A thin Cu_2S film was then deposited on top of the Al_2O_3 layer. The Al substrate, Al_2O_3, and Cu_2S were utilized as the gate electrode, gate oxide, and channel of the MOSFET, respectively. Unlike the MESFET case, there was virtually no leakage current between the Al gate and Cu_2S channel even under very high gate bias. A triangular depression was present at the center of the Cu_2S film. The film thickness at the depression was set to a molecular scale, allowing the vertical electric field at the depression to be significantly enhanced, thereby controlling the channel current.

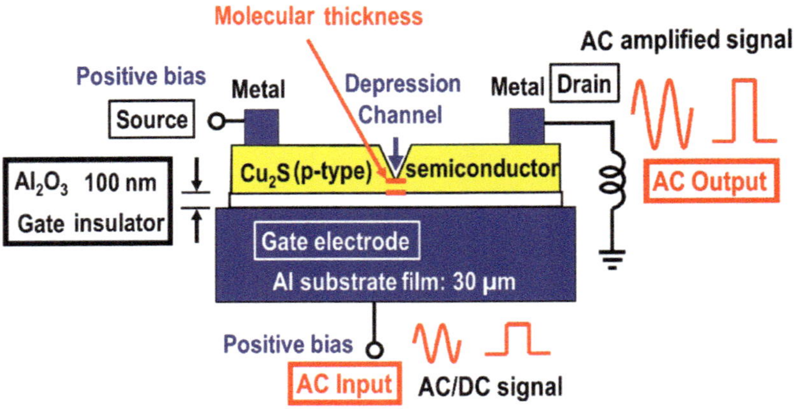

Figure 1.8 First MOSFET idea by J. Lilienfeld in 1928.

Cu_2S typically behaves as a high-resistivity p-type semiconductor, and Lilienfeld's FETs were typically p-channel FETs of the normally-on type, controlled by a positively biased gate electrode. These devices can also be classified as thin-film transistors (TFTs). However, unfortunately, there are no experimental records of Lilienfeld's MESFET and MOSFET, and it is assumed that his proposed devices did not function as amplifiers. This can be attributed to the unsuitable choice of semiconductor material (Cu_2S) and poor material quality, as well as the absence of technologies to adequately control the interfaces between materials.

During the 1920s, the concept of holes and pn junctions had not yet been developed. Additionally, there was no established theory for semiconductor physics, such as the semiconductor energy band theory, which was introduced by Alan Herries Wilson in 1931, employing quantum mechanical theory [18, 19]. Consequently, there was no fundamental understanding of the rectifying effects at the metal-to-semiconductor interface.

After Julius Lilienfeld's MOSFET concept in 1928, numerous attempts were made to realize the MOSFET, but the successful operation of the MOSFET was not achieved until 1960, as described later. Lilienfeld's MOSFET was a TFT-type transistor, assuming an ultra-thin semiconductor film on an insulator for the channel. However, in order to effectively control the carriers throughout the film using a vertical electric field, the film thickness needed to be extremely thin, on the order of 10 nm. Unfortunately, the technology available during that period was not advanced enough to fabricate usable films for these devices.

In the late 1930s, the analysis of the Schottky junction using band theory revealed the existence of ultra-thin surface inversion and accumulation layers in the semiconductor [20]. It was observed that under certain conditions, the high carrier-concentration carriers could exist in these ultra-thin surface layers, and it was proposed that MOSFETs could be realized by controlling the carrier current in these ultra-thin surface layers, even if the semiconductor film itself was not ultra-thin. These devices are referred to as bulk-type MOSFETs.

Figure 1.9 illustrates the five necessary conditions for the successful operation of bulk MOSFETs.

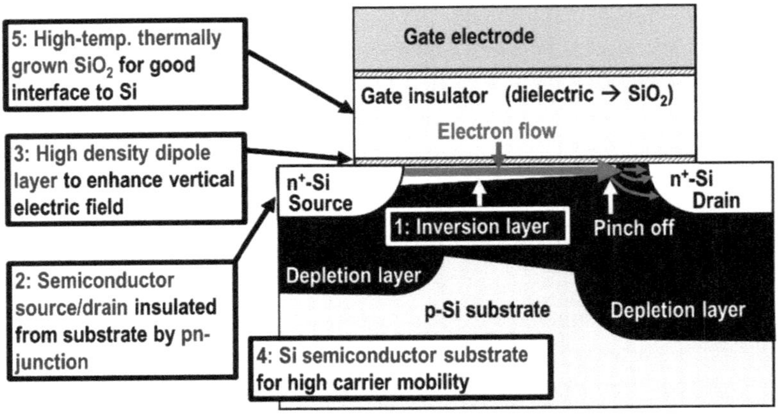

Figure 1.9 Five necessary conditions for bulk MOSFET successful operation.

The five conditions are as follows:

1. **Surface inversion channel:** As the channel of the carriers, the surface inversion layer should be used. The surface inversion layer has a thin enough thickness for the gate electric field to fully control the carriers within it. The inversion layer is separated from the substrate by a depletion (or space charge) region, allowing the channel carrier to flow exclusively within the inversion layer. The surface accumulation layer is not separated from the substrate and is therefore unsuitable for use as the channel layer.
2. **Semiconductor source and drain:** The semiconductor source/drain should establish an easy electrical connection with the channel inversion layer and be effectively insulated from the substrate This can be achieved by monolithically forming source and drain areas at the surface region of the semiconductor substrate through doping with an impurity of the opposite type to that of the substrate. This process establishes a pn junction between the source/drain and the substrate. Using a metal source/drain makes it difficult to establish an electrical connection solely with the ultra-thin channel inversion layer, resulting in a potential leakage current path from the metal source to the metal drain through the substrate.

3. **Adequate dipole density at the interface between the gate insulator and semiconductor:** High-density interface states exist at the semiconductor's surface, where charges trapped at these states can shield the electric field generated by the gate. To overcome this shielding effect, the electric field at the interface needs to be enhanced, which can be achieved through a dipole layer at the interface. To achieve a sufficiently high dipole density, the gate insulator should be a dielectric material with a suitably high dielectric constant.
4. **High-quality Si substrate:** For MOSFET applications, it is important to have a Si substrate with high carrier mobility and a long carrier lifetime. In the 1940s, Si and Ge were identified as suitable candidates in this regard. However, Ge was found to be unsuitable due to its poor interfacial properties with Ge oxides.
5. **SiO_2 gate insulator grown through high-temperature oxidation of Si:** The interface between the semiconductor and the gate insulator should exhibit good properties, characterized by low surface (interface) state density and surface roughness. By the late 1950s, numerous combinations of semiconductor substrates and gate insulators were investigated. Ultimately, it was discovered that SiO_2 film grown through high-temperature oxidation of the Si substrate provided the optimal solution at that time, exhibiting low density of interface states. Additionally, thermally grown SiO_2 possesses desirable properties, such as low leakage current, a dielectric constant value of 3.9 with sufficiently high density of interfacial dipoles, and resistance to dissolution in water.

Figure 1.10 shows the MOSFET/metal insulator semiconductor FET (MISFET) development trend from 1945 to 1960 [4].

Heinrich Welker's contributions to the development of the bulk MOSFET, as illustrated in Figure 1.10, were indeed significant. In his patent submission in April 1945 [21], he proposed a MISFET structure with a depletion layer at the surface region of the semiconductor, separating the surface channel carriers (electrons) from the majority substrate carriers (holes). He did not refer to this surface carrier layer as the 'inversion layer', but it enabled the control and modulation of the channel current. One notable aspect of Welker's proposal was his suggestion to use germanium (Ge) and silicon (Si) as the materials for the MISFET. He recognized their high carrier mobility, which is an important factor for achieving efficient device operation. Prior to his patent, no one had specifically mentioned Ge and Si as suitable materials for FETs. Furthermore, Welker emphasized the significance of using a dielectric film as the gate insulator in the MISFET structure. He noted that the interfacial dipoles between the dielectric and the semiconductor could enhance the gate electric field, enabling better control over the channel current. The structure proposed by Welker closely resembles the bulk MOSFET used today, with the exception that the source/drain was made entirely of metal. Consequently, there was a leakage path in the deep bulk region between the metal source and the drain. However, Welker suggested that this leakage could be suppressed by using a low-impurity substrate. Welker derived an accurate formula for the inversion channel current as a function of drain voltage (V_d) and gate voltage (V_g) for low V_d (triode mode operation) [22], although he was unable to conduct sufficient experiments due to the constraints of World War II [23, 24].

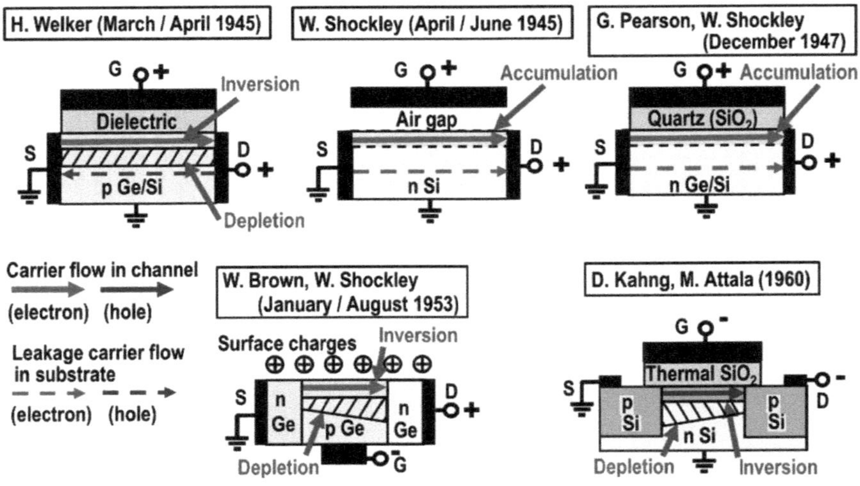

Figure 1.10 Various approaches toward MOSFET [4], copyright 2023 JSAP.

Overall, Heinrich Welker's patent submission laid the foundation for the bulk MOSFET, introducing key concepts such as the surface inversion layer, the use of Ge and Si as materials, and the importance of a dielectric gate insulator. However, his patent had not been recognized because the original document was lost because of World War II and recovered in 1952 [24]. The patent was granted in 1973.

In April 1945, William Shockley proposed a Si MOSFET with an air gap as the gate insulator [25]. His proposed design involved using a surface accumulation layer as the channel. However, there was a leakage path between the source and drain in the bulk region because the accumulation layer was not insulated from the bulk. Furthermore, the source/drain regions were composed solely of metal. Shockley conducted experiments until June 1945 but abandoned the project due to the lack of channel current modulation caused by the high density of surface charges trapped at the surface states in the Si, which shielded the gate electric field.

Meanwhile, W. Shockley's group at Bell Labs continued their research in the pursuit of FET operation. In December 1947, John Bardeen and Walter Houser Brattain discovered the operation of the point contact transistor (PCT), which marked the world's first transistor operation as shown in Figure 1.11 [26]. In January 1948, W. Shockley invented the junction bipolar transistor (BJT) as shown in Figure 1.11 [27], and its operation was experimentally confirmed in 1951 [28]. However, it is important to note that these devices were not FETs but rather junction transistors. The PCT modulated electron current injection at the collector Schottky junction (metal-collector and Ge-base) by altering the Ge-base potential, which was influenced by the emitter injection current. On the other hand, the BJT modulated electron injection at the pn junction (Ge-emitter and Ge-base junction) by applying a Ge-base potential change from the base terminal.

In December 1947, Gerald Pearson and W. Shockley of Bell Labs conducted experiments with Si and Ge MOSFETs using a quartz (SiO_2: dielectric) substrate as the gate insulator [29]. A Si or Ge film was deposited on one side of the quartz substrate as the channel, and an Au film was deposited on the other side as the gate electrode, as shown in Figure 1.10. The modulation of channel current by the gate electric field was observed, but it was only one-tenth of the value calculated [30] by the band theory due to the high density of interface states.

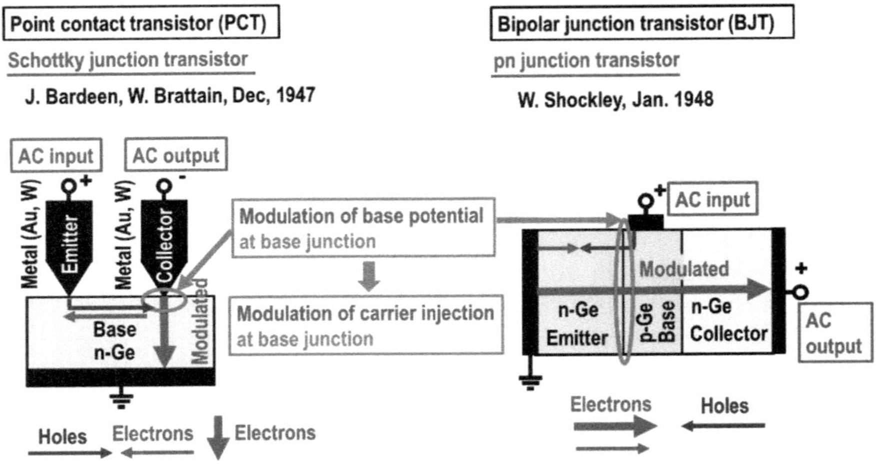

Figure 1.11 Point contact transistor and bipolar junction transistor.

The first FET operation with an inversion channel and semiconductor source/drain with pn junctions was presented by Walter Lyons Brown and W. Shockley in January 1953 [31, 32]. They conducted aging tests on Ge npn bipolar junction transistors by applying a reverse bias to the emitter-base and base-collector junctions. They discovered that a surface inversion conduction layer was formed at the base region due to the surface positive charges caused by the aging test. This inversion layer was electrically connected to the emitter (source) and collector (drain). They demonstrated that the conduction of the inversion layer could be controlled by changing the base (gate) potential through the back base electrode. The measured channel conductance was 1/300 of the theoretical calculation, presumably due to the low carrier mobility at the Ge surface caused by surface charges. This device was not a MISFET/MOSFET but rather a type of "one-sided" JFET (Junction FET) [33] with only a backside gate, which is the base.

The final requirement for achieving a successful MOSFET was to establish a high-quality interface between the gate insulator and the semiconductor, characterized by low densities of surface states and fixed charges. In 1958, Martin Attala and Eileen Tannenbaum of Bell Labs discovered that SiO_2 gate films grown through high-temperature oxidation (e.g., 900 °C) of Si could stabilize the Si interface, resulting in a significantly reduced density of slow states [34]. In 1960, Dawon Kahng and Martin Attala of Bell Labs published the first successful MOSFET operation using a Si substrate and a thermally grown SiO_2 gate insulator, as depicted in Figures 1.10 and 1.12 [35–37].

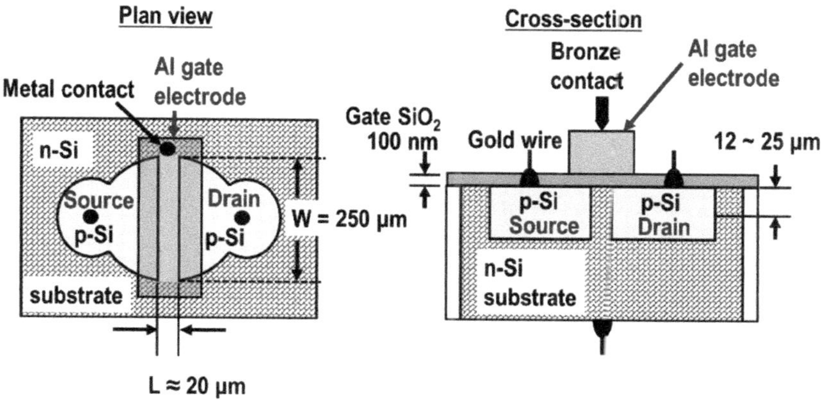

Figure 1.12 The first MOSFET with successful operation in 1960.

For the fabrication of MOSFETs, mechanically polished n- and p-type Si slices with [111] orientation were used as substrates. The slices were 1.5 mm square and 250 μm thick. The p-type substrates were doped with boron at a concentration of 10^{13} cm^{-3}, and the resistivity of the substrate was 1000 Ω cm. However, there is no available information on the n-type substrate in the literature. To create a 1 μm thick mask oxide for the source/drain impurity diffusion, the substrates were oxidized in 1-atmosphere steam for 2 hours at 1200 °C. The vapor phase diffusion was carried out using the "closed box technique" at 1250 °C for phosphorus and 1300 °C for boron. The resulting junction depth ranged from 12 to 25 μm. After etching the mask oxide, the Si surface was etched to a depth of about 3 μm in a 10 to 1 solution of nitric and hydrofluoric acids for 20–30 seconds, excluding the contact portion of the source/drain region. Gate oxides of 100 nm thickness were grown using the 'steam bomb oxidation technique' [38] at 650 °C, with a pressure of 45–55 atmospheres.

A 150 nm thick Al gate electrode was then evaporated. Initially, high-pressure oxidation was used for the growth of the gate oxide, but later on, other companies adopted normal atmospheric-pressure oxidation at higher temperatures above 1000 °C as the standard. Contact holes were drilled through the oxide to the source/drain regions, and a gold lead with a diameter of 38 μm was bonded to the Ge surface within the contact holes. A spring-loaded bronze point contact was used to establish contact with the Al gate electrode. The channel width-to-length ratio was 14, with a channel width of 250 μm according to ref [39], suggesting an estimated channel length of approximately 20 μm. P-channel MOSFETs exhibited good characteristics, whereas n-channel MOSFET characteristics were less favorable due to the high density of surface states.

1.4 MOSFET to LSI

It took 30 years from J. Lilienfeld's initial idea of the FET to the successful operation of the MOSFET by D. Kahng and M. Attala. The key factor behind the success of D. Kahng and M. Attala was their choice of using a Si substrate and a thermally grown SiO_2 gate insulator produced through high-temperature oxidation of the Si substrate. The Si/SiO_2 interface exhibited an exceptionally low interface state density. Over the course of 35 years, numerous researchers made significant contributions to the advancement of semiconductor device technologies, ultimately leading to the realization of MOSFET operation.

The discovery of the pn junction by Russell Ohl of Bell Labs in 1940 [40–42] was one of the most crucial milestones toward achieving the MOSFET. Another important milestone was the commercial production of the BJT and the development of relevant semiconductor process technologies in the 1950s. At the time when W. Shockley invented the BJT, there were very few semiconductor process technologies available. The technologies developed for BJT production included Ge/Si single crystal pulling, photolithography, vapor phase impurity diffusion, SiO_2 mask for impurity diffusion, and significant improvements in the quality and purity of Si crystals. It is certain that without the development of process technologies and the improvement in Si quality achieved through BJT mass production, the MOSFET would not have been realized in 1960.

BJT integrated circuits (ICs) were invented and fabricated by Jack Kilby in 1958 [43] and Robert Norton Noyce in 1959 [44], respectively. Then, small-scale BJT ICs (SSIs) became widely used for electronic applications, including the Apollo moon project, starting in the mid-1960s. Meanwhile, MOSFETs had been recognized as the most suitable transistors for large-scale integration due to their surface current conduction path, which allowed metal-line contacts to be made to all-transistor terminals from the chip surface. However, it took nearly 10 years for the first MOSFET developed by D. Kang and M. Atalla to evolve into LSIs, which involved integrating almost 1000 or more transistors on a single chip. This delay was primarily due to gate instability issues caused by various unknown charges present on the silicon surface and within the SiO_2 gate insulators (Figure 1.13) [45]. Additionally, production yield problems arose from overly ambitious integration attempts beyond the capabilities of the technology at that time [46].

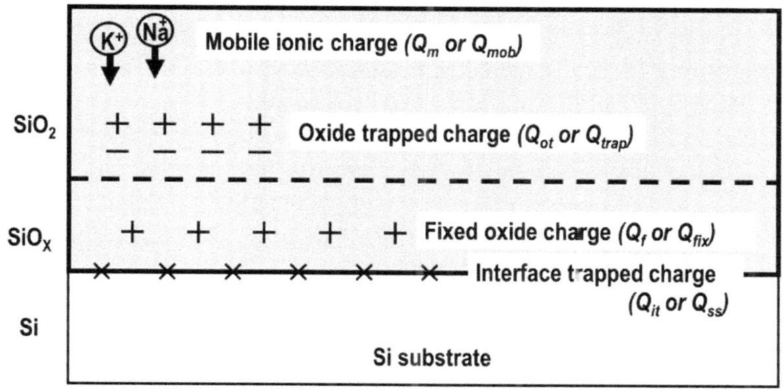

Figure 1.13 Various mysterious charges for Si MOS structure.

Then, the first generation of commercial MOS LSIs was fabricated between 1969 and 1971, following the resolution of the instability and yield issues. Notable examples of these LSIs include Intel's '1101' (256-bit SRAM) in 1969, '1103' (1k-bit DRAM) in 1970, and '4004' (4-bit microprocessor) in 1971.

Figure 1.14 summarizes the development trend of electric and electronic devices. Electric devices for ICT began with the use of relays in the 1830s to the 1840s. Relays were utilized in electric telegraph amplifiers and telephone exchange switches. By combining these switches, logic circuits could be constructed. Therefore, relays were also employed in early-stage computers and machine controllers before vacuum tubes were introduced. In fact, relays remained a key device for elevator controllers well into the 1970s due to their ability to handle high currents.

Subsequently, triode and pentode vacuum tubes [47] started to gain popularity in the 1910s for various electronic applications, including wireless communication, machine controllers, and computers. Vacuum tubes continued to be important electron devices even during the 1960s. However, with the advent of transistors and LSIs, vacuum tubes gradually shifted toward high-power and high-voltage applications, while transistors and LSIs became the mainstream electron devices from micro-PMOS LSIs to nano-CMOS VLSIs.

Once the MOS LSI era began with the introduction of 10–8 μm PMOS LSI technology between 1969 and 1971, the progress of LSIs advanced at an astonishing pace due to the scaling scheme [4, 48] (Figures 1.15 and 1.16).

In the ideal scaling scheme, both the horizontal and vertical dimensions of the MOSFET are decreased by the same factor of K, while the substrate's impurity concentration increases by a factor of $1/K$. This scaling ensures that the depletion layer width is also scaled by a factor of K. As a result, the I_d-V_d characteristics with V_g as a parameter are precisely scaled by a factor of K, as shown in Figure 1.15 [4]. Generally, the value of K is around 0.7. The clock frequency and transistor density of the MOS LSI chip increase by a factor of $1/K$ and $1/K^2$, respectively. However, the power consumption density remains unchanged, as shown in Figure 1.16.

Figures 1.17 and 1.18 [4] illustrate the actual scaling trend of MOSLSI since its inception. On average, the dimensions of MOSLSI have been reduced by a factor of 0.7 every 2.3 years, resulting in the development of 23 generations over a span of approximately 50 years. Additionally, the technology has evolved from PMOS LSI to NMOS LSI/VLSI (very large-scale integrated circuits), CMOS VLSI, BiCMOS VLSI, and the more recent nano-CMOS VLSI.

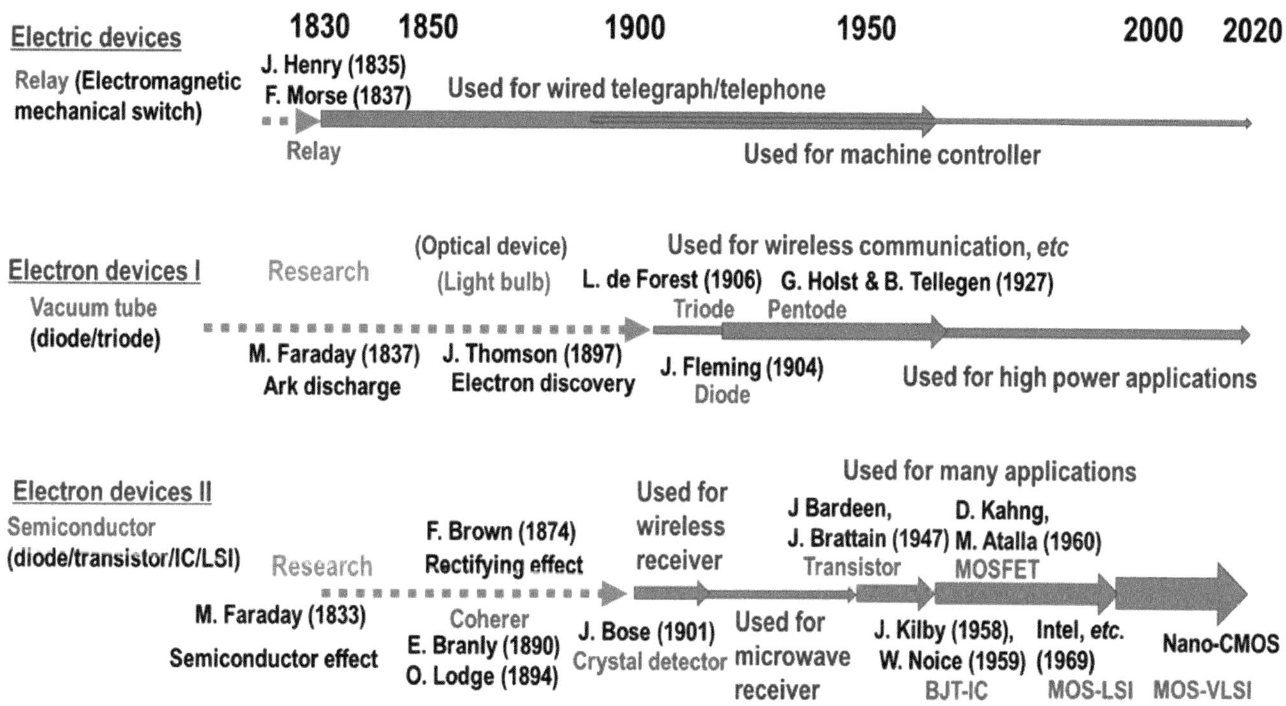

Figure 1.14 Trend of electric and electron devices.

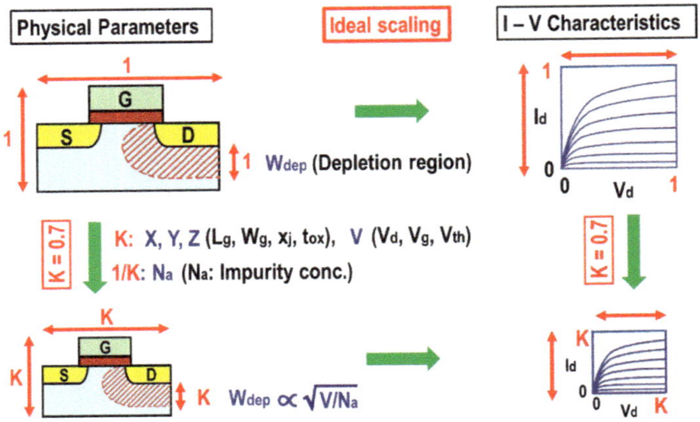

Figure 1.15 Ideal scaling of MOSFET [4], copyright 2023 JSAP.

Geometry & Supply voltage	L_g, W_g T_{ox}, V_{dd}	K	Scaling K : K=0.7 per generation 1 generation 2 ~ 3 years
Drive current in saturation	I_d	K	$I_d = v_{sat} W_g C_o (V_g - V_{th})$ C_o: gate C per unit area $\rightarrow W_g (t_{ox}{}^{-1})(V_g - V_{th}) = W_g t_{ox}{}^{-1}(V_g - V_{th}) = KK^{-1}K = K$
I_d per unit W_g	I_d/µm	1	I_d per unit $W_g = I_d / W_g = 1$
Gate capacitance	C_g	K	$C_g = e_o e_{ox} L_g W_g / t_{ox} \rightarrow KK/K = K$
Switching speed	t	K	$\tau = C_g V_{dd} / I_d \rightarrow KK/K = K$
Clock frequency	f	1/K	f = 1/t = 1/K
Transistor density	N	$1/K^2$	N $\rightarrow 1/K^2$
Power density	P	1	$fNCV^2 \rightarrow K^{-1}(K^{-2})K(K^1)^2 = 1$

Figure 1.16 Device parameter change by the ideal scaling scheme [4], copyright 2023 JSAP.

23 generations in 50 years: 2.3 year per generation

— PMOS — | — NMOS —

10 µm (1970) → 8 µm → 6 µm → 4 µm → 3 µm → 2 µm →

— CMOS —

1.2 µm → 0.8 µm → 0.5 µm → 0.35 µm → 0.25 µm → 180 nm →

(Micro-Electronics)

130 nm → 90 nm → 65 nm → 45 nm → 32 nm → (28 nm →)

22 nm(2012) → 14 nm (2014) →10 nm (2016) → 7 nm (2018) →

5 nm (2020) → 3 nm (2022)

(Nano-Electronics)

Figure 1.17 Actual scaling trend of logic MOS LSI [4], copyright 2023 JSAP.

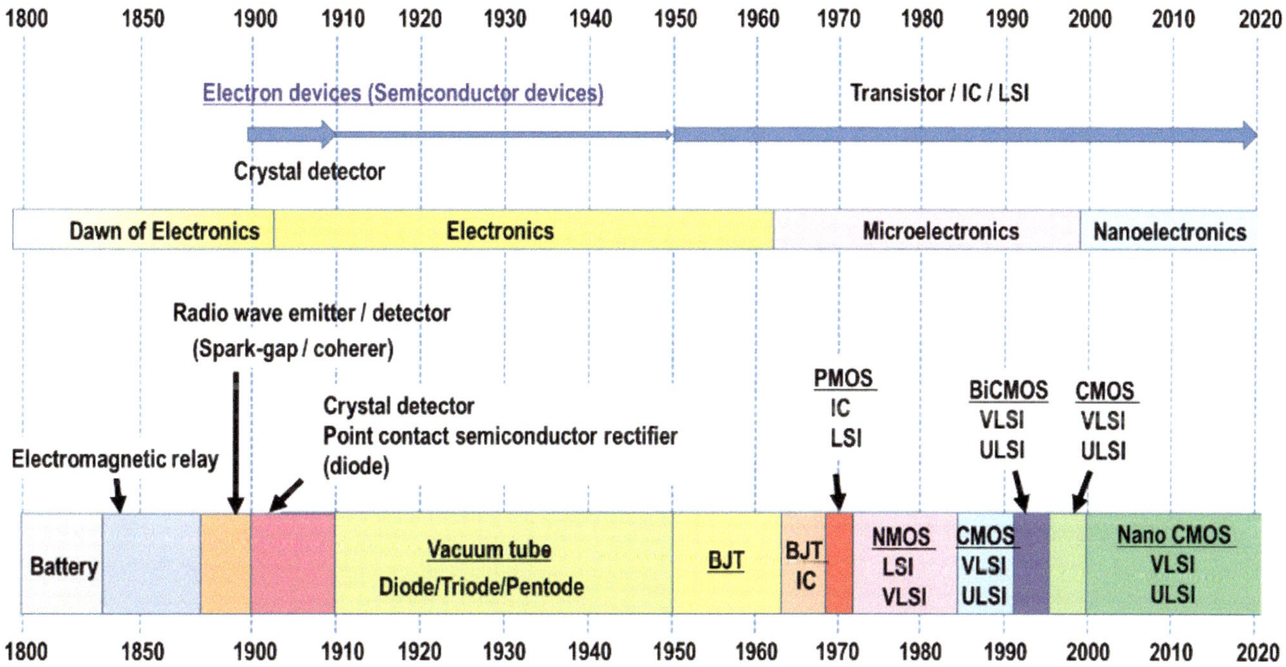

Figure 1.18 Trend of electron device technology.

Micro/nanoelectronics represents a new technological paradigm distinct from traditional electronics. In the early days of electronics, which utilized vacuum tubes, the minimum line width for electron devices was expected to be sub-millimeter. However, with the advent of photolithography techniques in the late 1950s, this dimension decreased to several tens of micrometers, marking the beginning of microelectronics with integrated circuits in the 1960s. Currently, microelectronics has further advanced to the realm of nanoelectronics, where the minimum line width is approaching 10 nanometers. Over the course of 115 years, spanning from the inception of electronics (triode vacuum tubes) to the present, the minimum line width has shrunk by a factor of approximately 10^5. This reduction has led to significant improvements in the performance of electric systems and devices, including operational speed, energy/power consumption, volume, weight, and cost, which have increased by orders of magnitude, ranging from billions to trillions.

The tremendous increase in performance has been achieved through the significant downsizing of device dimensions, an unprecedented feat in human history, as depicted in Figure 1.19 [4]. Today, semiconductor-based SD cards with capacities of 1 terabit (or 128 gigabytes) are readily available in the market at affordable prices, such as 10 US dollars. Let us envision a world without micro/nanoelectronics, relying solely on outdated vacuum tubes, using the following assumptions – price: 100 Japanese yen, size: $5 \times 5 \times 10$ cm^3, weight: 100 grams, and power consumption: 50 W. In this scenario, a 1 terabit memory composed of 1 trillion vacuum tubes would cost 100 trillion Japanese yen (or 1 trillion US dollars), equivalent to Japan's annual national budget. It would occupy a volume of $0.5 \times 0.5 \times 1.0$ km^3, surpassing the size of the world's tallest building. The weight would amount to 100 million tons, equal to the weight of 20 million elephants. The power consumption would reach 5 terawatts, equivalent to the output of 5,000 nuclear reactors. The yearly energy consumption of 1 trillion vacuum tubes would be 4×10^{20} calories, which is 2,000 times the global human consumption in a year. Therefore, without the development of micro/nanoelectronics, today's smart society with internet connectivity, artificial intelligence, and other advancements, depicted in Figure 1.20, would not exist. Thus, micro/nanoelectronics represents the most significant technological revolution since the advent of electronics.

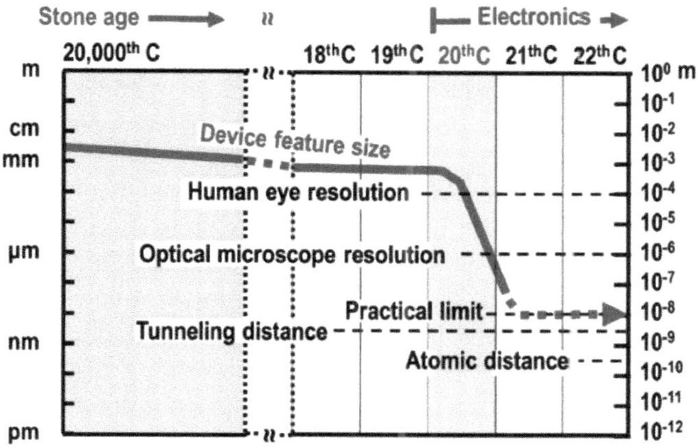

Figure 1.19 Trend of the minimum line width of the device in human history [4], copyright 2023 JSAP.

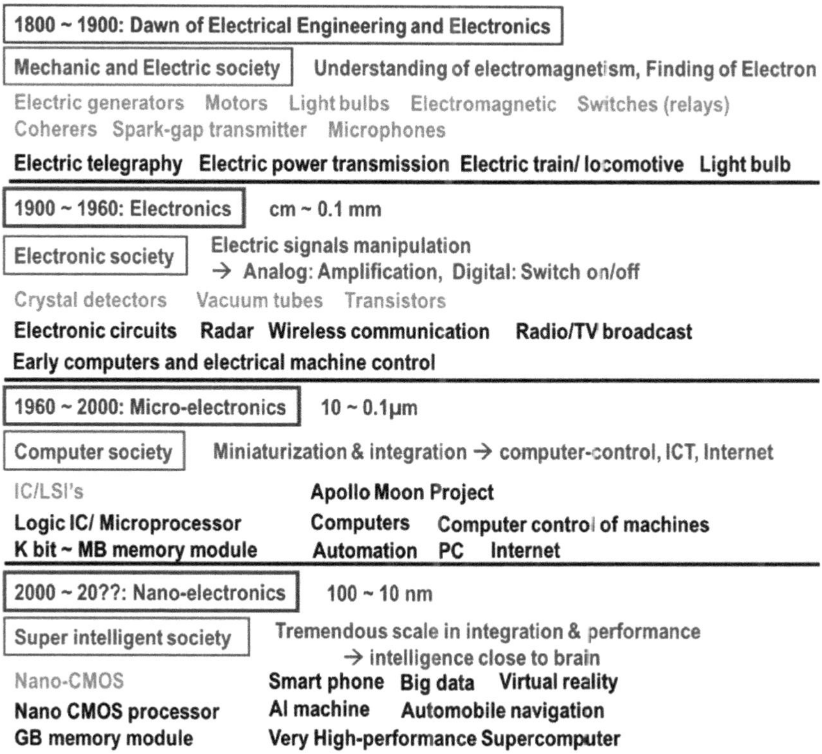

Figure 1.20 Progress of our society by electronics, microelectronics, and nanoelectronics.

1.5 Limit of Device Miniaturization

The progress of electronic devices has been achieved through device miniaturization. However, one may wonder what the limit of this miniaturization is.

The 'ultimate limit' of miniaturization is determined by the distance between atoms in the material, which is approximately 0.3 nm in the case of Si substrate, as shown in Figure 1.21 [4]. Changing the material does not significantly affect this limit. It is impossible to create device structures smaller than the atomic distance. Therefore, it is important to note that the term 'pico-electronics', referring to device sizes, is no longer applicable.

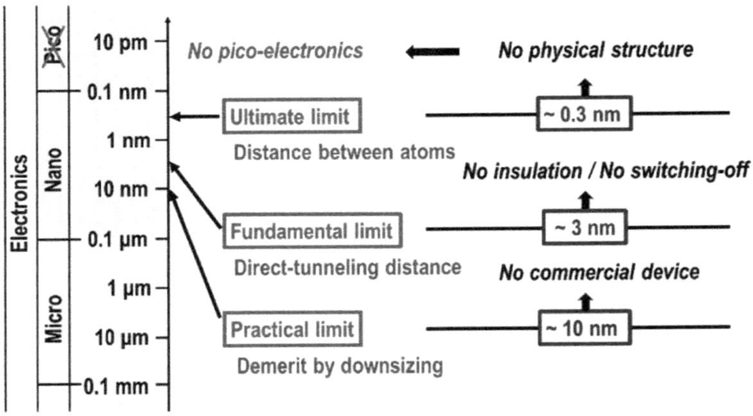

Figure 1.21 Miniaturization of the electron devices accelerated by MOSFET scaling scheme [4], copyright 2023 JSAP.

Before reaching the ultimate limit, there is another constraint known as the 'fundamental limit', which is defined by the distance required for direct tunneling. This distance is approximately 3 nm. When the spacing between structures becomes 3 nm, the wavefunction of the electrons penetrates the potential barrier, resulting in a significant direct tunneling current. As a result, there is no electrical insulation and devices cannot be switched off. Thus, 3 nm represents the limit for device operation.

The third limit is the 'practical limit', which is defined by the drawbacks associated with downsizing. When the disadvantages of downsizing outweigh the advantages, further miniaturization will cease. These drawbacks may include increased off-leakage current, higher production costs, reduced reliability, and greater variability in device characteristics, among others. According to the IEEE International Roadmap for Devices and Systems (IRDS) 2020, the practical limit is estimated to be around 10 nm, as explained later in this document.

Currently, there are logic devices marketed under the names '5 nm' or '3 nm' technology, but it is important to note that these numbers have little correlation with the actual physical sizes of the devices, as shown in Figure 1.22 [4]. In reality, the minimum line widths for these devices typically range from 30 nm to 15 nm. This discrepancy between the commercial names and actual sizes is a misleading practice within the logic device industry.

	2020	2022	2025	2028	2031	2034
Commercial name (Technology node)	5.0 nm	3.0 nm x 0.6	2.1 nm x 0.7	1.5 nm x 0.7	1.0 nm x 0.7	0.7 nm x 0.7
Gate length	18 nm	16 nm x 0.9	14 nm x 0.9	12 nm x 0.9	12 nm x 1.0	12 nm x 1.0
Metal wire pitch (width + space)	30 nm	24 nm x 0.8	20 nm x 0.8	16 nm x 0.8	16 nm x 1.0	16 nm x 1.0

Figure 1.22 Commercial names, gate length, and metal wire pitch of logic CMOS predicted by IEEE IRDS 2020 (International Roadmap for Devices and Systems) [4], copyright 2023 JSAP.

Before 2000, technology names (or commercial technology names) indicated sizes that represented gate length and metal line widths determined by photolithography capabilities. In the 2000s, gate length became significantly smaller than what photolithography could achieve through resist slimming with oxygen plasma etching. However, starting from the 2010s, due to various difficulties, the shrink rate of gate length and metal wire pitch began to exceed 0.7 while commercial technology names continued to shrink by a factor of 0.7 every two years. This has led to a substantial gap between commercial names and the actual minimum line width used in logic nano-CMOS VLSIs. According to IEEE International Roadmap for Devices and Systems (IRDS) 2020, as shown in Figure 1.22, 0.7 nm logic CMOS technology is predicted for 2034 but it is assumed that gate length and metal wire pitch will not decrease below 12 nm and 16 nm respectively. Hence, the practical limit for device miniaturization is estimated at around 10 nm. The causes of this practical limit will be explained later.

1.6 Miniaturization Limit by Power Consumption Increase

Based on the ideal scaling scheme, it was expected that the number of MOSFETs and clock frequency of the chip would increase by a factor of 2 and 1.4, respectively, every 2–3 years while maintaining the same power consumption. However, this expectation did not hold true in actual scaling.

Figure 1.23 shows the historical trend of scaling parameters from 1970 to 2000 [4, 49]. Over those thirty years, various device components such as gate length, gate oxide thickness, and junction depth decreased by a factor of 1/100. In contrast, power supply voltage only decreased by a factor of 1/10 starting from the early 1970s when NMOS LSI was introduced. Additionally, chip size increased by a factor of 10. This was due to maintaining supply voltage at 5 V for an extended period to ensure compatibility with TTL (transistor-to-transistor logic) standard which allowed customers to enjoy higher performance at 5 V with increased power consumption. However, due to issues like TDDB (time-dependent dielectric breakdown), hot carrier-induced degradation, and significant short-channel effects at 5 V, supply voltage had to be reduced to 3.3 V during the 0.5 μm generation in the mid-1990s. Since then, supply voltage has continued to decrease albeit at a slower rate, taking in account the tradeoff between the decrease in current drive and the increase in subthreshold leakage current as described later.

The insufficient reduction in supply voltage led to a dramatic thousand-fold increase in clock frequency over three decades. Unfortunately, this advancement also resulted in a hundred-thousand-fold rise in power consumption, leading to significant heat generation on the chip. During his presentation at ISSCC 2001 [4, 50], P.P. Gelsinger from Intel foresaw a remarkable surge in power consumption density for microprocessor chips. According to his predictions, in 2002, the power consumption of 10 W/cm^2 would be akin to the heat generation on a hot plate's surface.

By 2006, this value would increase to 100 W/cm^2, comparable to that of a nuclear reactor's surface. Further, in 2010, the figure would rise to 1000 W/cm^2, similar to the heat generated on a rocket nozzle's surface. The most astonishing projection was for 2016, with a staggering 10,000 W/cm^2, equivalent to the heat generated on the surface of the sun. This excessive heat generation raised concerns about the possibility of the chip melting.

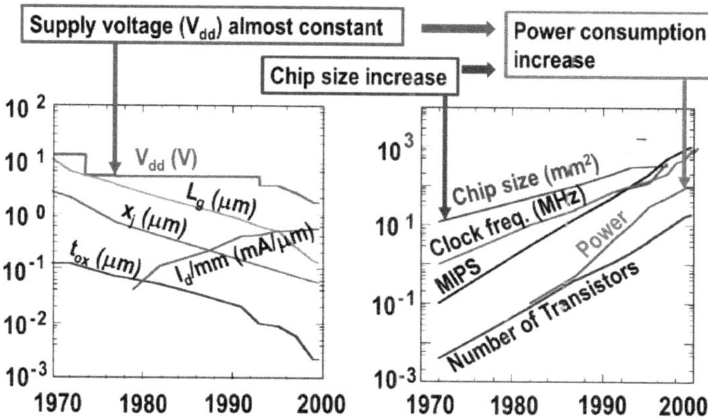

Figure 1.23 Actual past trend of scaling parameters of logic MOSLSI from 1970 to 2000 [4], copyright 2023 JSAP.

The solution to tackle the issue of excessive heat generation was to impose limitations on the increase in clock frequency. Consequently, the clock frequency of high-performance microprocessors has remained in the range of 3–5.5 GHz for the past 15 years. In Figure 1.24 [4], we can observe the average trend lines for microprocessor performance data in IRDS 2022. The number of transistors in a chip has continued to grow steadily, in line with Moore's law [51]. Initially, power consumption showed an upward trend until the mid-2000s when the increase in clock frequency nearly came to a halt, stabilizing at an average of around 3 GHz. By curbing the clock frequency, the growth in power consumption was significantly mitigated, although this did impact the rate of single-core performance improvement, causing it to slow down. However, to offset this drawback, the number of cores began to increase, leading to a continuous enhancement in the overall performance of microprocessors.

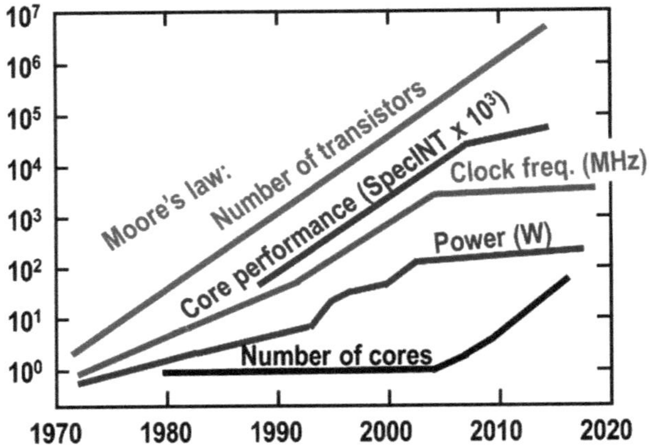

Figure 1.24 Forty years of microprocessor data published in ISSCC [4], copyright 2023 JSAP.

1.7 Miniaturization Limit by Leakage Current

Off-leakage current can be one of the most significant challenges when it comes to downsizing [4]. Off-leakage current refers to the current that occurs when the transistor is in the 'off' state or when the gate bias is set to 0 V. It is comprised of four components, as illustrated in Figure 1.25 [4]: punch-through, direct-tunneling, subthreshold, and gate insulator leakage.

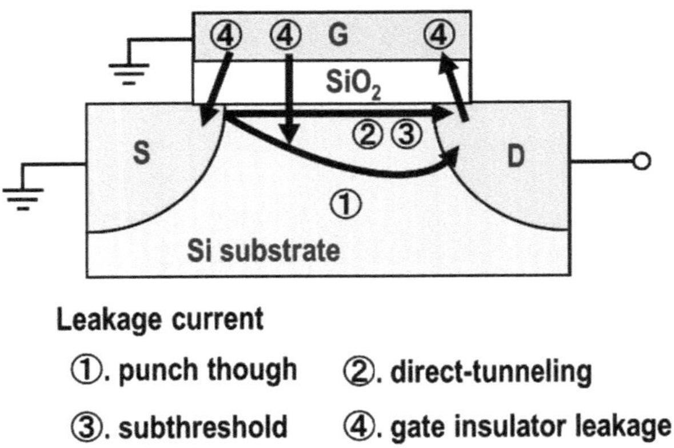

Figure 1.25 Off-leakage components of MOSFETs [4], copyright 2023 JSAP.

1.7.1 Punch-Through Current

When the gate bias (V_g) is set to 0 V, there is typically no leakage current between the source and drain of a transistor. However, as the gate length decreases, the depletion region near the drain touches the source. This causes electrons in the source region to flow into the depletion region due to the positive potential of the depletion region as shown in Figure 1.26 [4]. Eventually, these electrons move into the drain, resulting in a phenomenon known as punch-through. The punch-through current predominantly flows in the deeper region because the potential near the silicon surface is effectively controlled to 0 V by the gate bias.

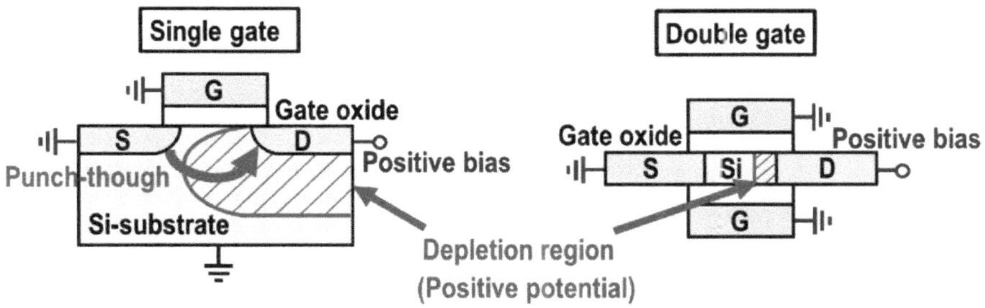

Figure 1.26 Punch-through current and its suppression by double and multiple gate structure [4], copyright 2023 JSAP.

To address the issue of punch-through, two solutions are commonly employed: reducing the supply voltage and decreasing the gate oxide thickness. Reducing the supply voltage helps shrink the depletion region, while decreasing the gate oxide thickness enhances the gate electric field, thereby improving the controllability of the silicon surface potential to 0 V. This approach aligns with the ideal scaling scheme.

Recently, reducing the supply voltage has become difficult, due to the rise in subthreshold leakage caused by the reduced threshold voltage, required under the reduced supply voltage, as discussed later. Furthermore, there are difficulties in reducing the gate oxide thickness because of long-term reliability concerns. Additionally, the substrate's impurity concentration, particularly in the high-impurity concentration region (referred to as 'halo' [52]) near the drain, has already reached 10^{18} cm^{-3} and cannot be further increased due to the reduction of the drain pn junction breakdown voltage [53].

As a result, an additional solution has been adopted, as shown in Figure 1.26 [4]. The concept of the solution involves cutting off the deep Si region where the punch-through current flows and utilizing a thin Si film as the channel. In this approach, the channel current is controlled not only by the top gate electrode positioned above the film but also by the bottom gate electrode situated at the backside of the film. This is known as a 'double-gate' MOSFET [54], which falls under the category of 'multiple-gate' MOSFETs. With this configuration, the potential of nearly the entire Si thin film region can be effectively controlled to 0 V, resulting in a significantly reduced depletion region.

In December 1992, D. J. Frank et al. from IBM confirmed the excellent operation of a 30 nm gate length double gate MOSFET through Monte Carlo simulation [55]. Additionally, it was verified using Monte Carlo simulation by C. Fiegna et al. from the University of Bologna/Toshiba that the double gate MOSFET operates reliably at room temperature with sub-25 nm gate length. In May 1993, C. Fiegna et al. published a new scaling scheme for future miniaturization, as depicted in Figure 1.27 [4, 56]. This scheme emphasized the necessity of the double gate MOSFET for achieving miniaturization into the sub-25 nm generation.

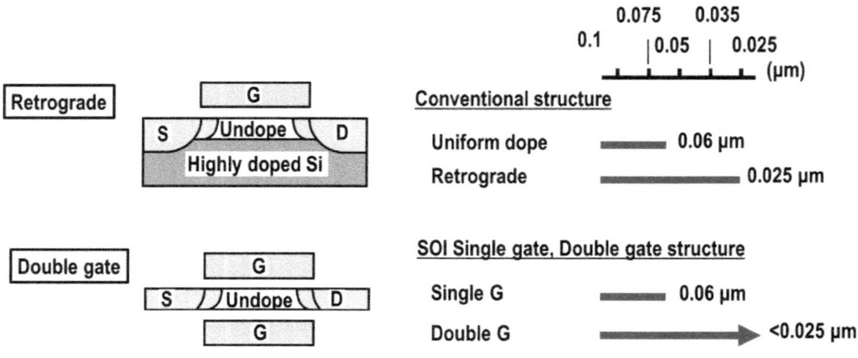

Figure 1.27 New scaling scheme proposed in 1993 [4], copyright 2023 JSAP.

Figure 1.28 [4] illustrates several types of double-gate or multi-gate structures. The first device utilized for production was a double-gate MOSFET with a vertical channel. This particular device showcased a fin-shaped vertical silicon film, which enabled the self-alignment of the top and bottom gate electrodes. In 1989, D. Hisamoto, E. Takeda et al. from Hitachi introduced the concept of the fin-FET, as documented in [57]. Subsequently, in 2000, D. Hisamoto, C. Hu et al. predicted the scalability of the fin-FET down to 20 nm [58]. At first, fin-FETs were produced on SOI wafers. A major milestone in production was achieved with the development of bulk fin-FET by C. Lee, K. Kim et al. of Samsung in 2004 [59]. Later, Intel introduced the bulk-type fin-FET, essentially a tri-gate FET [60], in their 22 nm logic CMOS products in 2011. For the past decade, Fin-FETs have been the dominant devices, but raising the fin height to achieve a higher drain current has presented challenges. Horizontal nano-sheets (or nano-ribbons) offer a straightforward solution to increase the channel width for enhanced drain current. The self-alignment technology for top and bottom gates was invented in 2002 [61] and demonstrated in 2003 [62], and 2004 [63] by S.-M. Kim's group in Samsung. It was also developed during the advancement of Si nanowire/ nanosheet FET devices by S. Deleonibus's group in Leti [64] and other organizations. Currently, nano-sheet channel MOSFETs [65] are replacing fin-FETs, and stacked nano-sheet structures or CFETs [66] are planned for the near future, as shown in Figure 111 [4]. In fact, the footprint area of SRAM cells has been reduced by half with the introduction of CFETs [66]. By utilizing multi-gate structures, the punch-through current is significantly reduced until the gate length reaches 10 nm.

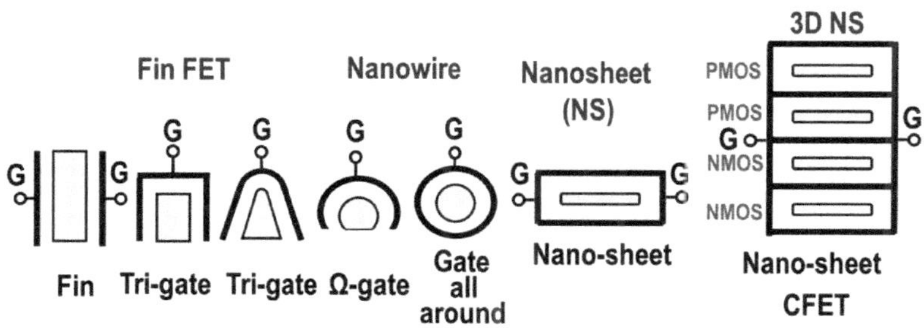

Figure 1.28 Various kinds of multiple gate structures [4], copyright 2023 JSAP.

1.7.2 Direct-Tunneling Current

For MOSFETs with a channel length of less than 3 nm, there is direct tunneling off-leakage between the source and drain, as illustrated in Figure 1.29 [4]. When the gate bias is 0 V, a potential barrier exists between the source and channel, which is the built-in potential of the pn junction (~0.7 eV). The presence of this barrier effectively suppresses leakage current in MOSFETs with long channels. However, as the channel length is reduced to less than 3 nm, quantum mechanical effects cause the wavefunction of electrons to penetrate the potential barrier, leading to a loss of insulation between the source and drain. This phenomenon represents the 'fundamental limit' discussed earlier. While MOSFETs may continue to function with slightly shorter channel lengths less than 3 nm, the off-leakage becomes excessively large, rendering it impractical to further reduce the gate length below 3 nm.

However, when the gate length is only slightly larger than 3 nm, the direct tunneling leakage remains considerably small. As a result, this type of leakage between the source and drain does not impose a limitation on downsizing since the limit is anticipated to be around 10 nm. The primary constraint at that point is mainly due to the subthreshold leakage current, as explained in the following section.

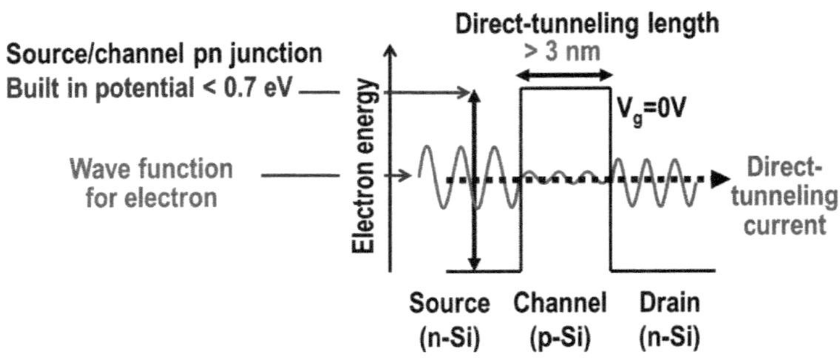

Figure 1.29 Direct-tunneling off-leakage between source and drain [4], copyright 2023 JSAP.

1.7.3 Subthreshold Leakage Current

The subthreshold current refers to the leakage current that occurs at V_g below the threshold voltage (V_{th}), as illustrated in

Figure 1.30 [4]. Ideally, there should be no leakage current below the threshold voltage. However, in reality, there is a leakage current that exhibits an exponential relationship with V_g.

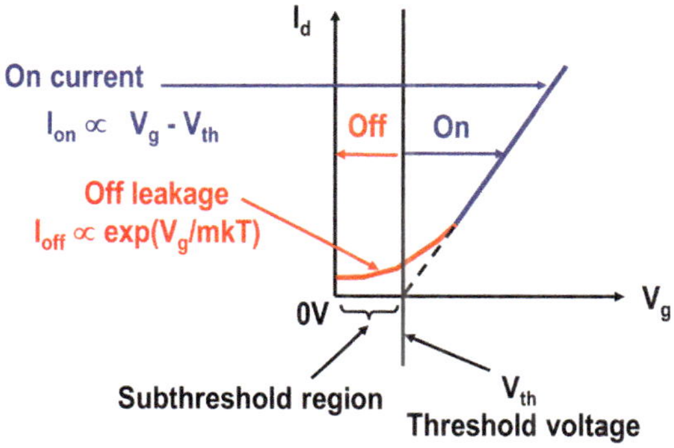

Figure 1.30 Subthreshold leakage current [4], copyright 2023 JSAP.

The flow of electrons in the subthreshold region from the source to the drain through the channel is akin to the flow of electrons in an npn BJT from the emitter to the collector through the base, as depicted in Figure 1.31 [4]. The key difference lies in the bias of the channel (or base), which is governed by the gate bias through capacitance coupling, while it is given direct contact from the base electrode in the case of the BJT. The subthreshold current is the pn junction current between the source (n-Si) and the channel (p-Si), and therefore, it is an exponential function of the voltage between the source and the channel. As a result, the subthreshold current can be expressed as $A \cdot \exp(\varphi_s/kT) = A \cdot \exp(V_g/mkT)$, where A represents a constant, φ_s denotes the channel potential, and m is defined as $C_{OX}/(C_{OX}+C_D+C_I)$, as shown in the figure.

Let us examine the scenario of scaling by two generations, as shown in Figure 1.32 [4]. In this case, the scaling factor is 0.7 × 0.7 = 0.5. Now, suppose we reduce V_{th} from 0.3 to 0.15 V. This change corresponds to shifting the I_d-V_g curve to the left, from the blue line to the green line. Consequently, at V_g = 0 V, there is a 1,000-fold surge in the off-leakage current for a subthreshold slope of 75 mV/decade. The reduction of the threshold voltage implies a decrease in the potential barrier φ_B, leading to an exponential rise in the number of electrons diffusing into the channel. As a result, the subthreshold off-leakage current is regarded as the most substantial limiting

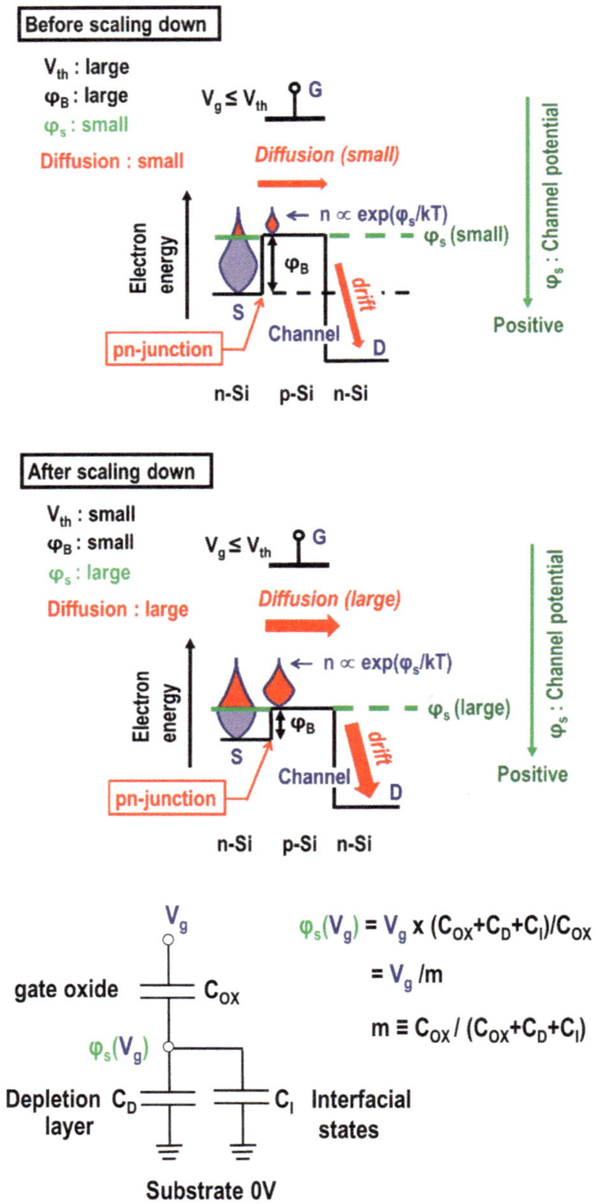

Figure 1.31 Subthreshold leakage current mechanism, it is just like the npn BJT current flow. The channel corresponds to the base, and the channel potential is given by the capacitance division of the gate bias [4], copyright 2023 JSAP.

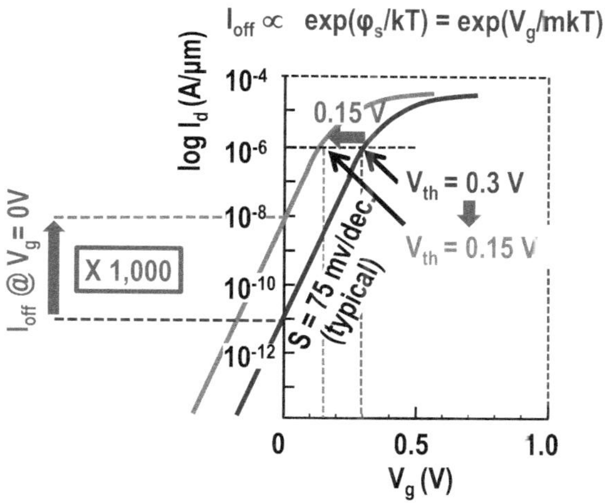

Figure 1.32 Significant increase of the subthreshold leakage by scaling [4], copyright 2023 JSAP.

factor, serving as the practical constraint before reaching the fundamental limit due to the direct-tunneling (3 nm). To alleviate the electron diffusion into the channel, one possible approach is to reduce the temperature to 77 K, according to the Boltzmann statistics. However, considering that our primary market is focused on room temperature applications, operating at 77 K is not as feasible as the mainstream devices. As per IRDS 2022, it is assumed that the limit for reducing the gate length is approximately 10 nm, primarily due to the subthreshold leakage.

1.7.4 Gate Insulator Leakage Current

When reducing the gate insulator thickness, not only does the insulator leakage current become a concern, but also the degradation of yield and reliability, due to defects in the thinner insulator. This posed a major challenge in the initial stages of MOS LSI development. Figure 1.33 [4] depicts the trend of gate oxide thickness and materials. During the 1960s and early 1970s, the high defect density of gate SiO_2 made it difficult to achieve thinning of the gate insulator without facing breakdown issues. From the mid-1970s onwards, significant progress was made in terms of dust and purity control, along with the optimization of the oxidation process. These improvements resulted in reduced defect density and increased reliability of thin gate SiO_2. As a consequence, continuous thinning of the gate SiO_2 was successfully achieved, reaching thicknesses of several nanometers.

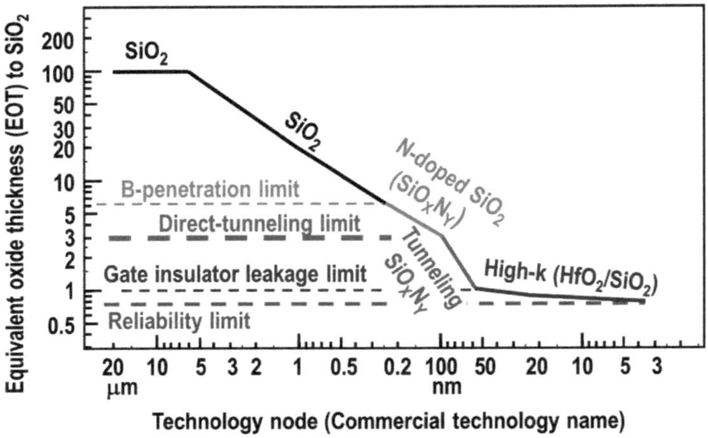

Figure 1.33 Trend of gate oxide thickness and material [4], copyright 2023 JSAP.

When the gate oxide thickness was reduced to several nanometers, a new issue emerged during the high-temperature LSI fabrication process. This is the B diffusion or boron penetration from the heavily B-doped p^+ poly Si gate into the Si channel through the gate SiO_2 film, as illustrated in Figure 1.36 [4]. To tackle this problem, T. Morimoto et al. from Toshiba introduced a solution in 1990 [67] by using a lightly N-doped SiO_2 gate insulator film, achieved through rapid thermal nitridation of SiO_2. It is essential to note that impurity penetration from the poly Si gate electrode only occurs for p-type impurities, while there is no such penetration for n-type impurities like P and As. Consequently, pure SiO_2 gate insulators were still commonly used for research on n-MOSFET miniaturization.

In 1972, Caver Mead et al. from Caltech predicted a limitation on reducing the gate length for MOSFETs to 0.25 µm, primarily because of the "tunneling current" through the 5 nm thick gate SiO_2 [68]. In his textbook published in 1979 [69], he described, "Finally, there appears to be a fundamental limit of approximately one-quarter micron channel length, where certain physical effects such as tunneling through the gate oxide and the fluctuations in the positions of impurities in the depletion layers begin to make the devices of smaller dimensions unworkable." Fortunately, his prediction turned out to be inaccurate. Experimental results continuously demonstrated the successful operation of MOSFETs with small gate lengths and thin gate SiO_2, disproving his initial hypothesis. For instance, in 1974, gate lengths (L_g), SiO_2 gate thicknesses (t_{ox}), and source/drain junction depths (x_j) were 1 μm, 20 nm, and 0.5 μm, respectively [48]. Similarly, in 1987, they were 0.1 μm, 4.5 nm, and 50 nm [70], and in 1992, they were 70 nm, 5 nm, and 40 nm [71], as shown in Figure 1.34 [4].

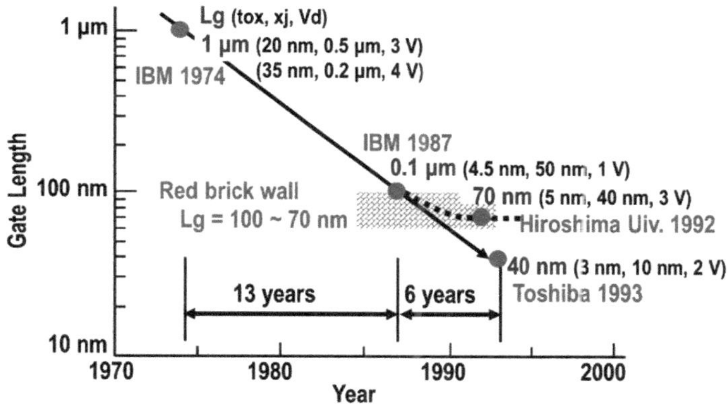

Figure 1.34 Trend of the miniaturization of n-MOSFETs at research level [4], copyright 2023 JSAP.

The paper published in 1974 [48] was the first comprehensive proposal and verification of an ideal scaling scheme, from a 5 μm gate length n-MOSFET to that of 1 μm. According to the ideal scaling derived from the 5 μm MOSFET, the combination of L_g, t_{ox}, and x_j should be 1 μm, 20 nm, and 0.5 μm, respectively. However, they concluded that a combination of 1 μm, 35 nm, and 0.2 μm (thicker t_{ox} but shallower x_j) would be a better choice, considering the 'reproducibility and reliability' of the gate insulator.

In the case of the 0.1 μm gate length n-MOSFET in 1987, the t_{ox} was 4.5 nm, which was slightly thicker than the ideally scaled thickness from the 1 μm gate length n-MOSFET (ranging from 35–20 nm), primarily prioritizing 'reliability'. Similarly, in 1992, the 70 nm MOSFET utilized a 5 nm thick gate SiO_2. Although the t_{ox} remained around 5 nm as predicted by *C.* Mead, they successfully operated MOSFETs with L_g less than 0.25 μm, down to 100–70 nm, by optimizing the impurity distribution design in the channel and source/drain edge. This achievement was possible even though the supply voltage, V_d, remained higher than the values expected by scaling, as explained earlier, since their focus was on achieving higher performance under a higher supply voltage.

To achieve MOSFETs with even smaller gate lengths, it became necessary to further reduce t_{ox} and x_j. However, uncertainty loomed over whether a thinner gate oxide would ensure good yield and reliability. Additionally, techniques to realize sub-20 nm-deep source and drain regions with sufficiently low sheet resistance applicable to MOSFETs were lacking. Consequently, in the early 1990s, the practical limit for downsizing was believed to be in the range of 100–70 nm, as indicated by the dotted line in Figure 1.34. This limitation was also suggested by NTRS (National Technology Roadmap for Semiconductors), which listed the minimum gate length as 70 nm [72, 73].

Over time, the quality and reliability of thin SiO_2 gate insulators improved. However, there was a widely held belief that 3 nm represented a hard limit for gate insulator thinning due to a sudden and substantial increase in leakage current, caused by 'direct-tunneling', as shown in Figure 1.35 [4]. Below a thickness of 3 nm, the electron's wave function in the gate electrode penetrates the potential barrier of the gate insulator and appears in the Si substrate. As a result, a SiO_2 film thinner than 3 nm was no longer considered an insulator but rather a conductor.

Efforts to surpass the 100–70 nm limitation and create sub-50 nm gate insulator MOSFETs have persisted. Scaling from the IBM 1 μm gate length MOSFET with t_{ox} = 35 nm and x_j = 0.2 μm, the corresponding parameters for a 50 nm gate length MOSFET would be t_{ox} = 17.5 nm and x_j = 10 nm. While reducing t_{ox} below 3 nm is not feasible, a special technique was developed to achieve a 10 nm-deep source and drain with sufficiently low resistance [74]. This technique entailed solid-phase diffusion of P (phosphorus) from a heavily P-doped PSG film using high-temperature RTA (rapid thermal annealing), obtaining a sheet resistance of 6 kΩ/sq and a junction depth of 12 nm. In 1993, by adopting the scaling scheme described above, M. Ono et al. from Toshiba successfully demonstrated the high-performance normal operation of a MOSFET at room temperature with the suppression of punch-through and short-channel effects [74]. This marked a significant achievement, as it was the first successful realization of a sub-50 nm gate length MOSFET, surpassing the expected limit of 100–70 nm depicted in Figure 1.34 [4].

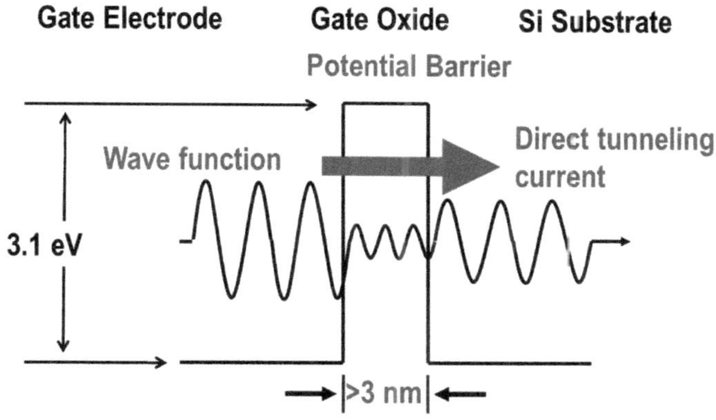

Figure 1.35 Direct-tunneling leakage current through the gate insulator [4], copyright 2023 JSAP.

In the realm of deep sub-50 nm gate length, reducing the SiO_2 thickness to less than 3 nm was considered crucial, as shown in the new scaling scheme depicted in Figure 1.36 [4, 56]. Consequently, the adoption of high-k (dielectric constant) materials became necessary. For instance, when using an insulator with $k = 16$, the capacitance value of a 12 nm thick high-k MOS capacitor would be equivalent to that of a 3 nm thick SiO_2 capacitor. Therefore, achieving nearly identical MOSFET electrical characteristics would be feasible under the condition that the interface between the Si and high-k gate insulator is as good as that of SiO_2, as illustrated in Figure 1.37 [4]. The greater physical thickness of the high-k film results in nearly complete suppression of direct tunneling current.

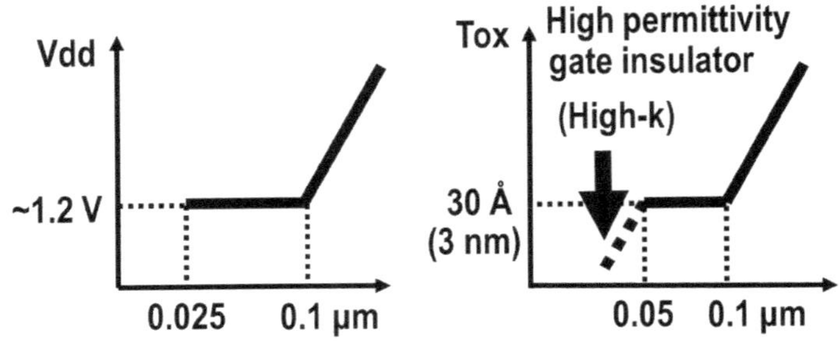

Figure 1.36 Trend of supply voltage and gate insulator thickness predicted by the new scaling scheme proposed in 1993 using Monte Carlo simulation [4], copyright 2023 JSAP.

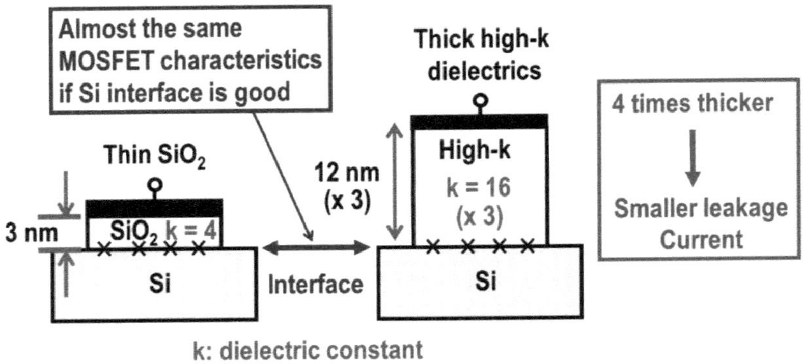

Figure 1.37 Concept of high-k gate insulator for MOSFET [4], copyright 2023 JSAP.

In 1993, there was limited understanding of which high-k materials were genuinely suitable for Si LSIs, and there was a lack of significant technological advancements in high-k/metal gate stacks targeted for their integration into products, although important pioneering works on high-k gate insulator Si-MOSFETs, such as those involving ZrO_2, had already been published in the late 1960s [75, 76]. The new scaling scheme provided a roadmap for incorporating high-k materials into sub-50 nm CMOS, and this milestone was eventually achieved in 2007 with the 45 nm logic CMOS, exactly 12 years later [77].

In the early 1990s, the study of high-k Si-MOSFETs was in its nascent stages, and no semiconductor company had achieved success in implementing high-k gate insulators. Consequently, it appeared unlikely that the gate insulator thickness could be further reduced below 3 nm, restricting the potential for further decreasing the gate length. However, even though the endeavor was challenging and almost reckless, the possibility of using ultra-thin SiO_2 films less than 3 nm was carefully investigated, despite the significant direct-tunneling current concerns.

From the experience of thin gate insulator MOSFET experiments, it was expected that the gate insulator leakage current (I_g) would become significantly smaller compared to the drain current (I_d) with decreasing gate length (L_g). This was because the gate area, and hence I_g, decreased proportionally to L_g, while I_d increased inversely proportional to L_g. In 1994, experiments were conducted by H. Momose et al. of Toshiba using MOSFETs with a 1.5 nm SiO_2 gate insulator [78]. The ultra-thin SiO_2 film was grown using Rapid Thermal Oxidation (RTO). The uniformity of the ultra-thin SiO_2 film was ensured by the strong oxidation of Si at high RTO temperatures during the early stages of oxidation, preventing local thinning, defects, or variations in the SiO_2 film. The standard deviation (σ) of the SiO_2 thickness across the entire wafer, evaluated through the gate leakage of $100 \times 110\ \mu m^2$ MOS capacitors, was only 0.008 nm (0.53%) [79].

Better-than-expected normal I_d-V_d characteristics were obtained with $L_g \leq 1\ \mu m$ because I_g became negligible compared to I_d. By properly decreasing the gate insulator thickness according to the scaling scheme, the reduction of the junction depth was relaxed to 30 nm with a sheet resistance of 1.4 kΩ/sq. Due to the ultra-thin gate insulator and lower source/drain resistance, a world-record transconductance value of 1010 mS/mm at V_d = 1.5 V and L_g = 90 nm was demonstrated [78].

After confirming their manufacturability and reliability, ultra-thin gate oxides in the direct-tunneling regime, ranging from 3 to 1 nm, started to be widely adopted in products. This breakthrough extended the limit of gate length reduction to 50 nm at the product level. The success of MOSFETs with a 1.5 nm SiO_2 gate insulator had a profound impact on the International Technology Roadmap for Semiconductors (ITRS), as illustrated in Figure 1.38 [4].

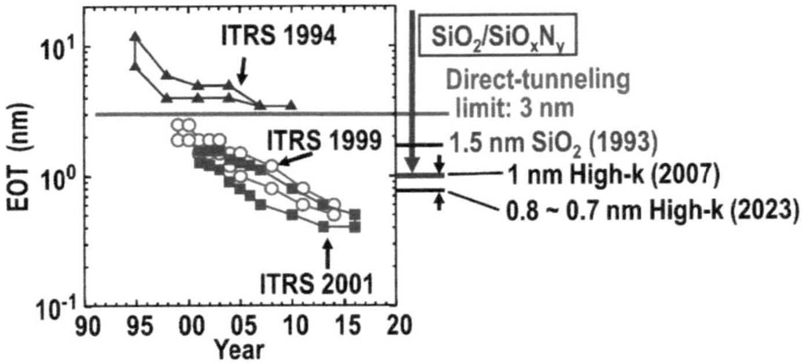

Figure 1.38 ITRS for the gate insulator [4], copyright 2023 JSAP.

The total gate leakage current of the direct-tunneling gate oxide in a chip cannot be ignored when the gate SiO_2 (or SiO_xN_y) thickness reaches 1.0 nm. Then, it took nearly 10 years until HfO_2 high-k/metal gate stack MOSFETs were introduced to 45 nm CMOS products by Intel in 2007, as mentioned earlier. However, according to IRDS 2020, the EOT value is constrained to approximately 0.7 nm due to long-term reliability concerns such as positive bias temperature instability (PBTI) and negative bias temperature instability (NBTI), as depicted in Figure 1.38 [4].

1.8 Miniaturization Limit by Other Reasons

In the previous subsection, we discussed the limitations of gate length reduction. It involves a trade-off between punch-through current and subthreshold leakage current in relation to the supply voltage. When maintaining a high drain voltage, subthreshold leakage can be suppressed by keeping the threshold voltage high, but it results in an increase in punch-through current, and vice versa. Lowering drain voltage is desirable for reducing power consumption, but it leads to subthreshold leakage becoming the limiting factor for gate length reduction. The thinning of the gate oxide also helps in suppressing subthreshold leakage. However, the thinning of the equivalent oxide thickness (EOT) is expected to be limited to approximately 0.7 nm due to long-term reliability concerns, as mentioned before and depicted in Figure 1.37. Taking these factors into consideration, IEEE IRDS estimates that the limit for gate length reduction is around 10 nm.

What are some other potential factors that could limit further miniaturization? Until approximately 10 years ago, there was a belief that escalating production costs, driven by expensive equipment like EUV lithography and the growing number of process steps, would hinder further downsizing. However, so far, these factors have not significantly limited progress, as the market has consistently demanded VLSIs with high performance and low power consumption. Nevertheless, in the long term, the continuous increase in process steps due to 3D integration may present challenges to the advancement of VLSI technology.

When considering photolithography, historical trends in resolution have demonstrated that resolution enhancement techniques can resolve features down to approximately one-tenth of the light wavelength used, as depicted in Figure 1.39 [4]. However, some distortions may occur in fine patterns. With the introduction of extreme ultraviolet (EUV) lithography, which provides a wavelength of 13.5 nm, it is anticipated that sub-10 nm patterns will become achievable through further technological advancements.

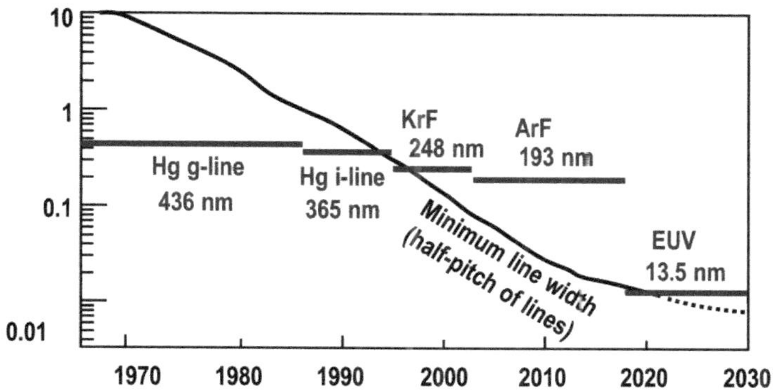

Figure 1.39 Trend of photolithography wavelength and resolution [4], copyright 2023 JSAP.

Regarding the resistance increase, significant improvements have been achieved in the resistance of narrow lines and long-term reliability through the use of new materials, as illustrated in Figure 1.40 [4]. The ongoing development of new materials will persist into the next generation to tackle these challenges.

The anticipated limit for narrow line widths is expected to be around 10 nm or even sub-10 nm.

Interconnects
Al → Al-Si → Al-Si-Cu→ Al-Si-Cu/TiN → Cu/TaN → Cu, Co, Ru?

Gate electrode
Al → Si → WSix/Si → $TiSi_2$/Si → $CoSi_2$/Si → NiSi/Si → W/TiN → ?

Gate insulator
SiO_2 → SiO_xN_y → HfO_2/SiO_2 → HfO_2/SiO_2 ?

Source/Drain
Si → $TiSi_2$/Si → $CoSi_2$/Si → NiSi/Si → Si ?

Channel
Si → Si ?

Figure 1.40 Trend of materials used for MOS LSI [4], copyright 2023 JSAP.

1.9 Future Prediction by IRDS 2022

According to IRDS 2022, it is projected that by 2028, the gate length and metal pitch will reach their limits at 12 nm and 16 nm, respectively, as depicted in Figure 1.41 [4]. The supply voltage is also expected to reach its limit at 0.6 V. The line width and supply voltage are likely to reach their limits in the near future. However, despite this, the pursuit of these limits remains important due to the uncertainty of the future. It is expected that the performance of individual MOSFETs (on-current, transconductance, CV/I, CV^2, f_T) will not experience significant improvements once downsizing reaches its limit (Figure 1.41). Similarly, the clock frequency is projected to plateau at 3–5 GHz.

When the device miniaturization approaches its limits, the adoption of 3D integration for MOSFETs (Figure 1.42 [4]) is expected to commence to uphold Moore's law. The main difficulties in advancing 3D integration are the continual rise in mask count and heat density (Figure 1.41 [4]).

Memory devices have already reached a milestone with the availability of a 100-layer 3D NAND flash memory in the market, as depicted in Figures 1.42 and 1.43 [4]. At the chip level, the integration of 3D chiplets has already begun (Figure 1.44), and the integration of chiplets with CAD-based design optimization is set to play a pivotal role in the future.

	2022	2025	2028	2031	2034	2037
Commercial Name (nm)	3	2	1.5	1.0	0.7	0.5
L_g (nm)	16	14	12	12	12	12
CET (nm)	1.0	0.9	0.90	0.9	0.9	0.9
M0 pitch (nm)	22	20	16	16	16	16
Mx pitch (nm)	32	24	20	16	16	16
# of tires	1	1	1	2	4	4
# of stacked NS	1	3	3	4	4	4
# of wiring layers	17	17	18	19	19	19
Footprint area scaling	1	0.74	0.55	0.26	0.13	0.08
# of CPU cores	12	20	33	55	92	155
# of GPU cores	36	49	66	136	286	477
Mask count	77	77	79	130	228	326
V_d (V)	0.7	0.65	0.6	0.6	0.6	0.6
Clock frequency (GHz)	3.18	3.28	3.36	3.42	3.47	3.57
Power density scaling	1	1.03	1.20	2.29	4.85	7.99

	2022	2025	2028	2031	2034	2037
Ion (μA/um)	874	787	851	763	737	753
Vt at Ioff = 10 nA/um (mV)	156	165	165	164	156	154
Subthreshold slope (mV/dec)	72	70	70	70	70	70
Transconductance (μS/um)	1605	1621	1751	1725	1653	1684
Effective mobility ($cm^2V^{-1}sec^{-1}$)	156	165	165	164	156	154
Ballistic Injection Velocity (cm/sec)	9E6	9E6	9E6	9E6	9E6	9E&
Rsd (Ohm/μm)	271	257	245	232	221	210
Mx resistance (Ω/μm)	300	475	920	1450	1450	1450
CV/I (ps)	1.06	0.96	0.84	0.88	0.90	0.88
CV^2 (fJ/switch) FO3 load	0.65	0.49	0.47	0.40	0.40	0.40
fT (GHz)	261	304	358	411	403	411

Figure 1.41 IRDS 2022 [4], copyright 2023 JSAP.

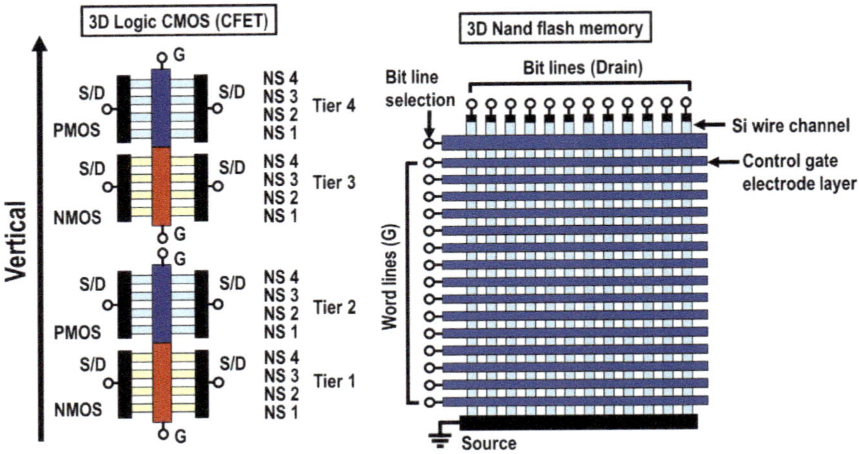

Figure 1.42 3D integration of MOSFETs (left) and NAND flash memory (right) [4], copyright 2023 JSAP.

Over the next 10–20 years, 3D integration is anticipated to encounter limitations arising from the increasing number of process steps, associated costs, and heat generation. Nonetheless, LSI remains a crucial component of our super-intelligent world, and technology optimization for each new application will continue even after reaching the limits of 3D integration.

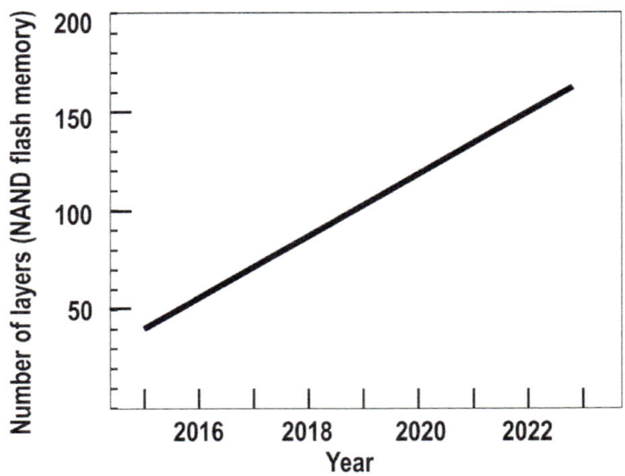

Figure 1.43 Trend of a number of layers for NAND flash memories [4], copyright 2023 JSAP.

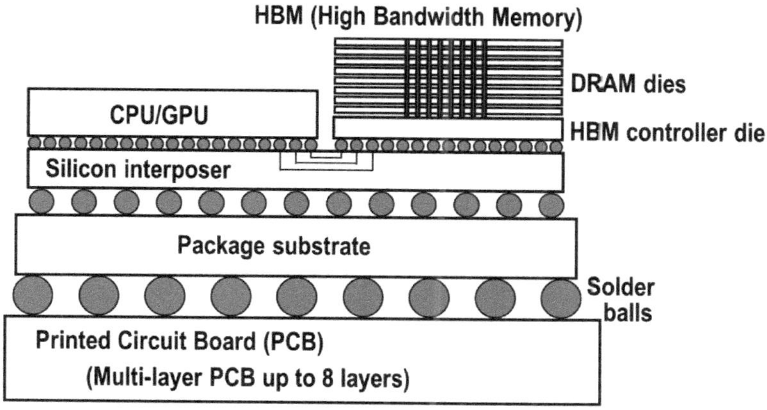

Figure 1.44 Example of 2D and 3D integration of chiplets [4], copyright 2023 JSAP.

1.10 Long-Term Future

In the long term, the integration of biological systems will be essential [4]. Neurons, in general, offer superior performance compared to MOSFETs. The most advanced MOSFET circuits currently operate at frequencies of 3–5 GHz with an internal supply voltage ranging from 0.7 to 0.6 V. However, due to the increase in subthreshold leakage and the degradation of the on/off ratio, the supply voltage cannot be reduced below 0.6 V. When billions of MOSFETs operate at 3 GHz on a single chip, they consume a significant amount of power. An artificial intelligence (AI) machine composed of MOSFETs consumes more than 100 kW of power for learning depending on the requirements.

Meanwhile, neurons operate at an impressively low voltage of less than 100 mV and a frequency below 1 kHz, leading to exceedingly low power consumption. The human brain's estimated power consumption is around 30 W. When a synapse receives an electrical pulse, it becomes activated and releases chemicals that trigger an electrical pulse in the next neuron. Consequently, when the synapse is deactivated, there is no leakage current between the two neurons, unlike the constant subthreshold leakage current observed in MOSFETs.

It is worth noting that a single neuron can establish connections with over 1000 other neurons through synapses, whereas a MOSFET can only connect to a few other MOSFETs. Neurons possess true 3D interconnects and a significantly greater degree of freedom in establishing connections with each other compared to MOSFETs, which rely on stacked layers of 2D interconnects. As a consequence, neurons are capable of creating much more intricate circuits than those achievable with MOSFETs. Additionally, MOSFETs face limitations in connecting to 1000 other MOSFETs due to the need for an extensive interconnect area, making it nearly impossible to drive the immense capacitance of these interconnects.

As a result, neuronal circuit systems offer significantly higher performance than that of MOSFETs. However, one might question why the human brain cannot consistently outperform AI in games such as 'go, chess, and 'shogi'. The reason lies in the fact that the human brain is not inherently designed to sustain prolonged focus on a single task for extended periods. The brain loses interest when engaged in the same activities for long durations, leading to fatigue and drowsiness. The human brain requires rest and sleep. Moreover, it cannot accurately retain vast amounts of data over extended periods, unlike AI systems.

On the other hand, AI based on semiconductor devices can tirelessly work for 24 hours and possess highly accurate memory. However, it consumes a significant amount of energy for a single task, which limits its ability to perform tasks that the human brain can effortlessly handle. Therefore, the ideal solution lies in the collaboration between AI and the brain. To enhance the reliability of the biological system's operation, we must develop direct communication between semiconductors and the brain. This implies that in the future, we may be able to directly communicate with the brains of animals such as dogs and cats. This is not a science fiction scenario! With machine learning that compares real images with brain blood flow, we are already capable of displaying images that humans see in their minds. In the long-term future, we might no longer need visual displays or audio speakers to watch movies if the data can be directly transmitted to our brains without the use of our eyes and ears with the progress of interface technologies. It might even be possible to transmit smell and taste directly to our brains.

In the future, birds, insects, rats, and fish could potentially be employed for surveillance and maintenance of tunnels, bridges, buildings, and more, utilizing their aerial, surface, terrestrial, and aquatic capabilities. They may prove to be more efficient, safe, and long-lasting compared to drones or robots. It is undeniable that machines cannot match dogs in tasks such as drug-sniffing, hunting, sledding, policing, rescuing, guiding, herding, and serving as pets and companions. Throughout history, animals have been utilized for food, labor, and companionship. As a society, we need a consensus on how to ethically treat animals and insects. Additionally, we must exercise great caution regarding AI's ability to control human minds and emotions through direct brain connections.

In the future, we may also have the ability to synthesize a simple artificial brain using neurons. Simultaneously, we should continue improving semiconductor devices by drawing inspiration from biological systems, such as implementing biological algorithms and replicating physical structures found in living organisms.

1.11 Conclusions

Rapid progress in ICT began in the 1830s with the invention of the electric telegraph. The demands of telecommunication have been the driving force behind the development of electric and electronic devices, including semiconductor devices.

The introduction of semiconductor devices, particularly transistors, marked the emergence of microelectronics, which eventually evolved into nanoelectronics. At the time of the transistor's invention by J. Lilienfeld in 1925, nobody could have predicted the commercialization of devices as small as 10 nm. The invention of the transistor laid the groundwork for today's nanoelectronics, enabling the development of a highly advanced society reliant on artificial intelligence (AI).

In the 18th and 19th centuries, mechanics played a pivotal role in information and communication technology (ICT). However, in the 20th and 21st centuries, a combination of electronics and mechanics emerged as the dominant technology. Looking ahead, the 22nd century and beyond will witness the convergence of bionics, electronics, and mechanics as the leading technologies. The future holds immense potential for the development of groundbreaking technologies.

References

1. H. Iwai, "Invention of the Transistor 75 Years Ago; The Origin of Device Miniaturization Towards Super-Intelligent Society," *IEEE EDS Newsletter*, vol. 29, no. 2, pp. 3–5, April 2022.
2. H. Iwai, "History of Transistor Invention: 75th Anniversary", ICSICT, October 2022.
3. H. Iwai and M. Durga, "The Transistor was Invented 75 Years Ago: A Big Milestone in Human History," Electrochemical Society, *Interface*, vol. 31, pp. 65–72, Winter 2022.
4. H. Iwai, "History of Junction Technologies, Commemorative Talk for the 75th Anniversary of the Transistor", *International Workshop on Junction Technology (IWJT)*, pp. 1–77, June 2023.
5. H. Iwai, "My LSI Development Experience at Toshiba (1973–1999), Hard Working, Exciting, and Splendid Days," *IEEE EDS Magazine*, vol. 1, no. 1, pp. 41–57, 2023.
6. H. Iwai, "The First MOSFET Design by J. Lilienfeld and its Long Journey to Implementation", pp. 65–88, in 75th Anniversary of the Transistor, Eds. A. Nathan, S. Samar, R. Todi, Willey-IEEE Press, 2023.
7. Samuel F. B. Morse, "Improvement in the Mode of Communication by Signals by the Application of Electro-Magnetism," US Patent 1647, filed April 7, 1938, issued June 20, 1840; Reissue No. 79, January 15, 1846; Reissue No. 117, dated June 13, 1847.
8. E. Branly, "Variations de conductibilité sous diverses influences électriques," *C. R. Acad. Sci. Paris*, vol. 111, pp. 785–787, 1890.
9. O. J. Lodge, "The History of Coherer Principle," Electrician, November 2, 1897.
10. Guglielmo Marconi, "Transmitting Electrical Signal," US Patent 586,193, filed July 13, 1897.
11. F. Braun, "Über die Stromleitung durch Schwefelmetalic (About the power line through sulfur metallic)", *Annalen der Physik and Chemie*, vol. 153, no. 4, pp. 556–563, 1874.
12. Arthur Schuster, "On Unilateral Conductivity," *The London, Edinburgh, and Dublin Philosophical Magazine and Journal of Science*, vol. 48, no. 318, pp. 251–258, 1874.

13. Jagadish C. Bose, "Detector for Electrical Disturbances," US Patent 755,840, filed September 30, 1901.

14. J. H. Scaff and R. S. Ohl, "Development of Silicon Crystal Rectifiers for Microwave Radar Receivers," *Bell System Technical Journal*, vol. 26, no. 1, pp. 1–30, 1947.

15. Lee de Forest, "Space Telegraphy," US Patent 879,532, filed January 29, 1907.

16. Julius E. Lilienfeld, "Method and Apparatus for Controlling Electric Current", Canadian Patent CA272437TA, filed October 22, 1925.

17. Julius E. Lilienfeld, "Device for Controlling Electric Current," US Patent 1900018, filed March 28, 1928.

18. A. H. Wilson, "The Theory of Electronic Semiconductors I," *Proceedings of the Royal Society of London*, vol. A133, pp. 458–491, October 1931.

19. A. H. Wilson, "The Theory of Electronic Semiconductors II," *Proceedings of the Royal Society of London*, vol. A134, pp. 277–287, November 1931.

20. W. Schottky and E. Spenke, "Zur quantitativen Durchführung der Raumladungs-und Randschichttheorie der Kristallgleichrichter (Quantitative treatment of the space charge and boundary-layer theory of the crystal rectifier)," *Wissenschaftliche Veröffentlichungen aus den Siemens-Werken*, vol. 18, pp. 225–291, 1939.

21. Heinrich J. Welker, "Halbleiteranordnung zur kapazitiven Steuerung von Strömen in einem Halbleiterkristall (Semiconductor arrangement for capacitive control of currents in a semiconductor crystal)," Deutshe Patentamt DE 980084, filed June 4, 1945, priority April 7, 1945.

22. Herbert Weiß, "Steuerung von Elektronenströmen im Festkörper (Teil I) (Controlling Electron Currents in Solids (Part I))," *Physikalische Blätter*, vol. 31, no. 4. pp. 156–165, April 1975.

23. Heinrich J. Welker, "From Solid State Research to Semi-Conductor Electronics," *Annual Review of Materials Research*, vol. 9, pp. 1–22, 1979.

24. M. Riordan, "How Europe Missed the Transistor," *IEEE Spectrum*, vol. 42, no. 11, pp. 52–57, Nov. 2005.

25. W. Shockley, "The Path to the Conception of the Junction Transistor," in *IEEE Transactions on Electron Devices*, vol. 23, no. 7, pp. 597–620, July 1976.

26. John Bardeen and Walter H. Brattain, "Three-Electrode Circuit Element Utilizing Semiconductive Materials," US Patent 2524034 field on June 17, 1948 (a continuation-in-part of application, filed February 26, 1948).
27. W. Shockley, "Circuit Elements Utilizing Semiconductor Material," US Patent, 2,569,347, filled June 26, 1948.
28. W. Shockley, M. Sparks, Morgan, and G. Teal, "p-n Junction Transistors," *Physical Review*, vol. 83, no. 1, pp.151–162, July 1951.
29. W. Shockley and G. L. Pearson, "Modulation of Conductance of Thin Films of Semi-Conductors by Surface Charges," *Physical Review*, vol. 74, p. 232–233, July 15, 1948.
30. W. Shockley, *Electrons and Holes in Semiconductors: With Applications to Transistor Electronics*, van Nostrand, 1950.
31. W. L. Brown and W. Shockley, "N-type Conduction on p-Type Germanium Surfaces," *Physical Review*, vol. 90, no. 2, pp. 336–336, 1953.
32. W. L. Brown, "N-Type Surface Conductivity on p-Type Germanium," *Physical Review*, vol. 91, no. 3, p. 518, August 1953.
33. W. Shockley, "A Unipolar Field-Effect" Transistor," *Proc. IRE*, vol. 40, pp. 1365–1376, 1952.
34. Martin M. Atalla, Edwin J. Scheibner, and Eileen Tannenbaum, "Fabrication of Semiconductor Devices Having Stable Surface Characteristics," US patent 2,899,344, filed April 30, 1958.
35. Dawon Kahng, "Electric Field Controlled Semiconductor Device," US Patent 3,102,230, filed May 31, 1960.
36. Martin M. Attala," Semiconductor Triode," US Patent 3,056,888, filed August 17, 1960.
37. D. Kahng, "Silicon-Silicon Dioxide Field Induced Surface Devices," Technical memorandum issued by Bell Labs., January 16, 1961 reprinted in S. M. Sze, *Semiconductor Devices: Pioneering Papers*, World Scientific Publishing Co., pp. 583–596, 1991.
38. Joseph R. Ligenza, "Method of Treating Silicon," US patent 2930722, filed February 3, 1959.
39. B. Lojek, *History of Semiconductor Engineering*, Springer, 2007.
40. Russell S. Ohl, "Alternating Current Rectifier," US Patent 2402661, filed March 1, 1941.
41. Russell S. Ohl, "Light-Sensitive Electric Device," US Patent 2402662, filed May 27, 1941.

42. Russell S. Ohl, "Thermoelectric Device," US Patent 2,402,663, filed April 11, 1942.

43. Jack S. Kilby, "Miniaturized Electronic Circuits," US Patent 3,138,743, filed February 6, 1959.

44. Robert N. Noyce, "Semiconductor Device-and-Lead Structure," US Patent 2981877, filed July 30, 1959.

45. B. E. Deal, "Standardized Terminology for Oxide Charges Associated with Thermally Oxidized Silicon," *IEEE Transactions on Electron Devices*, vol. 27, no. 3, pp. 606–608, March 1980.

46. Ross Knox Bassett, "To the Digital Age, Research Labs, Start-Up Companies and the Rise of MOS Technology," Johns Hopkins Studies in the History of Technology, Johns Hopkins University Press, 2002.

47. Gilles Holst and Bernardus Dominicus Hubertus Tellegen, "Means for Amplifying Electrical Oscillations," US Patent 1945040, filed November 25, 1927.

48. R. H. Dennard, F. H. Gaensslen, V. L. Rideout, E. Bassous, and A. R. LeBlanc, "Design of Ion-Implanted MOSFET's with Very Small Physical Dimensions," *IEEE J. Solid-State Circuits* vol. 9, no. 5, pp. 256–268, October 1974.

49. H. Iwai and S. Ohmi, " Silicon Integrated Circuit Technology from Past to Future", *Microelectronics Reliability*, vol. 42, no. 4-5, pp. 465–491, April–May 2002.

50. P. P. Gelsinger, "Microprocessor for the New Millennium: Challenges, Opportunities, and New Frontiers," *ISSCC*, pp. 22–23, February 2001.

51. G. E. Moore, "Cramming More Components onto Integrated Circuits," *Electronics*, vol. 38, no. 8, pp. 114 April 19, 1965.

52. C. F. Codella and S. Ogura, "Halo Doping Effects in Submicron DI-LDD Device Design," *IEDM Tech. Dig.*, pp. 230–233, 1985.

53. M. Ono, M. Saito, T. Yoshitomi, C. Fiegna, T. Ohguro, H. S. Momose, and H. Iwai, "Influence of High Substrate Doping Concentration on the Hot-Carrier and Other Characteristics of Small-Geometry CMOS Transistors Down to the 0.1 μm Generation', *Dig. Tech. Symp. on VLSI Technology*, Honolulu, Hawaii, USA, pp. 147–148, June 1994.

54. T. Sekigawa and Y. Hayashi," Calculated Threshold-Voltage Characteristics of an XMOS Transistor Having and Additional Bottom Gate," *Solid State Electronics*, vol. 27, no. 8/9, pp. 827–828, 1984.

55. D. J. Frank, S. E. Laux, and M. V. Fischetti, "Monte Carlo Simulation of a 30 nm Dual-Gate MOSFET: How Short Can Si Go?," *International*

Technical Digest on Electron Devices Meeting, pp. 553–556, December 1992, San Francisco, CA, USA.

56. C. Fiegna, H. Iwai, T. Wada, T. Saito, E. Sangiorgi, and B. Riccò, "A New Scaling Methodology for the 0.1–0.025 μm MOSFET," *Dig. Tech. Symp. on VLSI Technology*, pp. 33–34, May 1993, Kyoto Japan.
57. D. Hisamoto, T. Kaga, Y. Kawamoto, and E. Takeda, "A Fully Depleted Lean-Channel Transistor (DELTA): A Novel Vertical Ultra-Thin SOI MOSFET," *Tec. Dig. IEDM*, pp. 833–836, 1989, Washington, DC, USA.
58. D. Hisamoto, W. C. Lee, J. Kedzierski, H. Takeuchi, K. Asano, C. Kuo, E. Anderson, T.-J. King, J. Bokor, and C. Hu, "FinFET: A Self-Aligned Double-Gate MOSFET Scalable to 20 nm," *IEEE Transactions on Electron Devices*, vol. 47, no. 12, pp. 2320–2325, December 2000.
59. C. H. Lee, J. C. Yoon. C. Lee, H. M. Yang, K. N. Kim, T. Y. Kim, H. S. Kang, Y. J. Ahn, Donggun Park, and Kinam Kim, "Novel Body Tied FinFET Cell Array Transistor DRAM with Negative Word Line Operation for Sub-60 nm Technology and Beyond," *Dig. Tech. Symp. on VLSI Technology*, pp. 130–131, 2004, Honolulu, HI, USA.
60. Mark Bohr and Kaizad Mistry, "Intel's Revolutionary 22 nm Transistor Technology," May 2011. Available on line: https://download.intel.com/newsroom/kits/22nm/pdfs/22nm-Details_Presentation.pdf.
61. Sung-Min Kim, Dong-Gun Park, Chang-Sub Lee, Jeong-Dong Choe, Shin-Ae Lee, Seong-Ho Kim, "Field effect transistors having multiple stacked channels" US Patent US 7002207, field July 1, 2003, with priority October 1, 2002.
62. S.-M. Kim, S.-Y. Lee, C.-W. Oh, E.-J. Yoon, C.-S. Lee, J.-D. Choe, H.-J. Jo, S.-A. Lee, K.-C. Kim, B.-M. Yoon, H.-S. Lee, D. Park, and K. Kim, "A Novel MBC (Multi-Bridge-Channel) MOSFET: Fabrication Technology and Characteristics," in *Proc. Si Nanoelectronics Workshop*, p. 18, 2003.
63. Sung-Young Lee, Sung-Min Kim, Eun-Jung Yoon, Chang Woo Oh, Ilsub Chung, Donggun Park, and Kinam Kim, "Threedimensional MBCFET as an Ultimate Transistor," in *IEEE Electron Device Letters*, vol. 25, no. 4, pp. 217–219, April 2004.
64. T. Ernst, C. Dupre, C. Isheden, E. Bernard, R. Ritzenthaler, V. Maffini-Alvaro, J.-C. Barbe, F. De Crecy, A. Toffoli, C. Vizioz, S. Borel, F. Andrieu, V. Delaye, D. Lafond, G Rabille, J.-M. Hartmann, M. Rivoire, B. Guillaumot, A. Suhm, P. Rivallin, O. Faynot, G Ghibaudo, and S. Deleonibus, "Novel 3D Integration Process for Highly Scalable Nano-Beam Stacked-Channels GAA (NBG) FinFETs with HfO_2/TiN

Gate Stack," *IEDM Tech. Dig.*, pp. 1–4, December 2006, San Francisco, CA, USA.

65. S. Barraud, B. Previtali, C. Vizioz, J. M. Hartmann, J. Sturm, J. Lassarre, C. Perrot, P. Rodriguez, V. Loup, A. Magalhaes-Lucas, R. Kies, G. Romano, M. Cassé, N. Bernier, A. Jannaud, A. Grenier, and F. Andrieu, "7-Levels-Stacked Nanosheet GAA Transistors for High Performance Computing," *2020 IEEE Symposium on VLSI Technology*, pp. 1–2, June 2020, Honolulu, HI, USA.

66. C.-Y. Huang G. Dewey, E. Mannebach, A. Phan, P. Morrow, W. Rachmady, I.-C. Tung, N. Thomas, U. Alaan, R. Paul, N. Kabir, B. Krist, A. Oni, M. Mehta, M. Harper, P. Nguyen, R. Keech, S. Vishwanath, K. L. Cheong, J. S. Kang, A. Lilak, M. Metz, S. Clendenning, B. Turkot, R. Schenker, H. J. Yoo, M. Radosavljevic, and J. Kavalieros, "3-D Self-Aligned Stacked NMOS-on-PMOS Nanoribbon Transistors for Continued Moore's Law Scaling," *IEDM Tech. Dig.*, pp. 425–428, December 2020, San Francisco, CA, USA.

67. T. Morimoto, H. S. Momose, K. Yamabe, and H. Iwai, "Prevention of Boron Penetration from p^+ Poly Gate by RTN Produced Thin Gate Oxide," *ESSDERC*, pp. 73–76, September 1990, Nottingham, UK.

68. B. Hoeneisen and C. A. Mead, "Fundamental Limitations in Microelectronics – I. MOS Technology," *Solid-State Electronics*, vol. 15, no. 7, pp. 819–829, 1972.

69. C. Mead and L. Conway, *Introduction to VLSI Systems*, Addison-Wesley, USA, p. 37, 1979.

70. G. A. Sai-Halasz, M. R. Wordeman, D. P. Kern, E. Ganin, S. Rishton, D. S. Zicherman, H. Schmid, M. R. Polcari, H. Y. Ng, P. J. Restle, T. H. P. Chang, and R. H. Dennard, "Design and Experimental Technology for 0.1-μm Gate Length Low-Temperature Operation FET's," *IEEE Electron Device Lett.*, vol. 8, no. 10, pp. 463–466, October 1987.

71. T. Hashimoto, Y. Sudoh, H. Kurino, A. Narai, S. Yokoyama, Y. Horiike, and M. Koyanagi, "3 V Operation of 70 nm Gate Length MOSFET with New Double Punch Through Stopper Structure," *Ext. Abs. Int. Conf. SSDM*, pp. 490–492, August 1992, Tsukuba, Japan.

72. A. C. Diebold, "Overview of Metrology Requirements Based on the 1994 National Technology Roadmap for Semiconductors," *Proceedings of SEMI Advanced Semiconductor Manufacturing Conference and Workshop*, pp. 50–60, 1995, Cambridge, MA, USA.

73. W. J. Spencer and T. E. Seidel, "National Technology Roadmaps: The U.S. Semiconductor Experience," *Proceedings of 4th International Conference on Solid-State and IC Technology*, pp. 211–220, 1995, Beijing, China.

74. M. Ono, M. Saito, T. Yoshitomi, C. Fiegna, T. Ohguro, and H. Iwai, "Sub-50 nm Gate Length N-MOSFETs with 10 nm Phosphorus Source and Drain Junction," *IEDM Tech. Dig.* pp. 119–122, December 1993, Washington DC, USA.
75. T. Sugano and K. Hoh, "Interfacial Carrier Mobility of ZrO_2/Si Structure and n-Channel Enhancement FET Using This", *Annual Report of Engineering, Faculty of Engineering, University of Tokyo*, vol. 29, pp. 137–141, 1970 (in Japanese).
76. K. Hoh and T. Sugano, "Low Temperature Vapor Deposition of ZrO_2 Films on Si Surface," *Proc. Int. Conf. SSD, 1969, Suppl. JJAP.*, vol. 39, pp. 177–184, 1970.
77. K. Mistry, C. Allen, C. Auth, B. Beattie, D. Bergstrom, M. Bost, M. Brazier, M. Buehler, A. Cappellani, R. Chau, C.-H. Choi, G. Ding, K. Fischer, T. Ghani, R. Grover, W. Han, D. Hanken, M. Hattendorf, J. He, J. Hicks, R. Huessner, D. Ingerly, P. Jain, R. James, L. Jong, S. Joshi, C. Kenyon, K. Kuhn, K. Lee, H. Liu, J. Maiz, B. McIntyre, P.Moon, J. Neirynck, S. Pae, C. Parker, D. Parsons, C. Prasad, L. Pipes, M. Prince, P. Ranade, T. Reynolds, J. Sandford, L. Shifren, J. Sebastian, J. Seiple, D. Simon, S. Sivakumar, P. Smith, C. Thomas, T. Troeger, P. Vandervoorn, S. Williams, and K. Zawadzki, "A 45 nm Logic Technology with High-k + Metal Gate Transistors, Strained Silicon, 9 Cu Interconnect Layers, 193 nm Dry Patterning, and 100% Pb-free Packaging," *IEDM Tech. Dig*, pp. 248–251, December 2007, Washington DC, USA.
78. H. S. Momose, M. Ono, T. Yoshitomi, T. Ohguro, S. Nakamura, M. Saito, and H. Iwai, "Tunneling Gate Oxide Approach to Ultra-High Current Drive in Small-Geometry MOSFETs", *IEDM Tech. Dig.*, pp. 593–596, December 1994, San Francisco, CA, USA.
79. H. S. Momose, S. Nakamura, T. Ohguro, Y. Katsumata, and H. Iwai, "Uniformity and Reliability of 1.5 nm Direct Tunneling Gate Oxide MOSFETs", *Dig. Tech., Symp. on VLSI Technology*, pp. 15–16, June 1997, Kyoto, Japan.

Chapter 2

Towards Life Cycle Thinking and Judicious Ecodesign for the Internet of Things (IoT) Current Practices and Perspectives

Thibault Pirson,[a] Grégoire Le Brun,[a] Ernesto Quisbert-Trujillo,[b] Thomas Ernst,[c] Jean-Pierre Raskin,[a] and David Bol[a]

[a]*Université catholique de Louvain, ICTEAM, Louvain-La-Neuve, Belgium*
[b]*Université Grenoble Alpes, CNRS/Grenoble-INP, Grenoble, France*
[c]*Université Grenoble Alpes, CEA-Leti, Grenoble, France*

Thomas.ernst@cea.fr, jean-pierre.raskin@uclouvain.be

This book chapter aims at providing a clear overview of how life cycle assessment (LCA) can be used to support judicious ecodesign in the context of the Internet of Things (IoT). The main goal is to put interested designers on concrete tracks to integrate environmental considerations in the design of IoT systems over their entire life cycle. More especially, we target questions such as (i) what tools are currently available, (ii) where to find examples based on existing works, (iii) what barriers can be encountered, and (iv) what directions can be interesting to explore? We specifically focus on LCA, as it is a quantitative methodology based on already-existing tools that can be used by engineers in the process of designing technology. Given that

Outlooking beyond Nanoelectronics and Nanosystems: Ultra Scaling, Pervasiveness, Sustainable Integration, and Biotic Cross-Inspiration
Edited by Simon Deleonibus

ISBN 978-981-5129-32-8 (Hardcover), 978-1-003-50990-5 (eBook)
www.jennystanford.com

these considerations are still far from common in the design of IoT systems, we propose a streamlined overview of current practices and perspectives to foster life cycle thinking in this field.

2.1 Context and Introduction

2.1.1 The IoT: A Wide Ecosystem with Several Layers of Abstraction

Over the last decades, the rapid development of Information and Communication Technology (ICT) has deeply shaped our modern societies in what is called the "information age" [1]. This increasing dependency on ICT infrastructure and connectivity was also highlighted during the COVID-19 pandemic [2]. More specifically, a sub-part of ICT is significantly gaining momentum both in research and industry, namely the IoT. The IoT connects "things" (physical or virtual) to the Internet by embedding electronics in everyday-like objects. This resulted in a steep growth of IoT devices over the last years, with estimates projecting more than 50–200+ billion devices in 2030 [3].

Nowadays, IoT plays a key role in digitalization as it is seen as an efficient way to massively acquire data and actuate on the environment. In Western societies, IoT solutions are undoubtedly seen as important building blocks to realize projects such as Industry 4.0, connected cities, connected agriculture, connected mobility, and connected healthcare, among others. In addition, IoT solutions are often expected to support the so-called "circular economy" by ensuring better traceability and keeping added value "within the loop". Moreover, the data generated by IoT devices is key for promising applications including real-time decentralized monitoring, resource optimization, predictive maintenance, or remote control. Data is therefore associated with a financial value and the potential economic benefits of IoT are expected to reach about $11 trillion while contributing more than 11% to the global economy by the end of 2025 [4]. Massive investments are also supporting this emerging technology, including investments in sectors of semiconductors, artificial intelligence, and mobile communication networks.

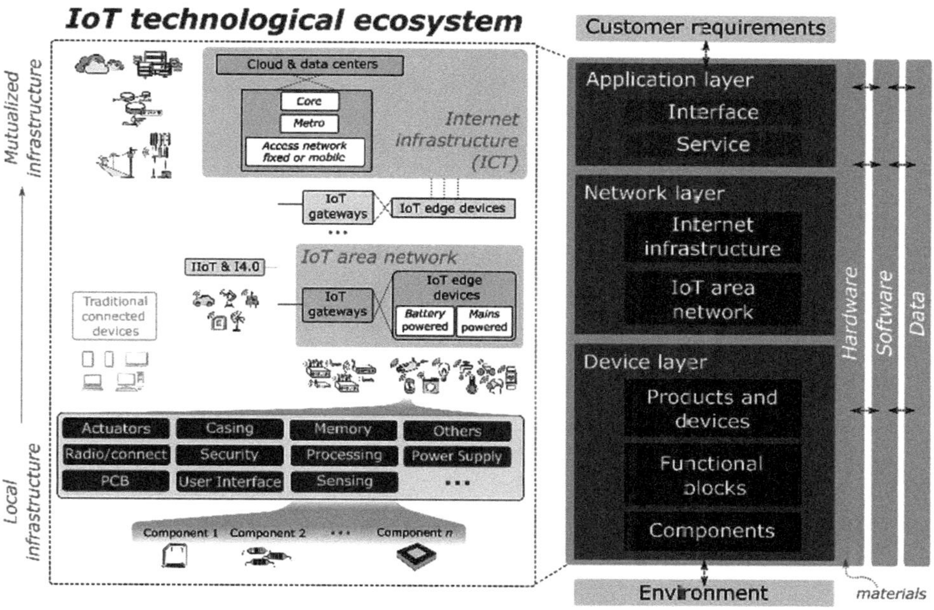

Figure 2.1 Schematic representation of the IoT technological ecosystem, with several layers spanning over multiple levels of abstraction. Figure inspired from [6–9].

Nevertheless, in order to ensure the proper operation of IoT services, a complex and vast ecosystem is required. It relies not only on a network of dedicated edge devices and gateways to acquire data or actuate on the environment, but also uses existing ICT infrastructures (i.e., the access network, clouds, and data centers) to ensure the data transfer, centralization, processing, and visualization. The IoT technological ecosystem therefore combines several layers of abstraction, where hardware, software, and data are involved. As shown in Figure 2.1, the IoT technological ecosystem gathers a wide diversity of layers ranging from components embedded in an IoT edge device to the network of IoT devices, the ICT infrastructure, and finally the IoT-M2M (machine-to-machine) applications. Data generated either by the edge devices or by the remote command (e.g., from cloud servers) flows from one point of the infrastructure to the other. This data also needs to be stored and processed in the cloud or directly at the edge. The complexity of each layer can vary significantly due to the large nature of applications targeted by IoT services and the vast offer of technology. Indeed, it is very important to be aware of the high heterogeneity within the IoT ecosystem, both in terms of hardware components and software [3]. Depending on the IoT device and the service considered, the hardware components can vary considerably, as illustrated in [3] through the concept of hardware profiles, or in [1, 5] through the concept of functions and capacities. For instance, an occupancy sensor is very light in terms of components; whereas, a drone, a connected smartwatch, or an industrial IoT gateway embeds several complex components to implement more advanced tasks, such as high data rate communication, sensing and processing, or memory storage. On the software side, many programming languages, operating systems, and protocols are involved in different layers, all allowing several advantages such as readiness, low-power consumption, or long communication ranges.

2.1.2 Positive and Negative Effects of IoT

Nowadays, ICT services are often presented as an effective solution for decreasing the environmental footprint of other sectors thanks to their optimization abilities. Indeed, several well-known industry reports claim that ICT services have a clear net environmental balance, as they enable reductions of environmental burdens in other sectors up to 10x their environmental footprint [10] (note that only carbon footprint is considered). In fact, ICT services also come with environmental burdens, due to their production, use, and disposal. For instance, the carbon footprint of ICT was evaluated at about 1000–2000 Mt CO_2-eq in 2020, which corresponds to about 2–4% of worldwide greenhouse gas (GHG) emissions [11]. In practice, it is far from trivial to obtain the net environmental balance of ICT and this is clearly out of the scope of this chapter. Nevertheless, the general framework popularized by Hilty, L.M. & Aebischer, B. [7] is useful for structuring the positive and negative contributions of ICT with respect to environmental impacts. Although the framework is by design normative and only focused on environmental impacts, it offers a simplified understanding of ICT being at the same time "part of the solution" and "part of the problem".

As IoT is part of the ICT, we propose to adapt this framework in the context of IoT, resulting in the classification done in Figure 2.2. Indeed, a similar distinction can be made between the technological level, i.e., direct effects, and the application level, i.e., enabling and systemic effects. The direct effects are the effects due to the production, use, and disposal of IoT systems. This is why they are presented as "part of the problem" with respect to environmental impacts. For instance, direct effects were considered for wireless sensor nodes in [12]. On the other hand, IoT can also be "part of the solution", as it can enable optimization and substitution effects at the application level. For instance, IoT could significantly help in remote control (thus reducing the need for human intervention), predictive maintenance, and efficiency improvements in essential management systems in the food, water, transportation, and energy sectors. IoT could also foster the transformation towards more sustainable patterns of production and consumption by supporting the tracking and transfer of information related to a specific product across value chains (through digital twins and digital product passports for instance). Yet, at the same time, the introduction of IoT devices can also create induction and obsolescence effects. In fact, the IoT relies by design on a massive deployment of devices. As soon as some devices have been deployed for a specific application, this likely induces the deployment of more IoT devices to get more data. In addition, the introduction of IoT systems could make older systems obsolete, if not directly compatible or less appealing. Eventually, this can lead to rebound effects, backfiring the environmental savings initially claimed and/or emerging risks such as multiplication of the vulnerable points for cyberattacks, therefore, increasing drastically the cybersecurity risk [13]. In the last report of the Intergovernmental Panel on Climate Change (IPCC), the rebound effects of ICT technologies are explicitly pointed out [14]. Taken as a whole, this raises questions about the roles of IoT in digital transformation. Nevertheless, this is a wide question, which cannot be addressed in a comprehensive manner in this chapter as we only focus on the direct effects of IoT. Up to now, environmental considerations for IoT systems are still far from being systematic and the evalution of environmental effects generated over the life cycle of IoT systems is still in its infancy, although it is gaining momentum in both academia and industry.

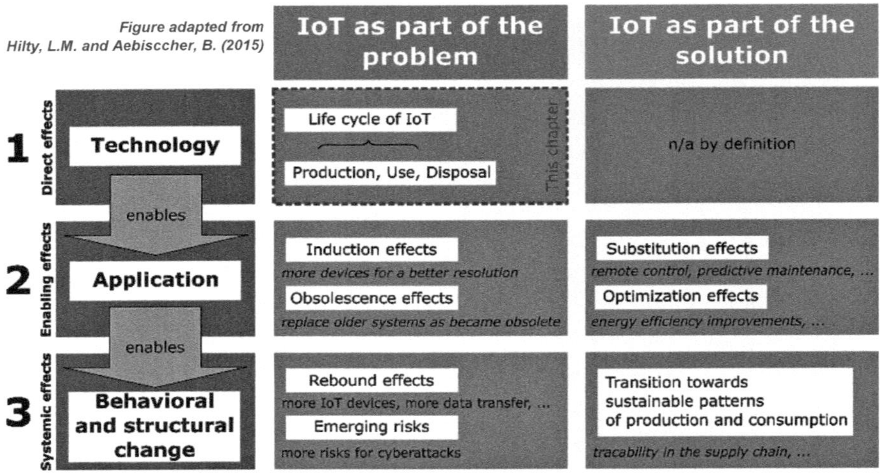

Figure 2.2 A streamlined matrix of the environmental effects of IoT, adapted from [7] focusing on ICT. This chapter focuses exclusively on the direct environmental effects.

2.1.2.1 Direct effects in the production phase

Before being used, every single IoT device must be produced. As IoT devices are predominantly made up of electronic components, the environmental impact of producing the IoT is closely related to the ones generated when producing electronics and semiconductors. This therefore includes a myriad of raw materials, complex supply chains all across the world, and high-tech advanced manufacturing processes. Although limited work has been done regarding the evaluation of the direct effects with respect to the production of IoT systems [3], several works are already available regarding the direct environmental effects generated by the production of electronics and semiconductors [15, 16].

2.1.2.2 Direct effects in the use phase

To include the impact of the use phase, the complex architecture of IoT systems requires considering not only the number and the energy used by end- and edge-devices but also all the mobilized resources within the Internet and cloud infrastructures.

Unfortunately, as designers from the local and cloud sides usually work in silos, both aspects are often addressed with high uncertainty (by means of hypotheses and scenarios concerning the potential number of end-devices or the potential electricity consumption of data centers, as pointed out in [3, 17]. Direct effects are also generated during the deployment, maintenance, and decommissioning [18–20].

2.1.2.3 Direct effects in the end-of-life phase

A major concern with IoT devices is end-of-life management [2]. Indeed, as the number of IoT devices is growing at a high pace, the potential amount of e-waste associated with such a massive deployment is likely to quickly grow accordingly [2]. In addition, very few legislations or constraints are in place to ensure proper disposal management when all these devices reach the end of their first life. Lessons from Waste Electrical and Electronic Equipment (WEEE) management point out that one of the main bottlenecks in electronic end-of-life management is the collection of devices, especially from consumer electronics [21]. For instance, an official report of the French government [22] shows that, from the 2.1 million tons of electronic devices commercialized in France in 2019, about 0.85 million tons were collected (~41%). Official statistics are often limited and more research should be done on this matter. Yet, the Global E-waste Monitor estimates that more than 80% of the WEEE are handled in an informal way [21], often being shipped towards developing countries, despite the Basel Convention, which prohibits such practices. Currently, the total amount of WEEE keeps on increasing, year after year [21]. As IoT is expected to penetrate aggressively the consumer market, industries, and infrastructures, it is critical to take into account their disposal scenarios to prevent the growth of WEEE with associated hazardous substances, leading to adverse impacts on human health and the environment [2]. In addition, IoT will embed electronics in everyday-like products that traditionally may not have included such components [2], which will likely lead to an important diversification of e-waste streams. Unfortunately, this will presumably complicate proper end-of-life management and worsen the current situation regarding WEEE.

To sum up, we have highlighted the existence of direct effects of IoT, during several phases of the life cycle. Identifying these impacts is important for two reasons. First, it prevents their omission when dealing with IoT systems. Too often, these impacts are not even mentioned, leading to a distorted picture of reality as the benefits of IoT do not come for free in terms of environmental impacts. Second, this can greatly help for future improvements of IoT systems in order to decrease their environmental footprint, without affecting their functionalities. Indeed, taking the environmental footprint into account when designing IoT systems is one of the key objectives of ecodesign approaches, which are therefore considered in this chapter.

2.1.3 Structure of this Chapter

The rest of this chapter is structured as follows. Section 2.2 provides an introduction regarding ecodesign and its challenges, while Section 2.3 focuses on LCA as a tool for ecodesigning the IoT. LCA fundamentals are therefore outlined, including definitions, goals, and an overview of current tools, software, and databases that can be used. Then, Section 2.4 provides concrete examples to illustrate how LCA can be useful for IoT designers in the context of judicious ecodesign approaches. Finally, take-home messages are provided in Section 2.5, and research perspectives and ongoing work regarding LCA and ecodesign for IoT are highlighted in Section 2.6.

2.2 Challenges of Ecodesign

In this section, we introduce the concept of ecodesign and provide readers with a pedagogical presentation of the main existing ecodesign tools and the phases of product development during which they can be applied. This will then help to better understand how it can be included in the traditional design of electronics and IoT systems to improve their environmental footprint.

Ecodesign can be defined as "the integration of environmental aspects into product design and development, with the aim to reduce adverse environmental impacts throughout a product's life cycle" [23, 24]. It is one of the levers that can be used to reduce one or several direct environmental effects of the IoT presented in the last section.

2.2.1 Product Design Process (PDP) and Ecodesign

Traditionally, the design of a product starts with the identification of a market need, or the customer requirements, i e., identifying the need for a function. Concepts (general working principles) are identified by the designer to fill the need during the conceptual design stage. The most promising concepts are elaborated in the configuration of a technical system with technical requirements, layout, materials, and scale (embodiment design). One or more of these are selected for detailed development and analysis of performance, safety, and cost (detailed design). The detailed design enables the construction of a prototype that, after testing and development, goes to manufacture, market launch, and, eventually, to the product customer verification. In project management, this is traditionally represented by a "V-model" [25] that allows the PDP to be represented linearly by its successive phases and their interactions. The V-model refers to the schematic view of this cycle, which takes the form of a V: a descending phase (customer requirements, and conceptual, embodiment, and detailed designs) and bottom (manufacture), followed by an ascending phase (testing, launch, and customer verification). The ascending phase mainly involves testing and verifying the choices made for the different design levels during the descending phase, right up to its final use, therefore, making it possible to adjust and optimize the designs to reach appropriate specifications (performances and costs). If we transpose this traditional process to the specific case of complete IoT systems, a similar PDP occurs in each IoT layer, as illustrated in Figure 2.3(a).

Figure 2.3(b) introduces ecodesign alongside the traditional PDP by referring to the common actions that aim at minimizing the environmental direct effects of a product over its whole life cycle, as stated by ISO standards 14006:2020 [23]. This standard fosters the co-optimization of environmental and technical performance by integrating environmental aspects into

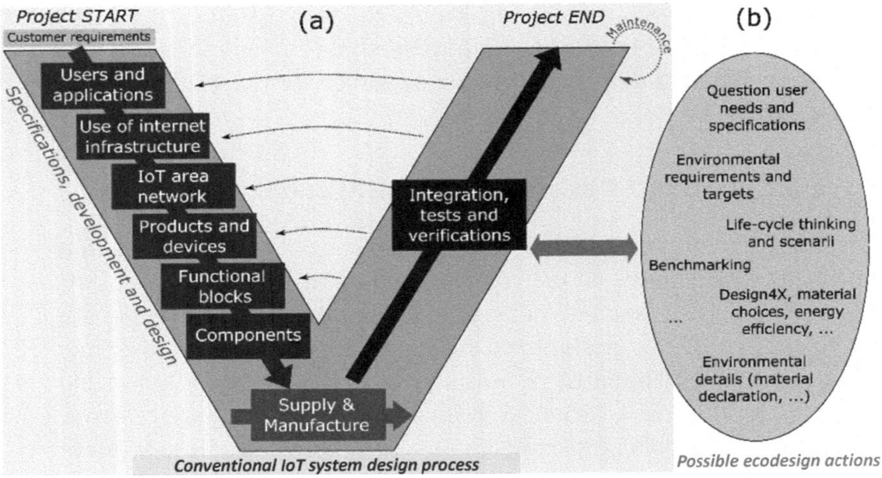

Figure 2.3 (a) Illustration of the V-model applied to IoT systems. Figure adapted from [30]. (b) Generic ecodesign actions alongside the traditional PDP applied to IoT systems [23].

the design process. The ecodesign must be integrated right into the early phases of the PDP, as it allows more flexibility for considering environmental aspects (e.g., meeting UE directives) together with the technical requirements. This is supported by the fact that about 80% of a product's environmental impacts are determined very early in the design stages, according to several authors [26, 27]. It is particularly critical for IoT as it is an emergent technology in constant evolution, with a high degree of heterogeneity regarding the devices and the services it provides. By applying these actions, the ecodesign process must highlight tangible outcomes useful to help IoT designers make informed choices during the PDP. Examples of tangible outcomes include:

- Minimization of the environmental risk, the exploitation of natural resources, the greenhouse emissions, and industrial pollution related to the product life cycle.
- Reduce the amount of raw materials used in the product and its constituent components and support the use of recycled materials in the product and components. Support Reduction, Reuse, Repair, Remanufacture, or Recycling (10'Rs scale [28]) strategies of products and components.

- Reduce product waste and amount of industrial waste and eliminate landfill disposal at product end of life.
- Minimize energy consumption at every stage of production and in particular during product use (duty cycle, ultra-low-power, data transfer, ...).
- Reduce the energy consumption of sensor systems (e.g., by changing awake/sleep patterns), edge devices (e.g., by reducing network size or designing efficient communication protocols), and cloud resources (e.g., by resource allocation policies).
- Minimize operating and maintenance costs of products.

Many of the abovementioned actions naturally emerge from the circular economy framework, which proposes an alternative production model to promote the responsible and optimized use of resources (waste prevention, recycling, ecodesign strategies such as modularity, and subsequent reuse of components and materials, minimizing the need of material loops), with the global goal of achieving increased sustainability [29].

Successfully applying ecodesign in the context of a specific product requires uncovering (i) how the initial PDP impacts the environment, (ii) determining the best way of addressing these impacts by making changes within the design, and (iii) validating the benefits over the whole product life cycle of the actions taken during ecodesign.

2.2.2 Ecodesign Tools

To support the ecodesign throughout different steps of the PDP, several ecodesign tools can be used. They differ mainly in the recommendation and assessment levels they can provide, as shown in Figure 2.4. The tools with high recommendation levels generally required low levels of environmental assessments, therefore missing quantitative data regarding the impacts of the product or service. On the contrary, precise assessment tools usually provide a low level of recommendations, as they rather provide specific estimates on the impacts than recommendations. As a result, much effort is required from the experts to translate the quantitative impact indications provided by the method into clear and useful recommendations for the targeted product or service.

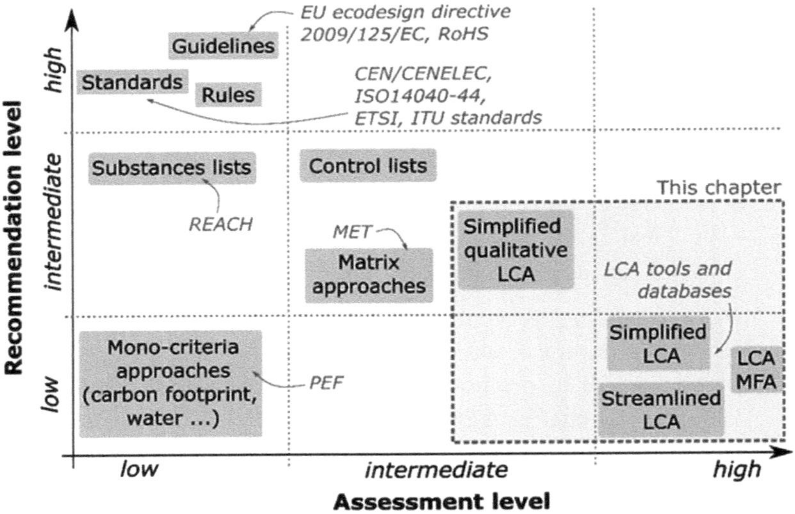

Figure 2.4 Overview of generic ecodesign tools with their associated recommendation level and assessment level. Figure adapted from [31].

In the context of electronics, the industrial and scientific communities have proposed different instruments at the recommendation and assessment levels to facilitate impact estimation and ecodesign in recent years. Rousseaux, P. et al. [32] record, until 2017, more than 70 instruments for enabling "green" electronics (among regulations, standards, methodologies, guidelines, and tools), and Ferrendier, S. et al. [33] facilitate design guidelines from the study of a considerable number of case studies (from small to big electronics). To illustrate these efforts, some important resources are presented below.

(i) **Ecodesign guidelines and rules**: Supportive research is currently conducted to increase the regulation on the design of electronic products, especially at the European level. Due to their very low assessment levels, the guidelines and rules can be operated during the conceptual and detailed design phases of the PDP to help generate alternative solutions:

- The EU Ecodesign Directive 2009/125/EC establishes a framework for the setting of ecodesign requirements for energy-related products (e.g., households, electronic displays, etc.) [34]. This ecodesign directive is currently extended to a new Ecodesign for Sustainable Products Regulation that will improve EU products' circularity, energy performance, and other environmental sustainability aspects. In particular, it will include new requirements on the carbon and environmental footprints of the products, as well as impose a Digital Product Passport that will improve transparency about products' life cycle impacts on the environment and help consumers and businesses make informed choices when purchasing products [35].
- The European RoHS Directive 2002/95/EC (Restriction of Hazardous Substances) aims to limit the use of ten hazardous substances in electrical and electronic equipment [36].

(ii) Standards: The European Committee for Standardization (CEN) and the European Committee for Electrotechnical Standardization (CENELEC) produce European standards, essential complementing tools that support the implementation of European Union (EU) Ecodesign and Energy Labeling legislation, providing dedicated methods for measuring the energy performance of various energy-related products and their material efficiency against the compulsory values and thresholds laid down in the directives [37]. In addition, standards are also issued by the European Telecommunications Standards Institute (ETSI) and the International Telecommunication Union (ITU) Study Group 5, dedicated to the environment, climate action, sustainable digitalization, and circular economy [38].

(iii) Substance lists: Substance lists such as the Registration, Evaluation, Authorisation and Restriction of Chemicals (REACH) regulates the use of chemicals, in order to favor the development and the use of components that are certified as less harmful for the human and the environment [39].

(iv) **Matrices**: They take the form of a table that integrates the three criteria (materials, energy, and toxicity (MET)) for each phase of the product life cycle. Matrices are quite simple to use in early design phases and the data to be filled in is qualitative, leading to average recommendation levels [40].

(v) **Mono-criteria approaches**: Among the first methods developed, they are generally based on material balances, bill of materials (BoM), or simply rely on energy efficiency or carbon balances of a specific product, process, or service. However, these approaches remain incomplete if one wants to identify actions to effectively reduce the environmental impacts of products. Indeed, undertaking changes without considering all the impacts can lead to an increase in some of the others (impact transfer) [41].

(vi) **LCA**: LCA is a standard quantification method [42] that aims to analyze flows (resource extraction and substance emissions) at the level of an individual, a product, a region, etc., in order to provide a better understanding of the environmental impacts on natural ecosystems. The main advantage of LCA over other methods in ecodesign is that it is quantitative, multi-indicators, and covers the whole life cycle of a product or service. LCA approaches usually require sector-specific data, making them particularly suitable to the detailed design stage of the PDP, especially for the comparison and the selection of the best design alternative based on their environmental performance. A great example of LCA applied to the ecodesign of electronics is the assessment of modularity as a circular strategy enabler for smartphones [43]. Although LCA tools offer a high level of assessment, they require time, expertise, and specific data [12]. Consequently, they usually come together with a non-negligible financial cost. This is why simplified LCA (S-LCA) tools have been developed, where the scope of impact categories is reduced and the study is often limited to a subset of life cycle phases [41]. They have the advantage of requiring less information, being quick to carry out, and easier to interpret, but the results are more limited in the conclusions that can be drawn.

In the context of IoT, it is clear that LCA-based methodologies can support ecodesign as they help to generate and compare design alternatives. However, although these methodologies allow to obtain quantitative evaluations, they are usually difficult to apply, since they require specific data about the life cycle of IoT systems. Moreover, they do not allow for a high level of recommendation. On the other hand, non-LCA-based approaches such as matrices, lists, or customized tools could be easier to use, but extremely limited regarding the high complexity and heterogeneity of full IoT systems (as observed in [44–46]). Finally, emerging guidelines [47–49] and standards [50] oriented to IoT may be practical to apply, but clearly insufficient to cover the whole PDP of IoT systems.

2.3 Fundamentals of Life Cycle Assessment

In this section, we will focus specifically on LCA-based methodologies as a key tool for ecodesign because (i) they are based on already-existing standardized ISO methodology, (ii) they provide quantitative results over the whole life cycle of the IoT products/services, (iii) they consider several environmental indicators in order to reveal impacts transfer, and (iv) they can help improving designs iteratively. Developing and using LCA is important to validate environmental gains and impacts expected from circular economy schemes, for instance. In practice, there is a recurrent use of "green buzzwords" unsupported by specific environmental analysis. This is specifically problematic for emerging technologies such as IoT.

2.3.1 LCA Methodology

According to ISO14040 [42], LCA is a standard quantification method that facilitates the estimation of multiple impact categories related to a product. Specifically, an LCA study aims at modeling several life cycle phases (i.e., modeling of extractions and acquisition of raw materials, production, use, and waste treatments), including all the energy and additional materials, products, and emissions involved in these phases. The transfer of potential impacts between life cycle phases, impact categories, or particular processes can therefore be identified and limited (or avoided).

The iterative steps for conducting an LCA study are presented in Figures 2.5 (a) and 2.5 (b). The fundamental concepts of LCA are introduced in the rest of this subsection. Nevertheless, for a more in-depth dive into the LCA methodology, we highly recommend the reader to read more comprehensive resources such as the general reference book written by Hauschild et al. [50a] and the International Reference Life Cycle Data System (ILCD) handbook from the European Commission [50b].

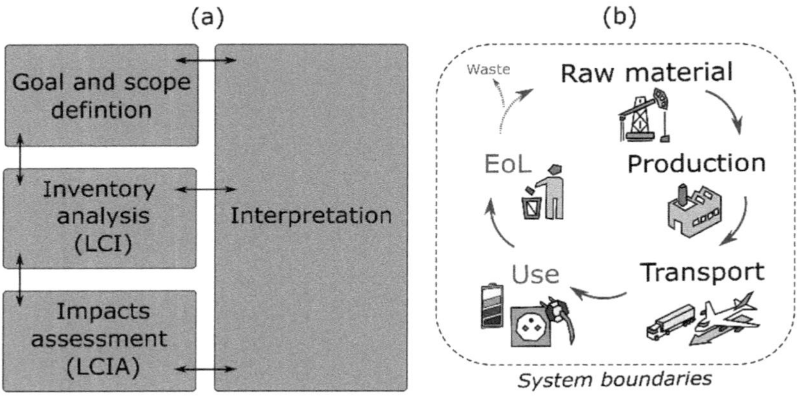

Figure 2.5 (a) LCA framework (adapted from ISO14040 [42]). (b) Generic representation of the different life cycle phases and the system boundary of a cradle-to-grave LCA study. It considers unit processes from the raw material extraction to the production, transport, use phase, and end-of-life.

The goal must define clearly the intended application of the study because it will frame the further steps of the methodology. The goal may be showing the environmental impact of a product based on an individual or comparison analysis, evaluating it based on a norm or a standard, evaluating its redesign alternatives, or facilitating the construction of ecological policies around it. On the other hand, the scope definition or the system boundary includes the limits of the product system (the life cycle model of a product in terms of unit processes, and elementary, product, and intermediate flows) and the level of details of the environmental assessment, which depends on the functional unit.

The functional unit defines the quantification of selected functions in an LCA study and guides the definition of a product system. The definition of functional units is generally given as a sentence and needs to be detailed enough to ensure appropriate impact estimation. For instance, "monitoring the domestic water consumption over a year". Functional units quantify the performance characteristics of a product to provide a reference to which the inputs and outputs are related. The definition of the functional unit is critical, as it allows the identification of the reference flow, which is the number of product systems, materials, energy, or even additional products needed to fulfill the functional unit as it is expressed.

The inventory analysis step involves the description and the data collection processes needed to model a product system, according to the functional unit evaluated by an LCA study. It includes the calculation procedures to quantify relevant inputs and outputs of unit processes of a product system and its outcome catalogs the flows crossing the system boundary. The inventory analysis includes key calculations of material and primary energy inputs for a product system and key calculations of its output emissions to air (e.g., CO_2 emissions), earth (e.g., WEEE), and water.

The impact assessment step consists of applying a life cycle impact assessment (LCIA) method to estimate the environmental impacts of a product system by associating the outcomes of the inventory analysis with specific impact categories and impact characterization factors. LCA software generally include several existing LCIA methods such as ReCiPe [51], CML-IA [52], or IPCC [53].

Starting from these fundamental concepts, LCA practitioners adopt different approaches to deal with daunting challenges involved in IoT systems such as estimating the impact related to the use of mutualized infrastructure or resources. Different approaches known as "top-down" and "bottom-up" can be used in this context. The main advantage of the first approach is available data, but the main disadvantage is the lack of granularity, preventing the identification of specific hotspots. Regarding the "bottom-up" approach, it usually provides a better granularity but suffers from the truncation error and may lack generalization [3].

2.3.2 LCA for IoT: Overview of Existing Tools (Software and Databases)

Although LCA benefits from this international standardization, it mainly remains focused on the description of the methodology at a theoretical level. Even if it helps regarding the question of "what to do", it provides very few details on "how to do" in practice. The LCA standard is defined for general products or systems, whereas the LCA for electronic devices and systems is very specific. In order to provide more recommendations to carry out the LCA of ICT products or services, additional guidelines, and specifications [54, 55] have been proposed by the European Telecommunications Standards Institute (ETSI) and the International Telecommunication Union (ITU) over the last decade. These can help ease the implementation of LCA on ICT products or services. Yet, to the best of the authors' knowledge, there are no official guidelines regarding LCA for IoT. Consequently, the LCA of IoT systems is generally carried out by using the existing guidelines, tools, and databases dedicated to electronics and ICT. This subsection therefore provides some practical insights to apply LCA in the context of IoT.

Different tools (mainly software and databases) can be used, even if trade-offs clearly exist with respect to resolution, complexity, and price. The choice should be made depending on (i) the goal of the LCA, (ii) the expertise of the designer regarding LCA, and (iii) the time and resources available for the assessment. In order to assist with this choice, an overview of the currently most used LCA software and databases for electronics is provided in Tables 2.1 and 2.2. In both tables, the qualitative assessment has been done based on the authors' prior knowledge and experience. Criteria used for the comparison are defined below each table.

Table 2.1 LCA software (based on the authors' knowledge and experience)

	Software						
	GaBi[1] [56]	Sima Pro [57]	EIME [58]	Open LCA [59]	Bright-way [60]	Excel	Other[2]
Expertise needed	high	moderate	moderate	moderate	moderate	low	n.a.
Custom modeling	high	high	high	high	high	high	n.a.
Open source	no	no	no	yes	yes	yes	no
Maintenance	yes	yes	yes	yes	yes	yes	n.a.
Price	high	high	moderate	free	free	(free)	high moderate
Goal	LCA	LCA	LCA	LCA	LCA	S-LCA	S-LCA
Sensitivity analysis	yes	yes	no	yes	yes	yes[3]	n.a.
Cloud-based	no	both	yes	no	no	no	n.a.
LCIA methods compatibility	high	high	high	high	high	mod	n.a.
Scriptable	no	yes[4]	no	yes	yes	no	n.a.

Criteria

Expertise needed: Which level of expertise is needed by the user regarding LCA and electronics in order to carry out the modeling?

Custom modeling: How much flexibility is given to the user to build custom processes and models?

Open source: Is the software open source?

Maintenance: Is the software maintained through time?

Price: How expensive is the software (qualitative)?

Goal: For which goal is the software intended?

Sensitivity: Is it possible to carry out sensitivity analysis (such as parametric, scenarios, and Monte Carlo analysis)?

Cloud-based: Is the software running remotely on a cloud-based interface?

LCIA methods compatibility: How compatible is the software with several LCIA methods?

Scriptable: Is it possible to write a script within the software?

n.a.: Not available

---

[1]GaBi is now called Sphera LCA

[2]Other LCA software such as PAIA, Eco-Audit, iPoint

[3]Limited/basic

[4]Depending on the license type

Table 2.2 LCA databases (based on authors' knowledge and experience)

	Databases						
	GaBi [61]	**Nega Octet [62]**	**CODDE [63]**	**Eco Invent [64]**	**Ansys Granta [65]**	**Base Carbone [66]**	**Eco System [67]**
Resolution	high	high	high	moderate	low	low	moderate
Documented	high	moderate	moderate	high	moderate	moderate	moderate
Open source	no	no	no	no	no	yes	yes
Up-to-date	yes	yes	yes	no[5]	yes	yes	yes
Maintenance	yes	yes	yes	yes	yes	yes	n.a.
Multi Indicator	yes	yes	yes	yes	yes	no	yes
Price	high	high	moderate	high	moderate	free	free
Specific to electronics	yes	yes	yes	no	no	no	yes

Criteria

Resolution: How detailed are the entries in the database?

Documented: How detailed is the documentation attached to the database?

Open source: Is the database open source?

Up-to-date: Is the database representative of cutting-edge manufacturing technologies and processes for electronics and semiconductors?

Maintenance: Is the database maintained through time?

Multi-indicator: Does the database allow for multi-indicator environmental analysis?

Price: How expensive is the database (qualitative)?

Specific to electronics: Is the database specific to electronics rather than generic?

Table 2.1 mainly points out that some software is more suited for simplified LCA thanks to the lower expertise needed; whereas, other software is usually more used in the context of detailed and specific LCA. Table 2.2 highlights that almost all available databases allow for multi-indicator analysis, apart from the Base Carbone database. However, only a few are specific to electronics down to different levels in terms of resolution and documentation. To the best of the authors' knowledge, the best state-of-the-art databases regarding LCA of electronics are GaBi Sphera LCA (Extension XI), CODDE, Ecoinvent, and Negaoctet.

[5]Regarding electronics

Finally, both tables clearly show that although some LCA software is open source, all advanced state-of-the-art and specific databases for electronics remain proprietary. In practice, this significantly limits the development of consolidated models shared among the community. Yet, confidentiality concerns due to the considerable amount of work needed to come up with such data and the need for regular updates, as rapid innovation cycles in the electronics sector (at least partially) explain the cost of these databases. The results that can be expected from these tools and databases will be exemplified in the next section through different use cases.

2.4 Illustrate LCA and Ecodesign in Several Layers of the IoT Ecosystem Through Specific Examples

This section aims at presenting existing use cases to illustrate ecodesign and LCA approaches in the different layers of the IoT technological ecosystem. Figure 2.6 is a graphical overview of several examples regarding LCA tools, studies, and ecodesign practices addressed previously in this chapter. The rest of this section will detail these examples and provide more insights for each layer in the IoT technological ecosystem.

2.4.1 Customer Requirements and Application Layer

Usually, the design process of IoT systems starts with a functional analysis, which gathers specific actions that local and mutualized equipment are supposed to do in order to cover the requirements of end-users. In general, there are three types of functions in functional analysis: (i) main functions — those that justify the creation of a device (e.g., "sensing water flow"); (ii) secondary functions — those that support main functions (e.g., "connecting to the Internet to transmit sensor data"); and (iii) constraint functions (e.g., "RoHS certified"). Normally, IoT designers and engineers align previous or up-to-date technologies to some functions and additionally, they list specific electronic components according to technical requirements i.e., "specifications".

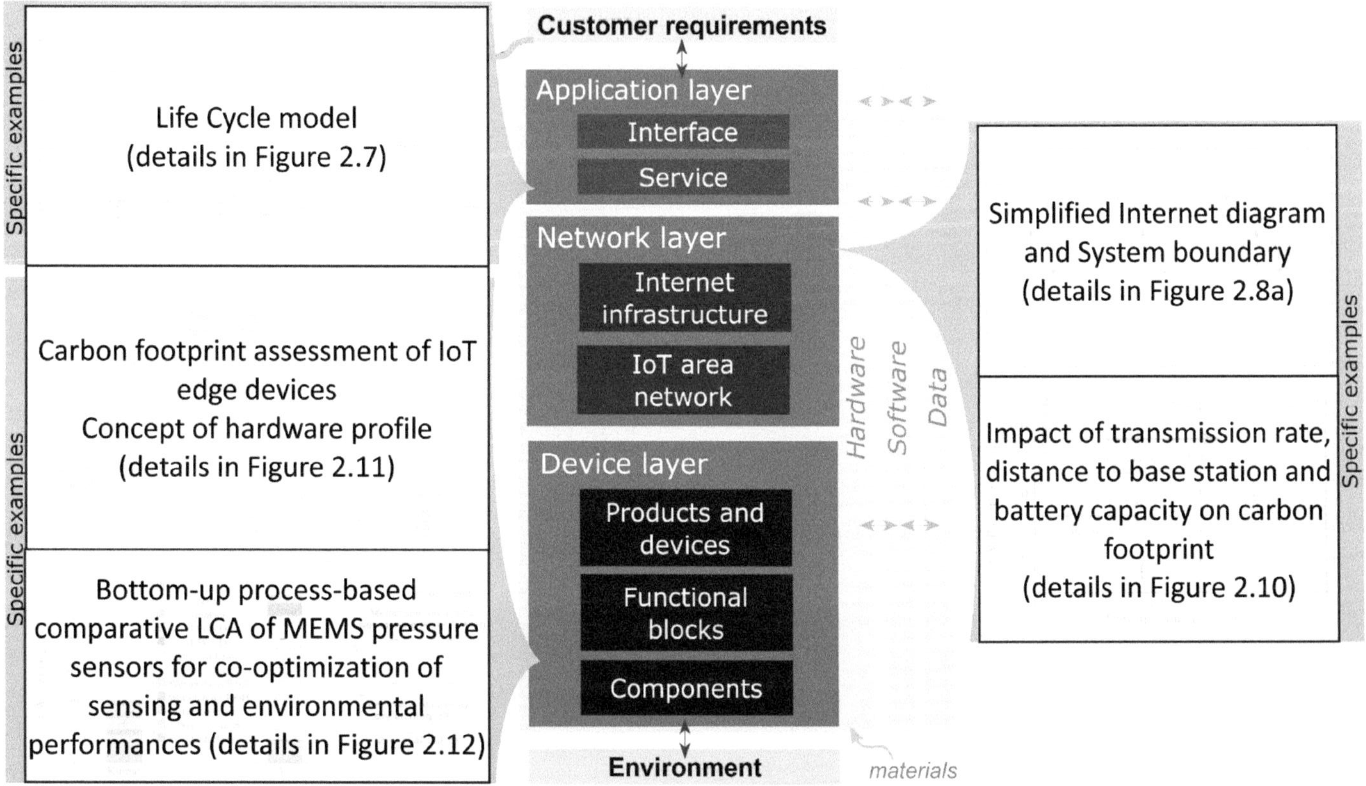

Figure 2.6 Graphical overview of this section focusing on examples of ecodesign practices, LCA tools, and studies mapped on the different layers of the IoT technological ecosystem.

Specifications are useful, yet insufficient to construct proper definitions of functional units of IoT systems in LCA studies. Indeed, functions usually describe the task that sensor systems, edge devices, or cloud resources do but they hardly describe the global service that IoT systems provide. This problem is exacerbated if multi-services, e.g., "smart monitoring" and "smart watering", are provided by only one IoT system [1]. Regarding ecodesign, putting a lot of emphasis only on high-quality services is also counterproductive, because it tends to oversize specifications concerning devices and mutualized infrastructure [1], as evidenced in Heinis, T. et al. [44].

To deal with these issues, sensor, edge, and cloud designers need to consider two aspects. Firstly, they have to define together the appropriate functional unit or units of full IoT systems, according to the most relevant aspects of services that these systems provide, as shown in [68]. Secondly, they have to adopt an integral ecodesign and life cycle modeling of all mobilized resources, based on (i) electronic components' functions and capacities for local equipment; and (ii) data volume and traffic for shared infrastructure, as shown in [1, 5].

2.4.2 Ecodesign Actions in Mutualized Infrastructure (Access-, Core-Networks, and Cloud)

The evaluation of the environmental impacts generated when transferring and treating data on the Internet infrastructure depends on the type of infrastructure that is used. In practice, IoT systems use remote servers and either common access networks (mobile/fixed) or a parallel infrastructure (private network, long-range technology (LoRa), SigFox, narrowband-IoT (NB-IoT), among others). The tendency among researchers working on reducing the ecological burden of local plus mutualized infrastructures is (i) measuring the impact of networks and data centers by an LCA-based top-down (e.g., [68]) or bottom-up approaches (e.g., [69–72]); and (ii) reducing the impact on the basis of the network activity of local equipment or service mutualization.

Example 2.1 Focusing on the question of efficient waste glass management in cities, Lelah, A. et al. [68] proposed a technical solution using a product-service system (PSS) based on a machine-to-machine (M2M) subsystem. Their LCA-based methodology assesses the environmental impacts of their solution, which is then used to propose possible improvements in the context of ecodesign. This study reveals the interest in mutualization to reduce environmental impacts as well as the importance of limiting data exchanges. More importantly, they consider on-site requirements and try to capture behaviors during the use phase through several scenarios and real-life data.

Example 2.2 Starting from the global service (functional unit) "tracking remotely the water consumption of a terrain during a period of time", Quisbert-Trujillo, E. et al. [5] modeled the life cycle of mobilized resources in the sensor, edge, and cloud layers on an IoT system (blue, green, and gray boxes, respectively). As shown in Figure 2.7, although the main functions of all devices and electronic components differ, all their outcomes contribute to providing the service, as defined in the functional unit.

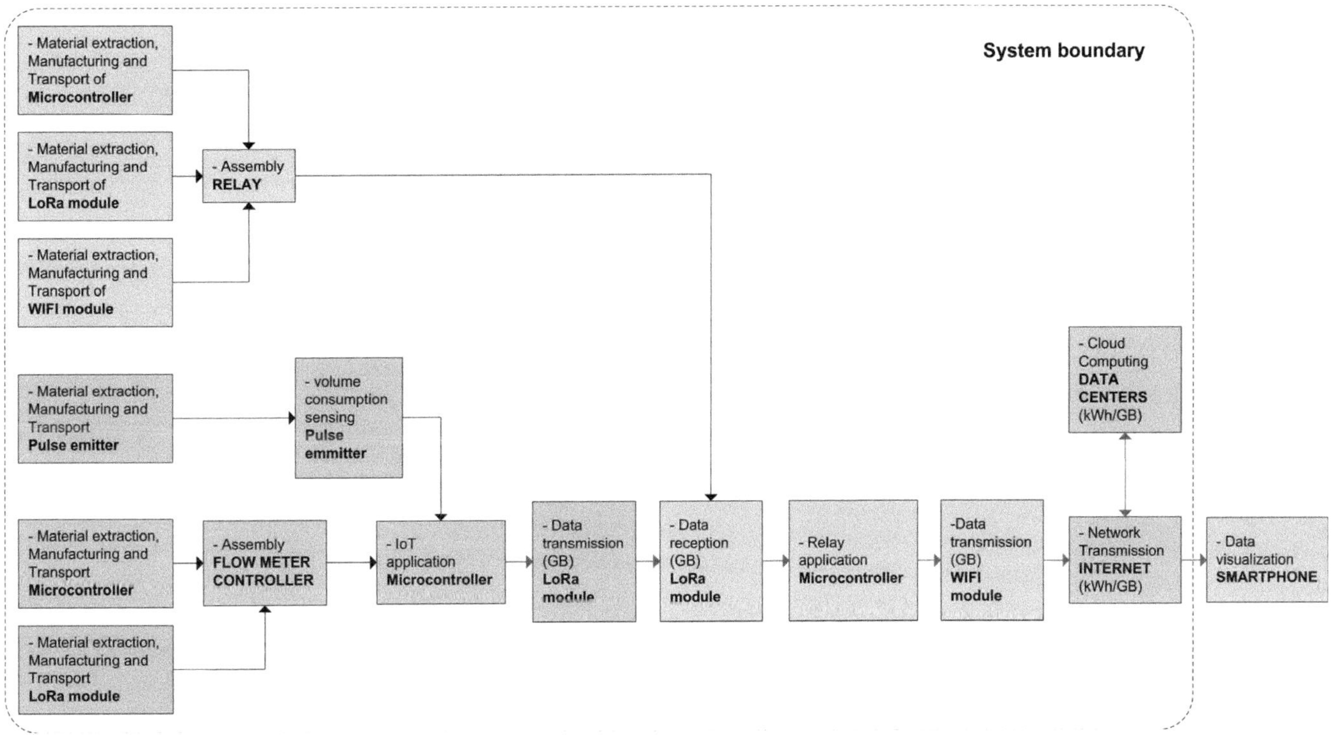

Figure 2.7 Life cycle model for an IoT system composed of a pulse emitter and flow meter (blue boxes), an internet relay (green boxes), and a cloud server connected to the internet (gray boxes). Black arrows represent physical flows and red arrows represent data flows. Figure adapted from [5].

Example 2.3 A great example of an intensity factor used to characterize the electricity consumption of data moving through the internet infrastructure is the study of Aslan. J. et al. [74]. The system boundaries are well defined (focusing on fixed access network and core network) and the evolution trend is provided over more than 15 years, as illustrated in Figure 2.8(a) and Figure 2.8(b), respectively.

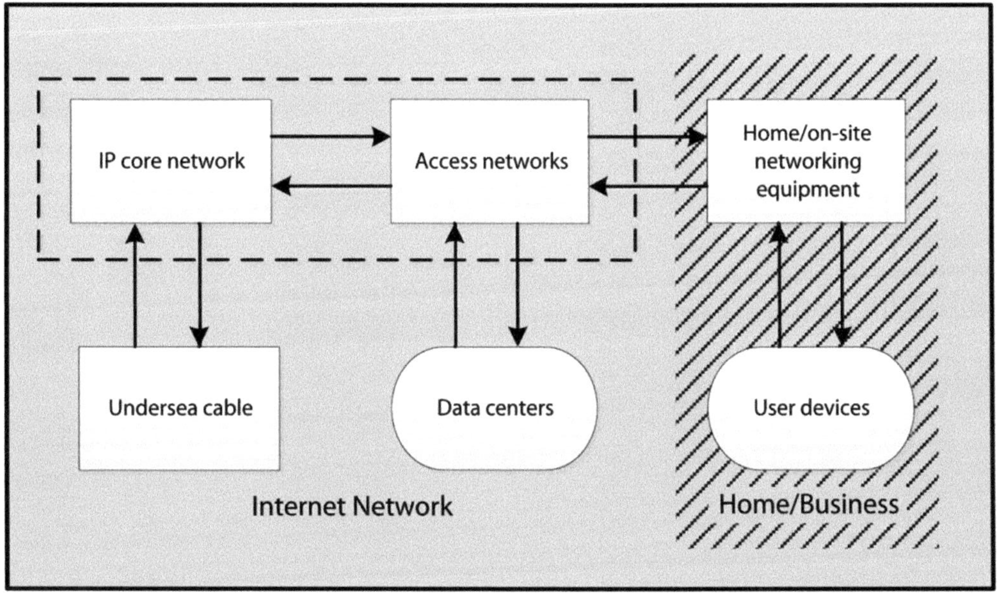

Figure 2.8 (a) Simplified Internet structure diagram and system boundary of the study focusing on data transmission.

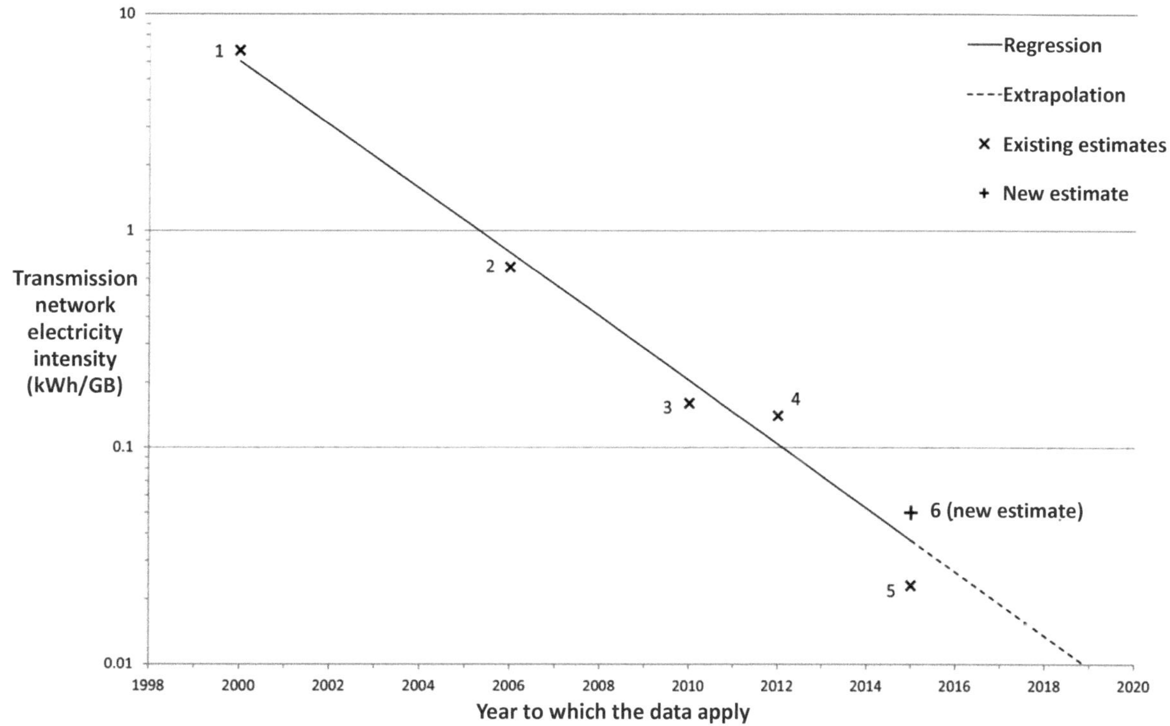

Figure 2.8 (b) Evolution of the estimates for the electricity intensity of the transmission network according to Aslan, J. et al. [74] (Both sub-figures are extracted from this work).

Regarding top-down postures, Malmodin, J. et al. [73] suggested that the impact of specific ICT sectors could be estimated by the allocation of statistical and economic data. Starting from the energy used in the French network Orange, Lelah, A. et al. [68] allocated, for example, the energy consumption data of necessary Internet services for running a wireless sensor network. Next, they found that the energy depletion impact of global telecom infrastructure (mainly GSM and Internet networks) may drastically decrease by more than 40% if hourly transmissions (between sensors and service servers) switch to daily transmissions.

Regarding bottom-up postures, environmental impacts are often evaluated based on the volume of data generated by IoT systems over specific periods of time. To convert data into environmental impacts, intensity factors are used (as observed generally in [74] and specifically to IoT in [1, 75]). These factors are generally focused on the electricity consumption per Gigabyte, i.e., kWh/GB, and improved quickly over time, together with the total amount of data transferred on mutualized networks [76, 77].

To couple either top-down or bottom-up approaches with the ecodesign of IoT systems, LCA practitioners and designers should consider the following aspects. Firstly, they should have a clear picture of the energy required and the volume of data generated by sensor systems and edge devices (for example in terms of size and transmission frequency). In this sense, an important aspect to take into account is the distinction between the energy intensity over a period of time (normally used to estimate the impact of local equipment) and the energy intensity per bit with the volume of data transferred (normally used to estimate the impact of the internet networks and cloud resources). Secondly, they should design (or redesign) the system guided by the essential information required by the service, as it was observed above in [1, 5]. Finally, they should be aware of the fact that service mutualization is highly related to multi-service IoT systems.

2.4.3 Ecodesign Actions in IoT, Local Area Networks

At the edge layer, Bonvoisin, J. et al. [18] demonstrated a more accurate estimation of impact when designers distinguish between the lifetime of physical devices and the lifetime of nodes, in terms of the place that these devices occupy in the network. This could be done normally on the basis of simulations of energy consumption of battery-powered sensors and edge devices, which is positively correlated to the number of replacements along the lifetime service of a network. Unfortunately, within this perspective, crucial aspects, beyond the energy consumption of local equipment (i.e., materials or circularity) are neglected. In fact, deployment, maintenance, and decommissioning should be systematically taken into account when performing LCA for an IoT network [18–20, 55]; together with specific ecodesign of end-devices, from the point of view of electronic design, data, and network activity. In addition, the communication protocol, the topology of the network (e.g., star vs. mesh), the number of IoT nodes in the network, the type of power supply (battery-powered vs. mains-powered), and the latency required by the final application can also significantly influence the environmental impacts of the whole IoT network [20, 78, 79].

In general, the topology of IoT area networks is designed to fulfill conventional metrics such as latency, power consumption, coverage, and price. Yet, environmental impacts generated over the whole life cycle of these networks are generally overlooked. They should be systematically considered ahead of the deployment of IoT area networks, thanks to simulations and modeling, as they can help in the identification of environmental hotspots. This should also be the case for IoT area networks using energy harvesting.

Example 2.4 An important work regarding the ecodesign of wireless sensor networks was done by Bonvoisin, J. et al. [18, 19]. By considering the replacement of end-devices, in [18], they showed that the initial impact of a wireless sensor network oriented to garbage collection (base case) could be reduced by more than 30% if overhearing of sensors is reduced to strict necessity (Alternative 1), and increasing battery capacities (Alternative 2) additionally. This sharp ecodesign was possible by considering the life cycle of physical devices and the life cycle of nodes (the representation of the device in the network, which considers energy consumption patterns and replacements according to the life cycle of the network). Unfortunately, the authors do not consider further ecodesign of electronic design of end devices, arguing (in a further work [19]) that they are already optimized to fulfill strong energy efficiency constraints.

Example 2.5 The choice of the communication strategy in an IoT network can have an important impact on the energy consumption of the network, and therefore on the lifetime of the IoT nodes if they are battery-powered. Figure 2.9 and Figure 2.10 both illustrate these two effects.

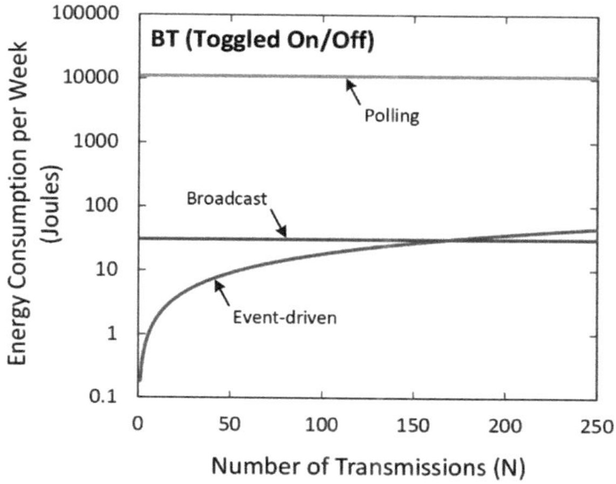

Figure 2.9 Comparison of the energy consumption for different communication strategies, i.e., broadcasting, polling, and even driving. Figure extracted from [78].

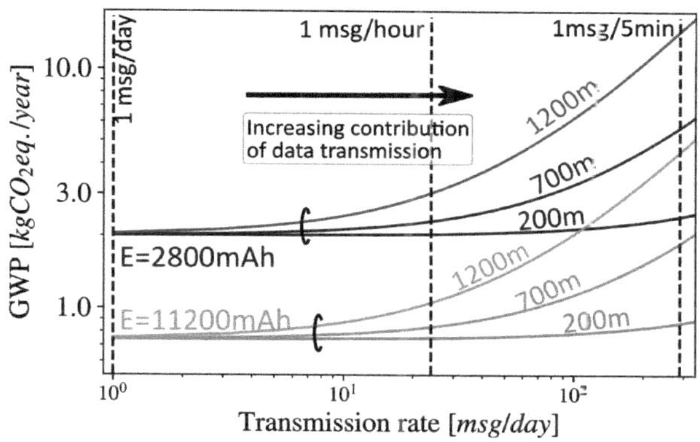

Figure 2.10 Impact of transmission rate, distance from node to base station, and battery capacity on the carbon footprint. Figure extracted from [20].

2.4.4 Ecodesign Actions in the IoT Product and Devices

This subsection reports a non-exhaustive list of actions that can be taken to ecodesign the IoT device level. It is common to see the investigation of highly integrated subsystems to reduce material consumption and the final size of devices. Probably, the main criticism of this perspective is the lack of a life cycle posture. For example, the more the integration of a sensor system, the more complex the production becomes, which mobilizes more resources (e.g., electricity) and increases impacts in the manufacturing phase (as pointed out by Wagner, E. et al. [12]). On the other hand, evidence in the literature shows that researchers are starting to consider more and more self-powered design, which is strongly coupled with software solutions such as intermittent computing. Self-powered subsystems (e.g., energy harvesters) might reduce the impact of local equipment (especially in terms of used batteries) and lower the need for maintenance of sensor networks. However, this may introduce intermittent functioning that easily degrades crucial components, as suggested in [80] and [81], or overhead in the production impacts the manufacture of energy harvesters. For instance, the environmental impacts of radio-frequency energy harvesting have been investigated in [82], highlighting the need to carry out specific LCA for the IoT even if it uses energy harvesting.

Designers working at a product/end-device level may also consider circularity as an ecodesign strategy but with certain prudency, because strategic design in this context (such as modularity-oriented design) may depend highly on exogenous factors (i.e., user behavior, inverse supply chain issues, etc.). Moreover, it may also provoke rebound effects. For example, the modularity design of Fairphone 3[6] is convenient only if a repaired phone has been used for more than 5 years [83]. In fact, there are extra impacts on the production as modular designs usually need more connectors or more sub-blocks to allow for modularity. Besides that, excessive reparations of modular screens may induce high consumption and losses of critical metals, as is shown for Yttrium and Indium [84].

On the other hand, an important issue when performing LCA for IoT products is the high heterogeneity existing in IoT devices [3]. For instance, this can be seen with sensors and actuators used in an IoT ecosystem (Figure 2.11). They range

[6]Smartphones usually act as edge devices in many IoT applications

from microelectromechanical systems (MEMS) (for temperature, movement, humidity, gas, and pressure sensing), to imagers, relays, and motors. In addition, they cover an ultra-wide range of applications from smart textiles, biomedical wearables, environment monitoring, predictive maintenance in industrial IoT or Industry 4.0, and smart buildings.

Ecodesigning the IoT sensor and actuator building blocks relies on concrete design choices based on environmental metrics during the early development steps, i.e., when the design is not completely fixed in its final configuration. The design choices include environmentally advantageous materials and manufacturing techniques, or the integration of the concept of design for X (X = reuse, disassembly, material recycling, ...), while keeping the device functionalities. Supporting sensors and actuators design with LCA metrics is still in its early stages and remains a crucial challenge, but recent works have begun to address it. For example, Delhaye et al. [85] introduced the co-optimization of technical and environmental performances (embodied energy, carbon footprint, and water consumption) during the early design of MEMS sensors by evaluating two different sensor scenarios (design configuration and process flow). They discuss the performances that could be reached by both processes and the resulting trade-offs in terms of environmental impacts (Figure 2.12). Also, Kokare, S. et al. [86] and Wan, Q. et al. [87] performed comparative LCA of stretchable and rigid electronics-based cardiac monitoring devices, highlighting the lower emissions of the former. Finally, Le Brun et al. [88, 89] investigated the technical feasibility of water quality sensors made of paper-based electronics to substitute their silicon-based counterparts, alongside their respective environmental burdens. Even if the functionalities are not fully substituted yet (lower performances in the case of paper-based sensors), they highlighted that the use of cellulose-based materials instead of silicon significantly reduces the impact per measurement in the case of single-use sensors. In similar ways but on other functional blocks, Zheng, L.R. et al. [90] used LCA to study the toxic emissions (air, seawater, freshwater, and industrial soils) generated during the manufacturing process of different antennas. The environmental assessments of printed radio-frequency identification (RFID) and printed circuit board (PCB) antennas are intended to advise the designer of an IoT system in selecting the most appropriate antenna.

Example 2.6 In order to ease the assessment and foster a more systematic evaluation of IoT edge devices, Pirson, T. & Bol. D. [3] introduced the concepts of 'hardware profiles' and 'functional blocks' as well as quantitative estimates to streamline the carbon footprint evaluation of IoT edge devices. If this is already an interesting step toward life cycle thinking for the IoT, this should be generalized to more environmental indicators.

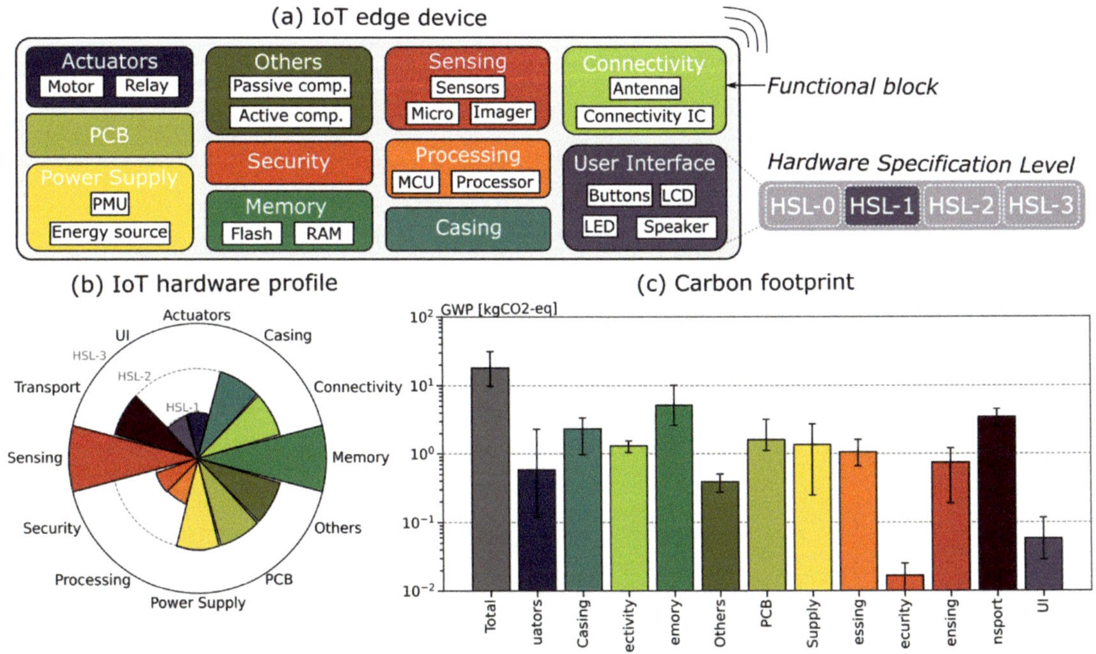

Figure 2.11 Streamlined carbon footprint assessment of IoT edge devices and illustration of the concept of hardware profiles. Figure adapted from [3].

Example 2.7 Delhaye et al. assessed and compared the environmental footprints of two microfabrication processes to support the ecodesign choices for silicon substrate types and the manufacturing process of MEMS pressure sensors based on silicon membrane release [85]. They conclude that dry etching releases approximately 10 times more global warming potential (GWP) emissions than wet etching. A co-optimization of technical and environmental performances was then performed by studying the pressure sensor sensitivities and GWP emissions of four typical scenarios (made of bulk silicon and silicon on insulator (SOI) substrates, and wet and dry etching techniques).

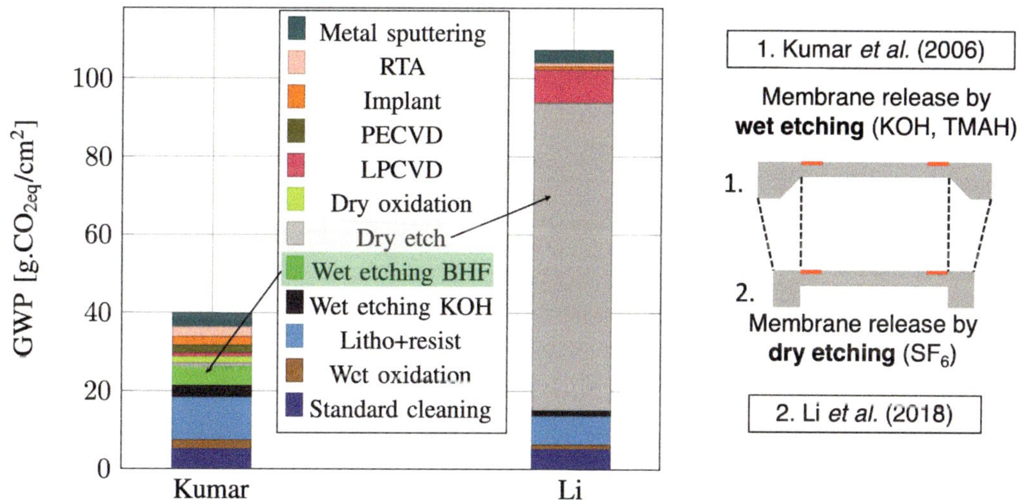

Figure 2.12 Bottom-up process-based comparative LCA of MEMS pressure sensors for co-optimization of sensing and environmental performances. Figure adapted from [85].

For specific studies, the use of simplified LCA can help designers as it can be applied early and in a short time frame during the product development process, overcoming the main limitation of a full LCA (which is time-consuming and requires the knowledge of final product specifications). Typically, simplified LCA can be used to screen materials and manufacturing techniques during an early-stage environmental assessment. Its early application can significantly influence the choice of materials and the development of devices, making it likely to avoid a significant amount of impact in later design stages. When material and process data are easily available (e.g., published generic data on nano-fabrication [91], or databases), it also has the advantage of being a versatile and rapid method, which could be adapted to any process (although it forces the designer to think about the essential data or the substantial information of the IoT system). Yet, a major limitation to introducing environmental assessment into the typical design workflow of device prototypes is that there is currently no data inventory automation, which is mainly to be performed with custom process-based life cycle inventory (e.g., build the process and material models in Excel sheets). In the future, such tools should be added to virtual prototype simulations, i.e., by increasing the common computer-aided design (CAD) and visualization tools with environmental data. For example, the integration of Ansys-Granta EduPack material databases (which contain environmental properties) with the Ansys multiphysics simulation tools (electromagnetism, mechanics, etc.) is a first step in this direction. Besides this, limits to the use of simplified LCA include the possible impact transfers due to the limited number of indicators considered, the general lack of life cycle inventory data in the IoT sector (because of the complex and opaque supply chain, or intellectual property), and the expertise required to identify appropriate end-of-life scenarios (very little overall knowledge about the actual end-of-life treatment of materials).

As a conclusion, at the product and device level, the bottom-up approach is usually preferred in order to better understand the contributions of each functional block in the overall environmental footprint of the device. In fact, they need a sufficient level of granularity in the modeling if the results are to be used for improving the design. Nevertheless, carrying out a complete LCA for each IoT device can be time-consuming and also requires some expertise.

The environmental impacts of an IoT device arise from its many functional blocks made of hardware components. In order to get a complete picture of the device's environmental profile, each of the functional blocks must be correctly modeled. Ecodesigning the device by assessing different scenarios both technically and environmentally (through comparative LCA) allows us to discuss device performances against its environmental burdens, support the development of new research practices or design guidelines, and investigate the most appropriate solutions in terms of environmental costs, design, and functionalities. Bearing in mind all of this, LCA practitioners need also to include in their analysis more dimensions (beyond energy consumption) of electronic components, sensor systems, and edge devices (as Quisbert-Trujillo, E. et al. [81] suggest), so that unintentional impacts can be spotted and avoided (or at least, reduced).

2.4.5 Environmental Considerations in the Manufacturing of IoT Devices

As pointed out throughout this chapter, the IoT technological ecosystem is mainly made up of electronic components, from passive components (e.g., capacitors, resistors, ...) to active ones (e.g., ICs, transistors, ...). Most of the IoT devices are currently fabricated with conventional microfabrication techniques, including subtractive complementary metal–oxide–semiconductor (CMOS) manufacturing, associated with low material efficiencies (e.g., high material losses during deposition processes [92]). From the LCA point of view, this allows us to build upon the already available knowledge regarding the environmental footprint of semiconductor manufacturing. In fact, several components such as integrated circuits, diodes, PCBs, or passive devices are already modeled in state-of-the-art databases specific to electronics. Yet, LCA for semiconductors still suffers from several gaps and blind spots, including high uncertainty in the upstream life cycle stages and supply chain impacts, as further discussed in subsection 2.4.6. More specifically, manufacturing processes evolve quickly to keep up with Moore's Law-like annual improvements [76], therefore introducing more and more advanced manufacturing techniques in the production lines [93]. The introduction of extreme ultraviolet (EUV) for advanced nodes is a telling example. This leads to an important increase in absolute energy consumption, even though the density of transistors increases exponentially at the same time. This is extensively discussed in a recent study from Pirson, T. et al. [94], which points out the critical need to consider both the increase in production volumes and the complexification of manufacturing processes for advanced CMOS technology nodes when analyzing environmental impacts of integrated circuits production [94–96]. Environmental considerations regarding the manufacturing of electronic devices remain limited, although massively produced. Moreover, most of the manufacturing occurs abroad (i.e., in Asia) therefore considerably limiting on-site measurements and LCA model updates. Corporate Social Responsibility (CSR) reports provide high-level aggregated values, but these are usually difficult to manipulate and too aggregated to have a proper understanding of the actual environmental impacts of specific components manufacturing [96].

In addition, the growing market for new IoT devices (e.g., conformal electronics or wearables) is pushing toward electronic components with new functionalities (e.g., flexible PCB) [48]. These roadmaps, therefore, stimulate interest in alternative paradigms (e.g., flexible or printed electronics), manufacturing techniques (e.g., additive manufacturing), and materials (e.g., biodegradable polymers, paper, textiles, functional inks, and nanomaterials), enabling the design of IoT building blocks with novel and relevant properties [97, 98], such as flexible printed PCB [99] or single-use paper-based physiological sensors [100] and batteries [101]. The awareness of the impacts generated by the mass-scale manufacture of sensors deployed within the IoT and the issues related to their end of life is leading to a trend towards integrating flexible substrates and (nano)materials that are simpler to process and to separate, with cost-effective, large-area additive manufacturing techniques (e.g., roll-to-roll processes [99]). Additive manufacturing consists of the additive deposition (or printing) of materials in a desired pattern with higher material efficiencies compared to the subtractive process. It involves fewer manufacturing steps per device and fewer harmful gases, operates at lower temperatures, and is less energy-intensive compared with conventional electronics microfabrication previously described [98].

It follows that printed electronics is often perceived as a "green" technology with a high potential of replacing established subtractive manufacturing methods [102]. However, even if studies show that additive manufacturing techniques and printed electronics are indeed less energy intensive [90, 103–105], there is a real lack of knowledge about the overall impacts of these technologies [102]: either LCA studies at the device level are not (at all) complete due to a lack of information on the components (large amount of possible materials), or a lack of knowledge in the field (early-stage LCA and monitoring of direct effects are very slightly integrated into design practices), or the indicators covered are not all relevant. For example, there is very little information on the actual and up-to-date impacts caused by the production of some functional nanomaterials (e.g., metallic nanomaterials or conductive polymers considered in functional inks for printed electronics), or the impacts that nanomaterials may have on ecosystems or human health during their degradation (e.g., emissions and toxicity potential), because the studies are very few in number, contradictory in their conclusions and not yet completed [102, 106, 107]. Indeed, we do not yet have the hindsight on the impacts generated by these technologies, which have sometimes only been developed in the last 10–15 years, leading to impacts potentially underestimated. Finally, regarding the end-of-life impacts of these new types of printed and flexible devices, although suppliers sometimes sell them as low-impact, recyclable (or "green"/"sustainable") alternatives to present-day electronics thanks to their additive manufacturing and their simple design, the end-of-life channels for the waste generated are absolutely not yet ready, mainly due to the contamination from electronics integrated into the flexible substrates (plastics, textiles, cardboard, papers, ...) that might make it into the recycling streams [104].

In order to leverage these new technology paradigms and judiciously ecodesign future IoT technologies, we argue for the importance of identifying and selecting technologically feasible and environmentally sustainable applications through multi-criteria design methodologies combining user acceptance, techno-economic feasibility, and environmental considerations as early as possible in the development process [102, 108]. Furthermore, the decision

Example 2.8 An extensive review and analysis of the environmental footprint of IC production was carried out by Thibault P. et al. [94]. This work points out the trade-off between technology downscaling and environmental impacts for three environmental indicators, as illustrated in Figure 2.13 for the GWP. It also provides a detailed qualitative analysis regarding the mismatch of system boundaries between existing key studies. Finally, it highlights the evolution of environmental impacts of IC production over 40 years, revealing historical trends and limits of technology downscaling in being an "effective solution" for environmental sustainability.

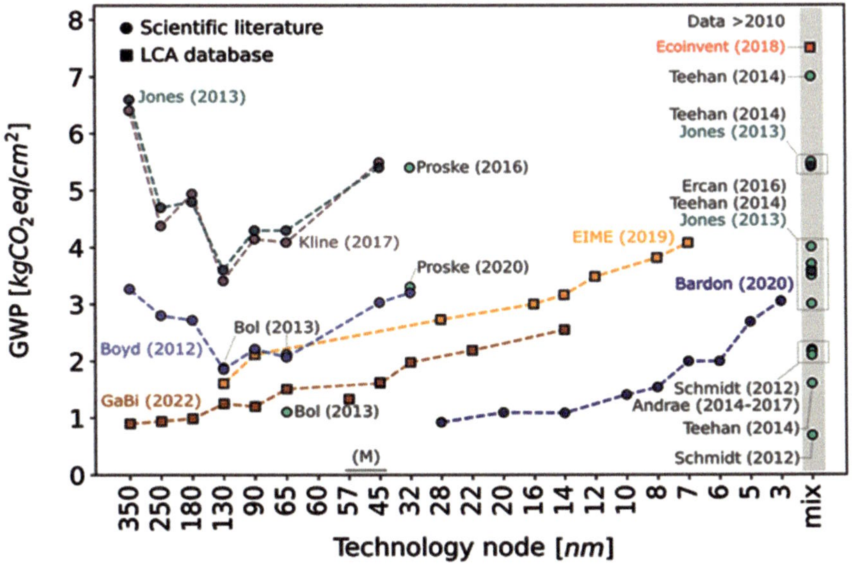

Figure 2.13 Carbon footprint of integrated circuit production per area, for several technology nodes. Figure extracted from [94]. *Disclaimer: there is a scope mismatch between studies. This Figure should be analyzed together with the qualitative analysis in the study* [94].

Example 2.9 Additive manufacturing refers to manufacturing processes that create devices in volume by adding material in successive layers (e.g., by melting plastics or metals), as opposed to conventional manufacturing of high-quality inorganic semiconductors that uses subtractive processes (e.g., by evaporating/sputtering and etching metals and semiconductors). The conventional processes of a significant number of semiconductors or other nanomaterials of modern electronic devices require up to six orders of magnitude or more energy than the energy required for processing a plastic or melting a metal component using additive techniques (typically 1 MJ/kg for aluminum or iron) [98]. Additive manufacturing is nowadays used in industrial processes (e.g., printed and flexible electronics), and is therefore envisioned to enable sustainability in the electronics sector through the usage of materials and technologies that have lower embodied energy (Figure 2.14).

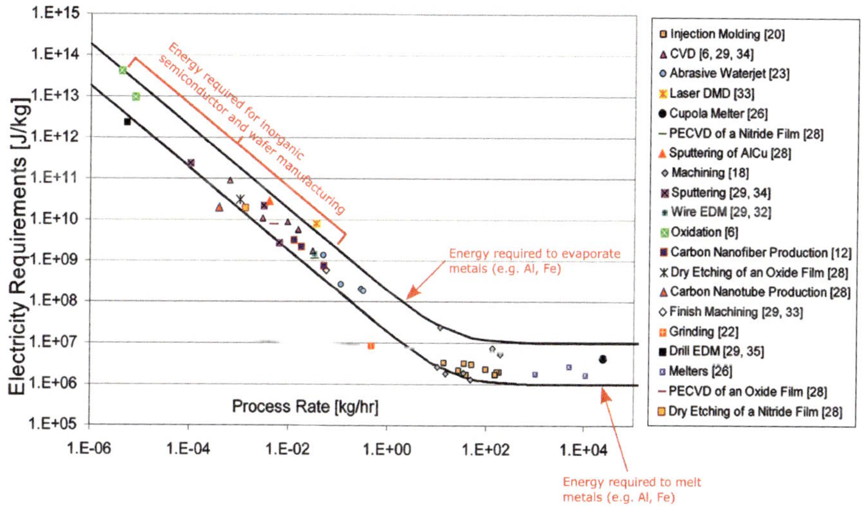

Figure 2.14 Work in the form of electricity used per unit of material processed expended in twenty typical manufacturing techniques, as a function of the rate of material processed (numbers in the brackets represent the reference cited therein). Adapted and reproduced from [98] with permission from the Royal Society of Chemistry.

to implement new electronic paradigms should be supported not only by quantitative proofs in the form of an LCA study (e.g., reducing energy or material consumption of the equipped product system), but also by qualitative analyses going beyond the sole focus on technological aspects.

2.4.6 Supply Chain Considerations in LCA for IoT Devices

The upstream life cycle stages such as semiconductor chemical production, raw materials extraction, and refining are generally poorly taken into account in the LCA of semiconductors and electronic devices, mainly because of the lack of available data. However, previous studies suggest that the production of ultra-pure chemicals could significantly change the hotspots from the IC fabrication to raw material production stages and therefore modify the interpretation of semiconductor LCA studies [93, 109]. In particular, the process of raw silica into semiconductor-grade silicon wafers is a highly energy-intensive process (>7200 MJ/kg Si out), even more if we consider high material losses at each step [110, 111]. In addition, the production of semiconductor chemicals (high-grade ultra-pure gases and chemicals used for microelectronics manufacturing) requires additional purification technologies, resulting in more energy-intensive chemicals. However, when available, the LCI data for these chemicals are usually representative of industrial grades (< 99% purity), while semiconductor grades require 99.9999% (6N) to 99.9999999% (9N) purity [109, 112]. The same reasoning also applies to the increasing number of specific minerals and metals used for IC manufacturing, e.g., germanium, ruthenium, cobalt, or tungsten for interconnects. These materials are associated with higher embodied energies and carbon footprints as their production (extraction and refinement) is very energy-intensive [113]. Being present in very small quantities in the ores, many minerals required for IC production are extracted as by-products from the production of major metals. Therefore, the energy of mineral extraction and refining is expected to increase with the progressive decrease of ore grade, becoming less concentrated in minerals

[114]. In addition, recent emerging issues linked to the availability and highly volatile prices of certain material elements like indium, ruthenium, platinum, and other precious metals, make them listed as critical materials by the European Union [115], and are questioning their future use in the microelectronics industry.

Geopolitics, the growing mineral demand to supply the high-tech and renewable energy industries supporting digitalization, and the rarefaction of high-grade mineral deposits (e.g., a lower number of sources of high-quality sand needed to process quartz for silicon wafers) have put the mineral and semiconductor material supply under pressure [116]. Therefore, the questions regarding the future availability of metals and industrial mineral resources have opened up new avenues of metal extraction. For example, the potential extraction of mineral resources from the deep-sea area can play a major role in this supply, e.g., exploitation of polymetallic nodules for coltan mining in the Clarion Clipperton Zone [117]. The extraction in such harsh environments could cause different damages than usual, onshore mining, and suggests considering LCA impacts on the biodiversity (e.g., sea) in addition to the commonly reported ones (e.g., GWP, acidification, eutrophication) [118, 119].

Aside from specific conventional materials, new IC fabrication processes still involve more and more ultra-pure gazes and exotic minerals (e.g., gadolinium, praseodymium, and holmium) in contact with the silicon. Altogether, this lack of knowledge about supply chain material impacts could lead to an underestimation of the actual environmental impacts of semiconductor production. Nevertheless, if it is already challenging to draw a relevant picture of the environmental impacts of the electronic manufacturing process, it only scratches the complexity of extending the scope to consider the whole IoT life cycle and more environmental indicators. In addition, it is necessary to anticipate future changes in the consumption and possible supply issues of raw materials in the microelectronics and IoT industries in order to prevent disruptions or high costs, either by material substitution or by reducing their consumption (by orders of decades) in a drastic way in the near future.

Example 2.10 The supply of raw materials needed by present and future IoT devices raises questions about its global sustainability. The production of many minerals and metals composing the IoT devices is highly energy intensive, and their refinement performed most of the time outside Occidental countries, also leads to emissions of greenhouse gasses (Figure 2.15).

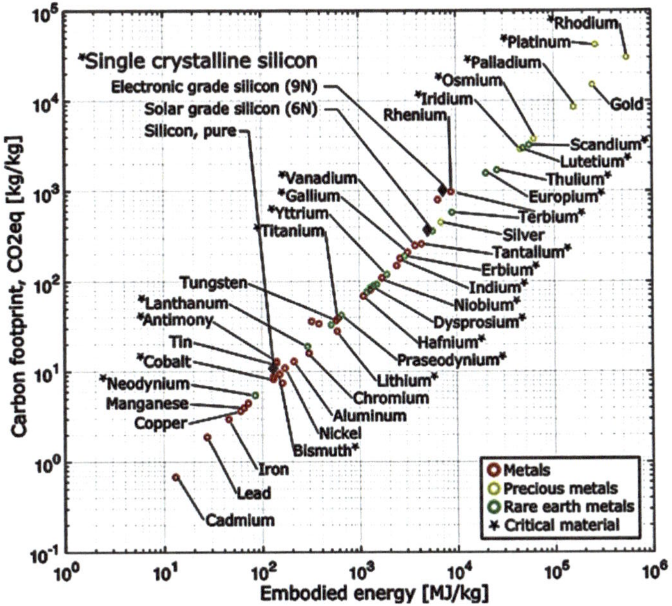

Figure 2.15 Embodied energy and global warming potential expressed in CO_2-eq of the extraction and refinement of standard industrial grade raw metals. Precious metals present the highest requirements in energy for production, followed by rare earth metals and metals. This can be directly correlated with their abundance and concentration in Earth's crust. Data adapted from [113].

The impacts of metal extraction are highly correlated with their concentration in the ores (Figure 2.16). Therefore, the rarefaction of many minerals (e.g., Cu, In, Zn, Au,...) widely used in the ICTs could only worsen the situation and call for re-thinking the IoT industry globally. Many ways could be explored, including the development of new technologies and materials to replace them, or to drastically limit their amount in some critical devices (sensors, memories,...).

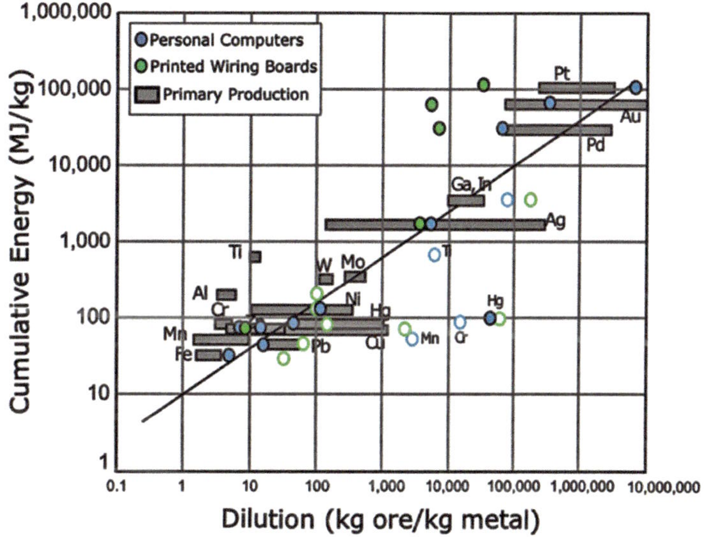

Figure 2.16 Graph of cumulative energy of metal production versus metal dilution (i.e., the inverse of concentration) in its original ore deposit (gray boxes), in personal computers (filled blue circles), and in printed circuit boards (filled green circles). The energy required for metal production is mainly driven by the original ore deposit abundance (concentration). Filled circles above the line are metals predominantly recycled; those metals below the black line (signified by empty circles) are too diluted compared to their concentration in ore deposits to be recycled at affordable price and energy cost. Graph reproduced from [114] with permission.

2.4.7 Limitations

Despite the results presented in the previous sections, examples of complete ecodesign of IoT systems are still scarce in the literature. Several factors limiting this complete ecodesign could be pointed out but the most important one identified in this work is clearly the lack of (vertical) cross-layer ecodesign and LCA-based methodology for the whole IoT technological ecosystem. In fact, if some isolated life cycle considerations are present within some layers of the IoT technological ecosystem as shown in this chapter, it is very rare to find studies taking into account the whole ecosystem at once and alongside the development of the IoT systems solutions, as depicted by red arrows in Figure 2.17. Here below, we provide more details about specific limitations preventing this cross-layer ecodesign.

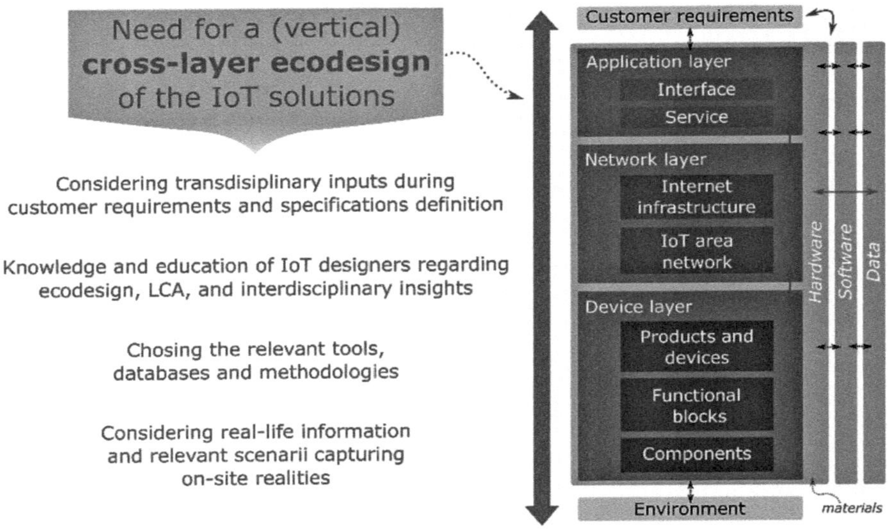

Figure 2.17 Need for a cross-layer integration of ecodesign practices in the design of IoT systems.

First, transdisciplinary inputs must be considered at the early stages of the design process. Indeed, this should be used as an opportunity to investigate both technical and non-technical solutions or a mix of both. Moreover, this would allow us to discuss some specifications that can have an important impact on the design of the other layers. A better understanding of the behaviors of users during the use phase will help to adapt the specifications accordingly. Moreover, it is necessary to start with a rational design from the proper definition of the services and main functions of IoT systems to effectively and efficiently satisfy the requirements from a sober treatment of information. Of course, this would help a lot to avoid oversizing the local equipment and mutualized infrastructure.

Second, due to a lack of general knowledge and education in the field, research and development teams often struggle to know where to start ecodesigning the highly heterogeneous field of IoT systems. In fact, most IoT designers (no matter the IoT layer) are not yet trained to transform environmental parameters into quantities related to human action, i.e., design decisions at the different IoT layers. Understanding environmental impacts is one thing, drawing conclusions on how to design technical solutions is another. They can find the apparent complexity of environmental issues daunting: the systemic approach needed to understand environmental burdens and their links with ecodesign requires a high level of effort, learning, documentation, and questioning to avoid the classic pitfalls (going beyond the energy-centered or "greenwashed" visions conveyed by many general communication channels, "carbon tunnel vision", understanding the interaction between indicators, targeting the correct hotspots, and avoiding the transfer of impacts from one dimension to another), which requires an important amount of time and expertise. In addition, IoT designers should be trained to include more interdisciplinary insights in the design process to better understand the trade-offs between the different layers and disciplines [120].

Then, choosing the relevant tools, databases, and methodologies remains a non-trivial task for designers [121]. There is a clear lack of methods and reliable data, and assessment tools are not adapted or integrated with existing design tools. Engaging in open data and open models in this sector could have a very strong impact, although this would require dealing with omnipresent confidentiality barriers. Most of the time, it is very difficult to develop common models that can be updated, challenged, and used as reference by the community. A significant step forward would be to recognize methodologically big impact contributors from basic inspection of electronic components (as proposed by [1]) or to provide life cycle details directly in electronic components' datasheets [16], which is far from the case today. For instance, even basic information such as die size and technology node are extremely difficult to obtain for IC. In addition, the very low transparency for industry data in public reports drastically complicates the understanding and the use of the results. Nowadays, Environmental Safety and Health (ESH) and CSR reports clearly fail to provide sufficient information to understand the actual environmental impacts of semiconductor production [94, 96].

Finally, real-life information and relevant scenarios should be considered in order to represent on-site realities, rather than models failing to provide a representative picture of the environmental impacts of IoT systems. Dynamic LCA models would be of high interest as current static LCA models fail to capture the evolution of IoT systems through time. This would be of high interest regarding the modeling of deployment, maintenance, and decommissioning phases, which are most often overlooked or not representative of reality.

2.5 Conclusions

Digital technologies are extremely recent in human history. The design tools and methodologies have changed considerably since their beginnings about fifty years ago. Overall, the historical evolution is all about efficiency improvement in metrics such as computing power per device and energy consumption per function, at the cost of exponential complexification. This has been made possible by a combination of automation tools and business stratification with increasingly specialized business units. The tools and methodologies are likely to continue to evolve in the future, for example, by adopting AI technologies, but hopefully also by taking environmental constraints into account. Nevertheless, the integration of environmental considerations and metrics in the conventional design process of technologies is still very scarce and far from systematic.

In this chapter, we therefore introduce existing tools, standardized methodologies, and inspiring examples to include environmental considerations in the design, manufacture, use, and end-of-life of IoT devices. The chapter offers a pedagogical kick-off summary for IoT engineers and designers willing to include environmental impacts and ecodesign considerations in their work, mainly based on the inputs from LCA. Practical examples are given for each layer of the IoT technological ecosystem and some known limitations are also pointed out to provide the reader with a streamlined yet complete overview of ecodesign and LCA of IoT systems.

Finally, we showed that cross-layers integration of LCA in the development of full IoT systems is urgently needed to support a judicious ecodesign of an end-to-end application, rather than localized optimization on sub-parts of a system. This also includes the need to integrate applications and users in the process of the impact evaluation to avoid rebound effects. Indeed, having access to simpler, cheaper, generalizable technological solutions is the open door to induction and rebound effects, which could backfire on the environmental advantages expected from such "solutions". Such changes call for new business models that highlight service and environmental impacts reduction on a global scale, rather than the traditional "product-oriented" current vision.

2.6 Perspectives and Further Work

Integrating LCA into current design flows and IoT design teams requires proper training and preparation. However, it is currently not a part of the classic development flow in the electronics sector, and engineers are rarely trained to integrate LCA and ecodesign into their every-day practice. To address this, future prototype simulations should be equipped with environmental data by increasing common CAD design and visualization tools with LCA methodology.

To support ecodesign, researchers must identify key IoT specificity whose impacts span several indicators and develop common procedures to transform these environmental parameters into quantities related to human action and design decisions at different IoT layers. This approach must be dynamic and agile enough to adapt to constraints and design changes over time to ensure sound decisions are made throughout the entire design stack. Ongoing work in this regard tries to expand this reflection, continuing the efforts already started by [5] and [1].

The ecodesign process requires appropriate environmental assessment tools to enable sound decisions from early research phases. As highlighted in this chapter, it is important to develop simplified LCA methods for each IoT layer that are coherent with full LCA approaches. Such simplified LCA approaches must be straightforward enough to allow regular checks and dynamic adjustments throughout the design process, and account for the increasing LCA modeling data quality (from conceptual design to real-scenario deployment). If homogeneously implemented for the different IoT layers, it will help facilitate communication about environmental aspects across all layers of the IoT system.

To address environmental concerns in the IoT sector, it is important to hold all actors in the chain responsible for the environmental damage or burden generated. However, this can be complicated by the long and complex electronics supply chain. To simplify this process, there should be an increase in reporting and accounting of actions, consumption, and damages of the building blocks of IoT networks, from raw materials to telecommunication networks.

By making all actors in the chain responsible for their actions and increasing accountability and innovation, the environmental impact of IoT products and processes can be better understood and mitigated, e.g., by the implementation of extended producer responsibility (EPR) environmental policies across all the IoT layers [122], and the introduction of extended consumer responsibility (ECR) within new business models.

In this sense, ongoing work is been made for developing and validating practical methodologies for sharp environmental assessment and ecodesign. These methodologies consider all stakeholders in real environments, such as Industry 4.0, and promising systems (e.g., long-distance/low-energy technology, intermittent computing, or energy harvesting). By considering different perspectives, this effort mainly considers the full life cycle and architecture of IoT systems and covers the mobilized resources, especially in terms of (i) new business models, (ii) energy used and data flow, (iii) the essential information of services that cover customers' needs, and (iv) critical dimensions of analysis (including physical, technical, and circularity aspects of electronic components), as it has been already undertaken in [1] and proposed in [81].

Finally, although we bring forward the need for ecodesign in this chapter, supported by the LCA methodology, it is important for LCA practitioners to be aware that LCA is mainly focused on improving the environmental performance of a product (or a service). This is therefore not sufficient to ensure that the absolute environmental impacts will decrease and that the environmental footprint of humanity will remain within planetary boundaries, no matter how advanced was the ecodesign process. Judicious ecodesign, therefore, calls for questioning the scale of deployment for the technological solution together with the real need for a technological solution. These are important aspects in the case of IoT as it pushes toward the massive deployment of electronics in our environment. In this context, increasing interdisciplinary and transdisciplinary interactions among designers, users, and policymakers would be of great interest.

Acknowledgments

- The work of Thibault Pirson was supported by the Fonds Européen de Développement Régional (FEDER) and the Wallonia within the Wallonie-2020 EU program.
- The work of Grégoire Le Brun was supported by the Fonds pour la formation à la Recherche dans l'Industrie et dans l'Agriculture (FRIA) through the Fonds de la Recherche Scientifique-FNRS.
- The work of Ernesto Quisbert-Trujillo was supported by the French National Research Agency in the framework of the "Investissements d'avenir" program (ANR-15-IDEX-02).

References

1. E. Quisbert-Trujillo, *Design methodology for sustainable IoT systems (Doctoral dissertation, University of Grenoble Alpes)*, Grenoble, 2022.
2. International Telecommunication Union (ITU), "Internet waste," International telecommunication union and the WEEE forum, Geneva, 2020.
3. T. Pirson and D. Bol, "Assessing the embodied carbon footprint of IoT edge devices with a bottom-up life-cycle approach," *Journal of Cleaner Production*, vol. 322, p. 128966, 11/2021.
4. McKinsey Global Institute, "The Internet of Things: Mapping the value beyond the hype," 2015.
5. E. Quisbert-Trujillo, T. Ernst, K. Samuel Evrard, E. Cor, and E. Monnier, "Lifecycle modeling for the eco design of the Internet of Things," in *Procedia CIRP*, 2020.
6. L. Shancang, D. X. Li, and Z. Shanshan, "The internet of things: A survey," *Information Systems Frontiers*, vol. 17, p. 243–259, 2015.
7. L. M. Hilty and B. Aebischer, *ICT innovations for sustainability*, vol. 310, Springer, 2015.
8. V. P. Kafle, Y. Fukushima, and H. Harai, "Internet of things standardization in ITU and prospective networking technologies," *IEEE Communications Magazine*, vol. 54, no. 19, pp. 43–49, 2016.
9. International Telecomunication Union (ITU), "Y.4113: Requirements of the network for the Internet of things," 2016.
10. Global Enabling Sustainability Initiative (GeSi), "SMARTer2030 ICT Solutions for 21st Century Challenges," 2015.
11. C. Freitag, M. Berners-Lee, K. Widdicks, B. Knowles, G. S. Blair, and A. Friday, "The real climate and transformative impact of ICT: A critique of estimates, trends, and regulations," *Patterns*, vol. 2, no. 19, p. 100340, 2021.
12. E. Wagner, C. Böhme, S. Benecke, N. N. Nissen, and K. D. Lang, "Advanced packaging for wireless sensor nodes in cyber-physical systems-impacts of multifunctionality and miniaturization on the environment.," chez *2017 IEEE International Conference on Prognostics and Health Management (ICPHM)*, 2017.
13. Y. Lu and L. D. Xu, "Internet of Things (IoT) cybersecurity research: A review of current research topics," *IEEE Internet of Things Journal*, vol. 6, no. 12, pp. 2103–2115, 2019.

14. Intergovernmental Panel on Climate Change (IPCC), "Climate change 2022: Mitigation of climate change," 2022.

15. M. Ercan, J. Malmodin, P. Bergmark, E. Kimfalk, and E. Nilsson, "Life cycle assessment of a smartphone," chez *ICT for Sustainability 2016*, 2016.

16. P. V. C. Louis-Philippe, Q. E. Jacquemotte, and L. M. Hilty, "Sources of variation in life cycle assessments of smartphones and tablet computers," *Environmental Impact Assessment Review*, vol. 84, p. 106416, 2020.

17. M. Koot and F. Wijnhoven, "Usage impact on data center electricity needs: A system dynamic forecasting model," *Applied Energy*, vol. 291, p. 116798, 2021.

18. J. Bonvoisin, A. Lelah, F. Mathieux, and D. Brissaud, "An environmental assessment method for wireless sensor networks," *Journal of Cleaner Production*, vol. 33, pp. 145–154, 2012.

19. J. Bonvoisin, A. Lelah, F. Mathieux, and D. Brissaud, "An integrated method for environmental assessment and ecodesign of ICT-based optimization services," *Journal of Cleaner Production*, vol. 68, pp. 144–154, 2014.

20. P. Maistriaux, T. Pirson, M. Schramme, J. Louveaux, and D. Bol, "Modeling the carbon footprint of battery-powered IoT sensor nodes for environmental-monitoring applications," chez *Proceedings of the 12th International Conference on the Internet of Things*, 2022.

21. C. P. Baldé, E. D'Angelo, V. Luda, O. Deubzer, and R. Kuehr, "Global Transboundary E-Waste Flows Monitor 2022," United Nations Institute for Training and Research (UNITAR), Germany, 2022.

22. E. Fangeat, A. Deprouw, M. Jover, M. Borie, and O. Tonsart, "Rapport annuel du registre des déchets d'équipements électriques and électroniques," In *Extenso Innovation Croissance and ADEME*, Angers, 2020.

23. ISO, "ISO 14006:2020 Environmental management systems — Guidelines for incorporating ecodesign," 2020. [Online]. Available: https://www.iso.org/standard/72644.html.

24. ISO, "ISO/TR 14062:2002 Environmental management — Integrating environmental aspects into product design and development," 2002. [Online]. Available: https://www.iso.org/standard/33020.html.

25. K. R. Fowler and C. L. Silver, *Developing and managing embedded systems and products: Methods, techniques, tools, processes, and teamwork*, Newnes, 2015.

26. K. Ramani, D. Ramanujan, W. Z. Bernstein, F. Zhao, J. Sutherland, C. Handwerker, J.-K. Choi, H. Kim, and D. Thurston, "Integrated sustainable life cycle design: A review," *J. Mech. Des.,* vol. 132(9): 091004, September 2010.
27. D. Su, *Introduction and sustainable product development,* D. Su, Éd., Springer, Cham, 2020.
28. P. Morseletto, "Targets for a circular economy," *Resources, Conservation and Recycling,* vol. 153, p. 104553, 2020.
29. E. MacArthur, "Circular consumer electronics: An initial exploration," 2018.
30. TPMB, "Using V-diagrams in engineering projects," 2019. [Online]. Available: https://www.theprojectmanagementblueprint.com/blog/procurements-management/the-v-diagram.
31. G. INP, "D.1.1: State of the art of ecodesign tools applicability in different product design stages," G.EN.ESI EU Project VII Framework Programme, 2015.
32. P. Rousseaux, C. Gremy-Gros, M. Bonnin, C. Henriel-Ricordel, P. Bernard, L. Floury, G. Staigre, and P. Vincent, "Eco-tool-seeker: A new and unique business guide for choosing ecodesign tools," *Journal of Cleaner Production,* vol. 151, pp. 546–577, 2017.
33. S. Ferrendier, F. Mathieux, G. Rebitzer, M. Simon, and D. Froelich, "Environmentally improved product design case studies of the european electrical and electronics industry," European Commission, 2002.
34. EU, "EU Ecodesign Directive 2009/125/EC of 21th october 2009 establishing a framework for the setting of ecodesign requirements for energy-related product," 2009. [Online]. Available: https://eur-lex.europa.eu/legal-content/EN/TXT/PDF/?uri=CELEX:02009L0125-20121204 from=EN.
35. EU, "Ecodesign for sustainable products," 2022. [Online]. Available: https://commission.europa.eu/energy-climate-change-environment/standards-tools-and-labels/products-labelling-rules-and-requirements/sustainable-products/ecodesign-sustainable-products_en.
36. EU, "Directive 2002/95/EC of the European Parliament and of the Council of 27 January 2003 on the restriction of the use of certain hazardous substances in electrical and electronic equipment," 2002. [Online]. Available: https://eur-lex.europa.eu/legal-content/FR/ALL/?uri=CELEX:32002L0095.

37. CEN-CENELEC, "Ecodesign, energy labelling and material efficiency," 2023. [Online]. Available: https://www.cencenelec.eu/areas-of-work/cen-cenelec-topics/ecodesign-energy-labelling-and-material-efficiency/.

38. ITU-T, "SG5 - EMF, environment, climate action, sustainable digitalization, and circular economy," 2022. [Online]. Available: https://www.itu.int/en/ITU-T/studygroups/2022-2024/05/Pages/default.aspx.

39. ECHA, "REACH," [Online]. Available: https://echa.europa.eu/fr/regulations/reach/understanding-reach.

40. R. v. Berkel, E. Willems, and M. Lafleur, "Development of an industrial ecology toolbox for the introduction of industrial ecology in enterprises—I," *Journal of Cleaner Production,* vol. 5, pp. 11–25, 1997.

41. F. Berthoud, C. Gossart, P. Balin, B. A. C. Charbuillet, E. Drezet, J.-D. Dubois, and M. Parry, "Impacts écologiques des Technologies de l'Information and de la Communication," *EDP Sciences,* 2020.

42. ISO, "ISO 14040:2006 Environmental management–Life cycle assessment–Principles and framework," 2006.

43. M. Proske, C. Clemm, and N. Richter, "Life cycle assessment of the fairphone 2," *Fraunhofer IZM,* 2016.

44. T. Heinis, C. Gomes Martinho, and M. Meboldt, "Fundamental challenges in developing internet of things applications for engineers and product designers," in *21st International conference on engineering design, ICED17,* Vancouver, 2017.

45. A. Middendorf, S. Deyter, J. Gausemeier, N. F. Nissen, and H. Reichl, "Integration of reliability and environmental aspects in early design stages of mechatronics," in *2009 IEEE International Symposium on Sustainable Systems and Technology,* Tempe, 2009.

46. J. Huang, Y. Meng, X. Gong, Y. Liu, and Q. Duan, "A novel deployment scheme for green internet of things," *IEEE Internet of Things Journal,* vol. 1, no. 2, pp. 196–205, 2014.

47. C. Zhu, V. C. M. Leung, L. Shu, and E. C.-H. Ngai, "Green internet of things for smart world," *IEEE Access,* vol. 3, pp. 2151–2162, 2015.

48. O. Gurova, T. Merritt, L. Papachristos, and J. Vajakaari, "Sustainable solutions for wearable technologies: Mapping the product development lifecycle," *Sustainability,* vol. 12, no. 120, p. 8444, 2020.

49. R. Arshad, S. Zahoor, M. A. Shah, A. Wahid, and H. Yu, "Green IoT: An investigation on energy saving practices for 2020 and beyond," *IEEE Access,* vol. 5, pp. 15667–15681, 2017.

50. I. T. U. (ITU), "ITU-T L.1370: Sustainable and intelligent building services," 2020.

50a. M. Z. Hauschild, R. K. Rosenbaum, and S. I. Olsen, *Life Cycle Assessment*. Springer, 2018, https://doi.org/10.1007/978-3-319-56475-3.

50b. European Commission - Joint Research Centre (JRC) - Institute for Environment and Sustainability, "International Reference Life Cycle Data System (ILCD) Handbook–General guide for Life Cycle Assessment-Detailed guidance," March, 2010, https://eplca.jrc.ec.europa.eu/uploads/ILCDHandbook-General-guide-for-LCA-DETAILED-GUIDANCE-12March2010-ISBN-fin-v1.0-EN.pdf.

51. M. Huijbregts, Z. Steinmann, P. Elshout, G. Stam, F. Verones, M. Vieira, and R. Van Zelm, "ReCiPe2016 v1.1. A harmonized life cycle impact assessment method at midpoint and endpoint level. Report I: Characterization," 2017.

52. J. Guinée, M. Gorrée, R. Heijungs, G. Huppes, R. Kleijn, A. D. Koning, L. V. Oers, A. Wegener Sleeswijk, S. Suh, H. Udo de Haes, H. D. Bruijn, R. V. Duin, and M. Huijbregts, *Handbook on life cycle assessment: Operational guide to the ISO standards*, Part III: Scientific background, Kluwer Academic Publishers, 2002.

53. Intergovernmental Panel on Climate Change (IPCC), "IPCC fifth assessment report. The physical science basis," 2013. [Online]. Available: http://www.ipcc.ch/report/ar5/wg1/.

54. European Telecommunications Standards Institute (ETSI), "ETSI TS 103 199 V1.1.1 – Environmental Engineering (EE) – Life Cycle Assessment (LCA) of ICT equipment – General methodology and common requirements," 2011.

55. International Telecommunication Union (ITU), "ITU-T L.1410: Methodology for environmental life cycle assessments of information and communication technology goods, networks and services," 2014.

56. Sphera, "Life cycle assessment lca product carbon footprint," 2023. [Online]. Available: https://sphera.com/life-cycle-assessment-lca-product-carbon-footprint-pcf/. [Accessed 17 March 2023].

57. PRé Sustainability B.V., "Simapro," [Online]. Available: https://simapro.com/. [Accessed 17 March 2023].

58. Bureau Veritas, "EIME - LCA and Eco-design Software," [Online]. Available: https://demo.bveime.com/login. [Accessed 17 March 2023].

59. Greendelta, "OpenLCA," [Online]. Available: https://www.openlca.org/. [Accessed 17 March 2023].

60. Brightway Developers, “Brightway life-cycle assessment framework,” [Online]. Available: https://documentation.brightway.dev/en/latest/index.html. [Accessed 17 March 2023].

61. Sphera, “Search managed LCA content (GaBi),” [Online]. Available: https://sphera.com/product-sustainability-gabi-data-search/. [Accessed 17 March 2023].

62. Bureau Veritas, “Negaoctet,” [Online]. Available: https://negaoctet.org/. [Accessed 17 March 2023].

63. Bureau Veritas, “CODDE,” [Online]. Available: https://codde.fr/evenement/eime-nouvelle-version-de-la-base-de-donnees-codde. [Accessed 17 March 2023].

64. Ecoinvent, “Ecoinvent,” [Online]. Available: https://ecoinvent.org/. [Accessed 17 March 2023].

65. ANSYS Inc, “Ansys Granta EduPack,” [Online]. Available: https://www.ansys.com/products/materials/granta-edupack. [Accessed 17 March 2023].

66. ADEME, “Bilans GES,” [Online]. Available: https://bilans-ges.ademe.fr/. [Accessed 17 March 2023].

67. soda4LCA 7.5.5, “WEEE LCI Database,” [Online]. Available: https://weee-lci.ecosystem.eco/. [Accessed 17 March 2023].

68. A. Lelah, F. Mathieux, and D. Brissaud, “Contributions to eco-design of machine-to-machine product service systems: The example of waste glass collection,” *Journal of Cleaner Production,* vol. 19, no. 19–10, pp. 1033–1044, 2011.

69. E. Ingemarsdotter, D. Diener, S. Andersson, C. Jonasson, A.-C. Mellquist, T. Nyström, E. Jamsin, and R. Balkenende, “Quantifying the net environmental impact of using iot to support circular strategies—The case of heavy-duty truck tires in Sweden,” *Circular Economy and Sustainability,* vol. 1, no. 2, pp. 613–650, 2021.

70. E. Dekoninck and F. Barbaccia, “Streamlined assessment to assist in the design of internet-of-things (IOT) enabled products: A case study of the smart fridge,” in *Proceedings of the Design Society: International Conference on Engineering Design*, 2019.

71. N. Ullrich, F. M. Piontek, C. Herrmann, A. Saraev, and T. Viere, “Estimating the resource intensity of the internet: A meta-model to account for cloud-based services in LCA,” chez *The 29th CIRP Conference on Life Cycle Engineering*, Leuven, Belgium, 2022.

72. L. Golard, J. Louveaux, and D. Bol, “Evaluation and projection of 4G and 5G RAN energy footprints: The case of Belgium for 2020–2025,” *Annals of Telecommunications,* pp. 1–15, 2022.

73. J. Malmodin, Å. Moberg, D. Lundén, G. Finnveden, and N. Lövehagen, "Greenhouse gas emissions and operational electricity use in the ICT and entertainment & media sectors," *Journal of Industrial Ecology,* vol. 14, no. 5, pp. 770–790, 2010.

74. J. Aslan, K. Mayers, J. G. Koomey, and C. France, "Electricity intensity of internet data transmission: Untangling the estimates," *Journal of Industrial Ecology,* vol. 22, no. 14, pp. 785–798, 2018.

75. O. Manz, S. Meyer, and C. Baumgartner, "Life cycle assessment of an internet of things product: Environmental impact of an intelligent smoke detector," in *11th International Conference on the Internet of Things*, St.Gallen Switzerland, 2021.

76. D. Bol, T. Pirson, and R. Dekimpe, "Moore's Law and ICT Innovation in the Anthropocene," chez *in Proc. IEEE Design, Automation and Test in Europe Conf.*, 2021.

77. Cisco, "Annual Internet Report (2018–2023)," 2020.

78. C. A. Gray, *Energy consumption of Internet of Things Applications and Services (Doctoral dissertation, University of Melbourne),* Melbourne, 2018.

79. R. Alvarenga, N. Préat, C. Duhayon, and J. Dewulf, "Prospective life cycle assessment of metal commodities obtained from deep-sea polymetallic nodules," *Journal of Cleaner Production,* vol. 330, p. 129884, 2022.

80. T. Daulby, A. Savanth, A. S. Weddell, and G. V. Merrett, "Comparing NVM technologies through the lens of intermittent computation," in *Proceedings of the 8th International Workshop on Energy Harvesting and Energy-Neutral Sensing Systems*, 2020.

81. E. Quisbert-Trujillo, T. Ernst, K. Samuel Evrard, E. Monnier, and M. Gallardo, "A methodology for supporting the sustainable future and eco design of the internet of things," in *International Conference on Sustainable Technology and Development*, 2021.

82. M. Gonzalez, P. Xu, R. Dekimpe, M. Schramme, I. Stupia, T. Pirson, and D. Bol, "Technical and ecological limits of 2.45-GHz wireless power transfer for battery-less sensors," *IEEE Internet of Things Journal, in press,* 2023.

83. M. Proske, D. Sánchez, C. Clemm, and S.-J. Baur, "Life cycle assessment of the Fairphone 3," *Fraunhofer IZM*, Berlin, 2020.

84. M. Ljunggren Söderman and H. André, "Effects of circular measures on scarce metals in complex products – Case studies of electrical and electronic equipment," *Resources, Conservation and Recycling,* vol. 151, p. 104464, 2019.

85. T. P. Delhaye, G. Le Brun, D. Flandre, and J.-P. Raskin, "Bottom-up life-cycle assessment of MEMS piezoresistive pressure sensors," chez *2021 Symposium on Design, Test, Integration & Packaging of MEMS and MOEMS (DTIP)*, Online, 2021.

86. S. Kokare, F. Asif, and G. Mårtensson, "A comparative life cycle assessment of stretchable and rigid electronics: A case study of cardiac monitoring devices," *Int. J. Environ. Sci. Technol*, vol. 19, p. 3087–3102, 2022.

87. Q. Wan, Z. Zou, and L. Zheng, "Life cycle assessment of paper based printed interconnections for ECG monitoring," *European Journal of Engineering Research and Science*, vol. 2, no. 16, 2017.

88. G. Le Brun and J.-P. Raskin, "Material and manufacturing process selection for electronics ecodesign: Case study on paper-based water quality sensors," *Procedia CIRP*, vol. 90, pp. 344–349, 2020.

89. G. Le Brun, M. Hauwaert, A. Leprince, K. Glinel, J. Mahillon, and J.-P. Raskin, "Electrical characterization of cellulose-based membranes towards pathogen detection in water.," *Biosensors*, vol. 11, no. 157, 2021.

90. L.-R. Zheng, H. Tenhunen, and Z. Zou, "Life cycle assessment (LCA) for printed electronics," chez *Smart Electronic Systems: Heterogeneous Integration of Silicon and Printed Electronics*, Wiley, 2018.

91. N. Krishnan, S. Boyd, A. Somani, S. Raoux, D. Clark, and D. Dornfeld, "A hybrid life cycle inventory of nano-scale semiconductor manufacturing," *Environmental Science & Technology*, vol. 42, no. 18, pp. 3069–3075, 2008.

92. A. Kotliarenko, O. Azzolini, G. Keppel, C. Pira, and J. Esposito, "Investigation of a possible material-saving approach of sputtering techniques for radiopharmaceutical target production," *Applied Sciences*, vol. 11, no. 119, p. 9219, 2021.

93. M. G. Bardon, P. Wuytens, L. Å. Ragnarsson, G. Mirabelli, D. Jang, G. Willems, and B. Parvais, "DTCO including sustainability: Power-performance-area-cost-environmental score (PPACE) analysis for logic technologies," chez *2020 IEEE International Electron Devices Meeting (IEDM)*, 2020.

94. T. Pirson, T. Delhaye, A. Pip, G. Le Brun, J. P. Raskin, and D. Bol, "The environmental footprint of IC production: Review, analysis and lessons from historical trends," vol. 36, no. 11, pp. 56–67, 2023.

95. T. Pirson, T. Delhaye, A. Pip, G. Le Brun, J. P. Raskin, and D. Bol, "The environmental footprint of IC production: Meta-analysis

and historical trends," chez *ESSDERC 2022-IEEE 52nd European Solid-State Device Research Conference (ESSDERC)*, 2022.

96. G. Roussilhe, T. Pirson, M. Xhonneux, and D. Bol, "From silicon shield to carbon lock-in? The environmental footprint of electronic components manufacturing in taiwan (2015–2020)," arXiv preprint arXiv:2209.12523, 2022.
97. K. Fu, Z. Wang, J. Dai, M. Carter, and L. Hu, "Transient electronics: Materials and devices," *Chem. Mater.*, vol. 28, no. 111, pp. 3527–3539, 2016.
98. M. Irimia-Vladu, ""Green electronics: Biodegradable and biocompatible materials and devices for sustainable future," *Chem. Soc. Rev.*, vol. 43, no. 1588, 2014.
99. T. H. Phung, A. N. Gafurov, I. Kim, S. Y. Kim, K. M. Kim, and T. M. Lee, "IoT device fabrication using roll-to-roll printing process," *Nature*, 2021.
100. S. Liu, D. S. Yang, S. Wang, H. Luan, Y. Sekine, J. B. Model, A. J. Aranyosi, R. Ghaffari, and J. A. Rogers, "Soft, environmentally degradable microfluidic devices for measurement of sweat rate and total sweat loss and for colorimetric analysis of sweat biomarkers," *EcoMat.*, vol. 5, no. 11, 2022.
101. R. Moumneh, R. Hanus, G. Le Brun, J.-P. Raskin, and L. Francis, "Paper-based stacked reverse electrodialysis cells for energy generation from salinity gradient," chez *Micro and Nano Engineering Conference*, Torino, Italy, 2021.
102. T. M. Prenzel, F. Gehring, F. Fuhs, and S. Albrecht, "Influence of design properties of printed electronics on their environmental profile," *Matériaux & Techniques*, vol. 109, no. 1506, 2021.
103. J. Liu, C. Yang, H. Wu, Z. Lin, Z. Zhang, R. Wang, and C. P. Wong, "Future paper based printed circuit boards for green electronics: fabrication and life cycle assessment," *Energy & Environmental Science*, vol. 7, no. 111, pp. 3674–3682, 2014.
104. K. Schischke, N. F. Nissen, and M. Schneifer-Ramelow, "Flexible, stretchable, conformal electronics, and smart textiles: Environmental life cycle considerations for emerging applications," *MRS Communications*, vol. 10, pp. 69–82, 2020.
105. A. Sudheshwar, V. Beni, N. Malinverno, R. Hischier, Y. Nevo, B. Dhuiège, M. Borras, A. Chbani, C. Aucher, and S. Martinez-Crespiera, "Assessing sustainability hotspots in the production of paper-based printed electronics," *Flexible and Printed Electronics*, vol. 8, no. 1015002, 2023.

106. R. Hischier and T. Walser, "Life cycle assessment of engineered nanomaterials: State of the art and strategies to overcome existing gaps," *Science of the Total Environment,* vol. 425, p. 271–282, 2012.

107. B. Salieri, D. A. Turner, B. Nowack, and R. Hischier, "Life cycle assessment of manufactured nanomaterials: Where are we?," *NanoImpact,* vol. 10, pp. 108–120, 2018.

108. A. Sudheshwar, M. N. R. Hischier, B. Nowack, and C. Som, "Identifying sustainable applications for printed electronics using the multi-perspective application selection approach," *Journal of Cleaner Production,* vol. 383, no. 1135532, 2023.

109. A. Plepys, "The environmental impacts of electronics: Going beyond the walls of semiconductor fabs," chez *IEEE International Symposium on Electronics and the Environment,* 2004

110. G. Phylipsen and E. Alsema, "Environmental life-cycle assessment of multicrystalline silicon solar cell modules," Netherlands Agency for Energy and the Environment (NOVEM), Utrecht, 1995.

111. E. D. Williams, R. U. Ayres, and M. Heller, "The 1.7 kilogram microchip: Energy and material use in the production of semiconductor devices," *Environmental Science & Technology,* vol. 36, no. 124, p. 5504–5510, 2002.

112. T. Higgs, M. Cullen, M. Yao, and S. Stewart, "Review of LCA methods for ICT products and the impact of high purity and high cost materials," chez *2010 IEEE International Symposium on Sustainable Systems and Technology*, 2010.

113. M. F. Ashby, *Materials and Sustainable Development,* Springer, 2016.

114. O. Vidal, F. Rostom, C. Francois, and G. Giraud, "Global trends in metal consumption and supply: The raw material–energy nexus," *Elements,* vol. 16, no. 15, pp. 319–324, 2017.

115. IEA, "The 2020 EU critical raw materials list," 2020. [Online]. Available: https://www.iea.org/policies/15274-the-2020-eu-critical-raw-materials-list.

116. G. Calas, "Mineral resources and sustainable development," *Elements,* vol. 13, pp. 301–306, 2017.

117. P. A. J. Lusty and B. J. Murton, "Deep-ocean mineral deposits: Metal resources and windows into earth processes," *Elements,* vol. 14, no. 15, pp. 301–306, 2018.

118. N. Santero and J. Hendry, "Harmonization of LCA methodologies for the metal and mining industry," *The International Journal of Life Cycle Assessment,* vol. 21, p. 1543–1553, 2016.

119. R. Alvarenga, N. Préat, C. Duhayon, and J. Dewulf, "Prospective life cycle assessment of metal commodities obtained from deep-sea polymetallic nodules," *Journal of Cleaner Production,* vol. 330, p. 129884, 2022.

120. N. Moreau, T. Pirson, G. Le Brun, T. Delhaye, G. Sandu, A. Paris, D. Bol, and J.-P. Raskin, "Could unsustainable electronics support sustainability?," *Sustainability,* vol. 13, no. 12, p. 6541, 2021.

121. V. Lofthouse, "Ecodesign tools for designers: defining the requirements," *Journal of Cleaner Production,* vol. 14, no. 115–16, pp. 1386–1395, 2006.

122. OECD, "Extended producer responsibility," 2023. [Online]. Available: https://www.oecd.org/environment/extended-producer-responsibility.htm.

Part II

Off the Beaten Tracks for Future Pervasive Nanosystems Technology

Chapter 3

Helium-Ion Microscopy and Its Applications in Nanoscale Devices

Shinichi Ogawa,[a] Gregor Hlawacek,[b] Fayong Liu,[c] Shu Nakaharai,[d] Nirjhar Sarkar,[e] and Shane Cybart[e]

[a]*Advanced Industrial Science and Technology, Tsukuba, Japan*
[b]*Helmholtz-Zentrum Dresden-Rossendorf, Dresden, Germany*
[c]*Ocean University of China, Qingdao, People's Republic of China*
[d]*National Institute for Materials Science, Tsukuba, Japan*
[e]*University of California at Riverside, Riverside, California, USA*

shin1.ogawa1@gmail.com

The term "Helium-Ion Microscope (HIM)" is still unfamiliar to many people. HIM is a type of gas field ion source (GFIS) microscope based on the principle of the projection microscope proposed by E.W. Muller of the University of Pennsylvania in 1951. It utilizes a needle-shaped metal tip, sharpened to the atomic level, as the ion source filament. In 1975, W.H. Escovitz et al. proposed a microscope using hydrogen gas for GFIS, and in 1978, J. Orloff et al. developed the current GFIS. Around the same time, a liquid metal ion source (LMIS) was also developed, enabling more stable beam extraction than GFIS. However, as the research, development, and commercialization of Ga Focused Ion Beam (Ga FIB) microscopes using LMIS progressed, the advancements

Outlooking beyond Nanoelectronics and Nanosystems: Ultra Scaling, Pervasiveness, Sustainable Integration, and Biotic Cross-Inspiration
Edited by Simon Deleonibus

ISBN 978-981-5129-32-8 (Hardcover), 978-1-003-50990-5 (eBook)
www.jennystanford.com

of GFIS were temporarily overshadowed. In 2006, more than a decade after the maturity of Ga FIB microscopy technology, HIM, which utilizes helium gas for GFIS, was commercialized. Since then, there has been steady progress in the research and development of HIM technology applications, as well as advancements in HIM equipment. These applications include surface imaging using secondary electrons emitted from the near-surface region of the sample irradiated by helium ions, which closely resembles observations made by secondary electron microscopy (SEM). Additionally, analysis techniques such as backscattering spectrometry are employed. Furthermore, HIM technology enables nano-level fabrication and modification of materials, leading to numerous new applications.

This chapter provides an overview of the HIM system and presents typical examples of its applications, focusing on electronic material processing and device formation from the perspective of nanofabrication or modification. For instance, it explains how the electrical conduction properties in graphene films can be controlled by creating nanopore arrays or point defects using helium-ion beam irradiation. It also discusses the formation of Josephson junctions in superconductors by manipulating the disorder of atomic arrangement through helium-ion beam irradiation. The aim of this chapter is to offer a clear and accessible explanation of the applications of HIM technology in the field of electronic device engineering, with a particular emphasis on the unique capabilities provided by helium-ion microscopy.

3.1 Helium-Ion Microscope Instrumentation

The HIM entered the market more than 15 years ago and its development has been delayed by the fact that the development and the stable operation of a three-atom source are demanding and require unique solutions, which have not been available during the first attempts in the 70s of the 20th century. Here, I will try to highlight the key components and instrumentation, which have enabled the operation of the gas field ion source (GFIS) and allowed researchers to obtain astonishing results in several fields including high-resolution imaging and high-resolution material modification as well as spatially resolved ion induced materials property engineering. However, due to the propriety nature of some of the key components, this description unfortunately has to stay superficial in some key aspects (e.g., details of the functioning of the GFIS source itself).

3.1.1 Gas Field Ion Source

The uniqueness of the HIM is rooted in its source. Like in other scanning beam methods, the resolution of the HIM is ultimately limited by the achievable spot size of the sample. To the smallest possible spot size, at least the following criteria of the below non-exhaustive list need to be fulfilled.

- Small beam divergence
- Low energy spread
- Small virtual source size

All of these have multiple consequences for the final beam size and I will try to discuss them in the following paragraphs.

A small beam divergence will contribute to increasing the brightness of the source by concentrating more current close to the axis of the beam. However, it will also allow to better focus the beam by reducing the spherical aberrations as high-angle ion trajectories can safely be cut away—by using suitable apertures—without losing current in the main beam.

A low energy spread will—in a similar fashion as the small beam divergence—facilitate the lens design as fewer chromatic aberrations will occur.

While a small virtual source size may seem trivial, it is hard to achieve in reality. In fact, in most other charged particle beam optics, the source needs to be demagnified in order to reach the ultimate spot size. While this can be achieved by using suitable electrostatic (best suited for heavy charged particles like ions) or electromagnetic (in the case of light particles like electrons) lenses, this will result in increased spherical and chromatic aberrations if strong lenses have to be used.

The GFIS source used in the HIM happens to fulfill all three requirements simultaneously, resulting in reported imaging resolutions as small as 0.35 nm [1].

3.1.1.1 A short history of focused ion beam sources

The first developments of ion sources for focused ion beam (FIB) applications date back to the 70s and 80s of the previous century with the earliest attempts dating as far back as 1943 [2]. While today LMIS-based Ga-FIBs are the standard their initial development [3] has been preceded by the first attempts to use a GFIS [4]. Escovitz et al. [5] demonstrated the use of GFIS-based imaging for the investigation of biological specimens, and new imaging modes such as scanning transmission ion microscope (STIM) have been explored quickly [6].

However, with the fast progress in the development of scanning electron microscopes (SEMs) and the realization that Ga-based LMIS can be used for spatially resolved material removal at the micro and nanometer scale and not only for powering satellites, interest in GFIS-based imaging and nanofabrication quickly declined. This changed when in 2006 Ward et al. [7] showed a helium-based realization of a GFIS that utilized only a single atom of a trimer at the apex of a tungsten tip.

3.1.1.2 Gas field ion source

The GFIS source shares its fundamental working principle with atom probe tomography (APT) [8] and the field ion microscope (FIM) [9]. The foundations for the latter have been laid by Muller et al. [10] and led to the development of both APT and HIM. In the FIM a sharp needle is placed inside a vacuum container and biased with respect to a screen a few centimeters away. Upon introduction of a suitable imaging gas, a magnified image of the atomic structure can be observed on the screen.

The underlying processes are explained in detail in monographs [11] and papers but can be summarized as follows. The point-like shape of the needle focuses the field lines and the geometric field enhancement increases the field strength at the tip apex. Once a sufficient field strength is reached the gas atoms will lose an electron via a tunneling process into the needle at the *pointiest* parts of the tip (e.g., single adatoms and small clusters [called trimer if it consists out of three needle-atoms], kink, corner, and edge positions). The positively charged ions are then accelerated away from the tip following the field lines toward the screen. If the field

Table 3.1 Ionization potentials and polarizability of common gases intentionally or unintentionally present in the GFIS source environment [12–14]

Gas	Ionization potential	Polarizability
	eV	1×10^{-24} cm^3
He	24.6	0.20
Ne	21.6	0.29
Ar	15.8	1.63
H_2	15.6	0.80
N_2	14.5	1.74
CO	14.0	1.97
O_2	13.6	1.57
H_2O	12.6	1.43

is increased even higher than needle atoms might be ionized and a process called field evaporation starts. The latter is used in APT and is called field evaporation. Here, the ionized needle atoms are counted and by pulsing the field their mass can be measured using time of flight (TOF) techniques. However, in the FIM and HIM, the needle material and imaging gas are selected in such a way that the best imaging field (BIF) for the needle is higher than for the gas. This restricts the choice of materials to elements, such as Ir or W, if He or Ne are to be used allowing to perform field ionization and avoiding field evaporation during normal operation (see Table 3.1).

In the HIM both techniques—field ionization and field evaporation—are used. After applying a high voltage to a sharp ⟨111⟩ oriented tungsten needle, an imaging gas—typically helium—is allowed into the ultra-high vacuum (UHV) source volume. Using field evaporation atoms from the tip apex are removed until a well-ordered structure is visible and three atoms are left in the topmost atom terrace. An example is presented in Figure 3.1. An add atom being the *pointiest* object in the field of view (FOV) ionizes most of the He as is evident from the intensity (see Figure 3.1(a)). Also, other *pointy* objects like terrace corners and edges are visible. The terraces themselves appear dark as the geometric field enhancement in this comparatively flat area is insufficient for efficient He ionization. The add atom is only weakly bound compared to tungsten atoms placed in crystallographic correct positions and can easily be field evaporated using a small voltage pulse. In Figure 3.1(b), the clean tungsten tip showing the expected hexagonal arrangement of the terrace edges is achieved. After adjusting (reducing) the field strength such that the He ionization is maximized at the trimer the source is ready for operation (see Figure 3.1(c)). Each of the three tungsten atoms in the trimer (as well as all other protruding edges, corners, and kink sites) creates a small He beamlet as schematically depicted in Figure 3.2.

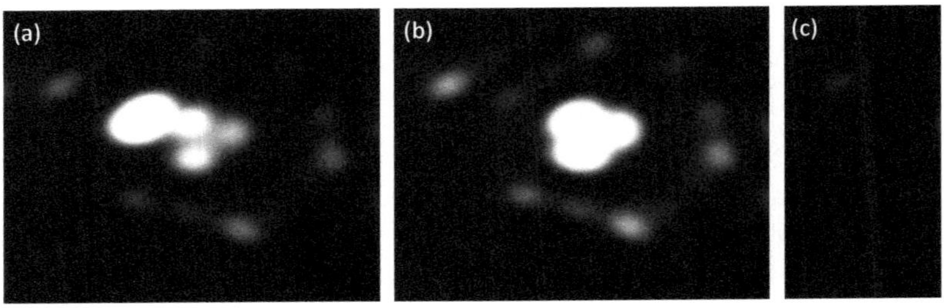

Figure 3.1 Sequence of SFIM images obtained in a HIM. An add atom (present in (a)) is removed by applying a voltage pulse to the source. Two additional terraces are visible in (b) after the adatom has been removed. After reducing the field strength, the three atoms can clearly be seen and the He ionization occurs predominantly at the trimer (c).

While a monomer would allow even higher currents, this trimer-named configuration provides a higher stability and longer source lifetime.

In order to provide a suitable trimer lifetime and low number of adatoms, it is necessary to keep the source at liquid nitrogen temperature and with the high voltage on at all times. The reason for the latter is related to the ionization potentials of the different elements present in the source environment. The ionization potential and polarizability of some selected gases, which are either used or are potentially present in the source vacuum, are listed in Table 3.1. Helium and neon stand out as the gases with the highest ionization potential.

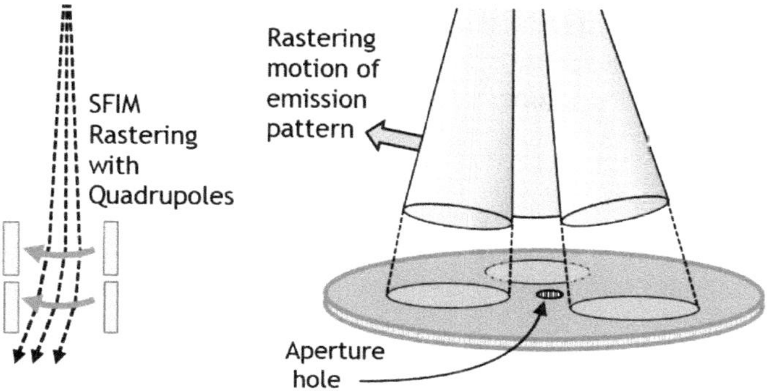

Figure 3.2 Working principle of SFIM mode in the HIM. Sketch of the GFIS source emitting three beamlets from the trimer. Reproduced with permission from [15].

While it may sound counterintuitive, this is one of the reasons that makes, in particular, He the best choice for a GFIS with respect to source operation. Once the trimer is formed and the best imaging voltage (BIV) has been selected—with the trimer at cryogenic temperatures—only He atoms can reach the trimer location and are subsequently ionized. The reason is related to the fact that only at the trimer location, the geometric field enhancement is strong enough to allow reaching the required BIF strength for helium or neon ionization (4.4 $V^{\circ}A^{-1}$ and 3.5 $V^{\circ}A^{-1}$, respectively). Other gases with a lower ionization potential will be ionized—and subsequently accelerated away—before reaching the trimer. This is particularly important for contaminates such as O_2, N_2, or H_2O, which could chemically react with the W atoms destroying the trimer.

The cryogenic temperatures present in the source volume support a stable and efficient GFIS operation in a number of ways. First, the additional cryogenic pumping helps remove contaminates from the source vacuum helping to prevent unwanted chemical or physical reactions. Secondly, it supports the collection and surface accommodation of the imaging gas. By reducing the temperature of both imaging gas atoms and the emitter itself, it also helps reduce the energy spread and supports the mechanical stability of the tip by suppressing tungsten atom diffusion. Too low temperatures (below ≈20 K) will hinder the transport of the imaging gas and result in unstable emissions. The used source temperature in the HIM of ≈85 K is a compromise between the above considerations and a vibration-free technical solution based on the usage of a solid nitrogen Dewar.

Table 3.2 Overview of the key performance and operating parameters the GFIS. After [16] based on [17–20]

Quantity	Value
Extraction voltage (He-BIV)	25–35 kV
Operating temperature	60–90 K
Base pressure	5×10^{-10} mbar
Operating pressure	1×10^{-7} mbar to 5×10^{-6} mbar
Brightness	5×10^{9} Acm^{-2} sr^{-1}
Reduced brightness	1×10^{9} Am^{-2} sr^{-1} V^{-1*}
Energy spread	<1 eV
de Broglie wavelength	≈80 fm
Virtual source size	<0.25 nm
typical max. current	≈150 pA

*Under optimal conditions using a BIV of ≈30 keV.

In Table 3.2, an overview of some of the key performance and operation parameters of the GFIS is presented.

3.1.2 Ion Opticsm

The ion optical column of the HIM is comparable to other FIB columns but possesses some distinct features that differentiate it from others. In Figure 3.3, an overview of the ion optical column of the HIM is given. While elements like condenser and objective lens, scanning octupole, and beam blanker are standard components, the upper scanning quadrupole is not often found on other ion optical columns. This upper quadrupole enables the creation of scanning field ion microscope (SFIM) images by scanning an image of the trimer magnified by the condenser lens over an aperture. The image is created by recording the intensity of the secondary electron (SE) signal from a conductive sample below the objective lens. If the intensity is present in the magnified trimer image a bright pixel with the corresponding intensity is recorded, and a dark pixel otherwise (for examples see Figure 3.1). SFIM imaging is used during trimer formation and instrument alignment. The low magnification of the ion optical column and high resolution necessitates and good mechanical alignment of the trimer with respect to the optical axis of the column. This is enabled by mounting the source in a swivel head, which can be tilted to compensate for any beam tilt and shifted to correct for lateral misalignment of the used He beamlet (see Figure 3.2).

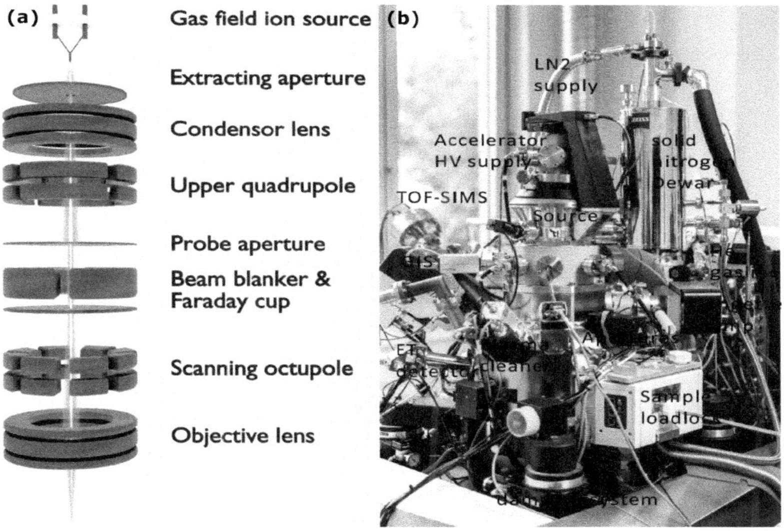

Figure 3.3 The ion optical column of the HIM. (a) An overview of the most important optical elements found inside the vacuum vessel is shown. (b) A picture of the ion optical column is presented. Various parts are labeled for easy identification.

The ion optical column of the HIM is operated with a crossover between the condenser and the objective lens. This enables precise beam current control by moving the position of this crossover with respect to a lower-lying aperture plan. This also trims larger angle components from the used beamlet and consequently reduces the spherical aberrations and increases the achievable lateral resolution. Due to the very low beam currents used in the HIM negative effects due to Coulomb repulsion of the ions in the crossover can be neglected.

Two different things need to be considered when trying to understand the reason for the surprisingly high lateral resolution of the HIM.

(1) Optical performance of the ion optical column.
(2) Physical interaction of the charged particles with the sample.

The optical performance of the ion optical column depends on a number of parameters, which depend on the properties of the source and the optical elements used.

Diffraction: Most importantly influenced by the de Broglie wavelength of the particles used at the given acceleration voltage (see Table 3.2).

Brightness: A property of the source (see Table 3.2).

Spherical aberration: Depends on the angular spread in the beam and the properties of the optical elements.

Chromatic aberration: Depends on the energy spread of the charged particle beam and the properties of the optical elements.

A more detailed analysis of the parameters for the HIM can be found in [16, 21, 22] and an in depth discussion on charged particle optics can be found in [23] and others. The above list of parameters is summarized in the plot presented in Figure 3.4. Here, the achievable resolution is calculated for a high resolution imaging application using a small current of only 500 fA.

However, in the same way such calculations can also be performed high current milling and implantation applications [16]. As can be seen from Figure 3.4, a spatial resolution of <0.5 nm should be achievable for the given column configuration.

The ion solid interaction between the accelerated charged particles and the sample atoms is the other important mechanism to consider when discussing the achievable resolution. It is important to consider this as the recorded signal is not only determined by the beam diameter but also by the sampled volume. This sampled volume is larger than the beam diameter, due to the scattering of the primary particles and the spread of secondary particles contributing to the signal. It is, however, smaller than the beam interaction volume due to the limited escape depth of the detected particles used for image creation. For the simple case of SE-based imaging in a scanning charged particle microscope, the comparison presented in Figure 3.5 can be used. Different Monte Carlo simulation codes have been used to calculate the interaction volume of electrons at different energies as well as various ions with a silicon target. The thin blue line in the figure indicates the escape depth of the secondary electrons. In fact, this quantity is smaller by approximately a factor of two for ion-induced secondary electron (i-SE) as compared to electron-induced secondary electron (e-SE). However, both are on the order of 1–5 nm depending on materials [24, 25].

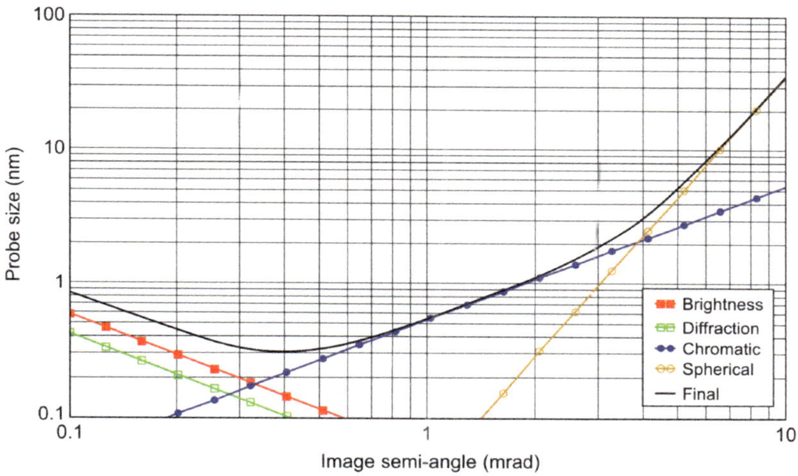

Figure 3.4 Optical performance of the HIM ion optical column. The calculated achievable resolution for a low current column configuration useful for high resolution imaging applications is presented. Reproduced from [22].

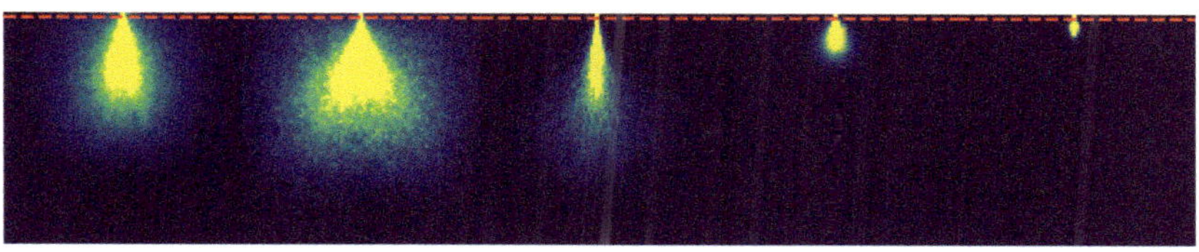

Figure 3.5 Monte-Carlo simulation results of the interaction of electrons (CASINO) and different ions (SRIM) with a silicon target. Plotted are the deposited energy density and the energy available for ionization events in the case of electrons and ions, respectively. The presentation is of qualitative nature only, but shows that the energy for secondary electron production is limited to a very small cylinder in the case of the light ion as compared to the other cases. The escape depth for the secondary electrons is indicated by the dashed red line close to the sample surface.

From these considerations it becomes clear that the narrow beam diameter and slow widening due to scattering makes the He ion beam an ideal candidate for high resolution surface sensitive imaging. Similar conclusions can be reached by comparing the trajectories of the primary and secondary particles directly (see for example [23] or [16]). This is further supported by the higher SE yield in the case of ions (2 to 8) as compared to electrons (≈ 1).

3.1.3 Detectors

This subsection introduces the available detectors. While the Everhart Thornley detector (ETD) and the backscattered ion detector (BSID)—discussed in sub-subsections 3.1.3.1 and 3.1.3.2—are available on all or on all older instruments, respectively, some of the following detectors are only available on prototypes or a small subset of instruments. The following sub-subsections summarize the various imaging methods. The emphasis lies on a brief description of the hardware for application examples and a description of the underlying physical processes the reader is referred to in the original works cited in the text.

In Figure 3.6, some of the detectors present in an Orion+ are visible.

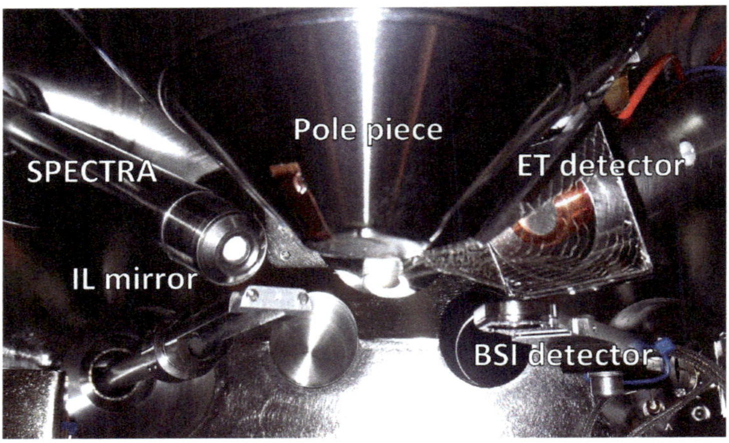

Figure 3.6 Inside of the vacuum chamber of an Orion+instrument. The following detectors are visible starting in the lower left corner in a clockwise manner: a parabolic mirror of Gatan MonoCL 4 used for IL, a SPECTRA detector for RBS, ETD for SE imaging, and a BSID used for BSI imaging. The pole piece containing the objective lens dominates the center of the image.

3.1.3.1 Secondary electron imaging

The Everhart Thornley detector (ETD) is the working horse in the HIM and more than 90% of all images are obtained using SE recorded by the ETD. The ETD was developed in the 60s of the last century [26] and is the dominating detector used in SEMs, FIBs, and HIM. The ETD is extremely efficient in detecting low-energy electrons by combining a scintillator with a light pipe and a photomultiplier.

An important difference to conventional ETDs found in SEMs is the delicate grid with high transparency. Together with a few other improvements the efficiency is exceeding 90% [16].

3.1.3.2 Backscattered ion imaging

For backscattered ion imaging (BSI) providing quick qualitative access to the sample composition, a BSID based on a micro-channel plate (MCP) can be used. The implementation used in the HIM and visible in the lower right of Figure 3.6 utilizes an annular MCP in combination with a biasing grid. The biasing grid enables the simultaneous recording of both BSI and SE images by deflecting the low-energy SE toward the ETD. The detector provides materials contrast [27] and has been used for subsurface imaging in semiconductor technology [28] and to reveal buried inorganic particles in biological specimens as well as in channeling experiments [29].

3.1.3.3 Rutherford backscattering in HIM

The methods promise a near damage-free depth-resolved analysis of the elemental composition. Due to the fact that the majority of the backscattered particles are neutral, detection schemes depending on the charge of the backscattered particles such as magnetic [30] or electrostatic analyzers [31, 32] would have an excellent energy resolution in the relevant energy range are of limited use [33]. Micro calorimeters [34] have an excellent energy resolution as well but their integration into a high-resolution imaging tool has been prevented so far by the challenging required mechanical and thermal decoupling.

Two detectors have been tested for their use in RBS-based imaging. In one implementation, a silicon drift detector (SDD) has been employed to measure the energy of the backscattered He atoms [35]. The energy resolution of only 4.5 keV and the fact that the backscattered ions will damage the detector reducing the sensitivity limited the usefulness of this approach. Nevertheless, it has been used to identify bulk materials as well as thin films of heavy elements on light substrates [35, 36].

Later, Klingner et al. [37] presented a different approach using TOF enabled by new blanker electronics, which could easily be retrofitted to existing instruments. Using a flight path of roughly 1 m, an ΔE of 1.6% for 25 keV He ions has been achieved [33].

However, all approaches will have to deal with an enhanced number of multiple and dual collision events, which will unavoidably result in an increased energy struggle of the detected particles [33]. This complicates the fitting of the obtained spectra and thus limits the applicability of the method to real-life problems in which no or only limited information is available on sample composition and structure.

3.1.3.4 Secondary ion mass spectroscopy in HIM

Enabled by the addition of neon as an imaging gas, secondary ion mass spectrometry (SIMS) started to be a viable method to obtain information on the elemental and chemical composition of the sample surface. In SIMS, the charged sputtered particles are extracted from the sample chamber into a spectrometer and then analyzed with respect to their mass. On the HIM, two approaches have been implemented and are commercially available from different companies.

Based on the above-mentioned fast blanker electronics [37], a TOF-SIMS add-on has been developed [38]. A resolution of 8 nm has been reported and the tool has been used successfully for the analysis of Li-ion batteries [39] and other materials [40].

A different approach utilizes a magnetic sector analyzer to separate the different sputtered ions and achieves a comparable high lateral resolution in the below 20 nm range [41, 42]. The latter instrument has recently been upgraded for the simultaneous recording of all masses [43] and is part of a new multi-method instrument called the npSCOPE [44].

The main difference between the two methods is that the former uses a pulsed primary beam and consequently only sputters the sample when a signal is recorded. The latter approach—in the initial implementation with which only four masses could be recorded simultaneously—needed several cycles to collect all potential masses bearing the risk that an interesting surface layer is destroyed before being analyzed. With the recent upgrade to a focal plane detector using a large MCP in combination with a delay line detector [43], this limitation is overcome and the analysis of all masses can be done quickly as no duty cycle as in TOF-SIMS is required.

3.1.3.5 Ion luminescence (IL) in HIM

IL is the ion-based implementation of cathode luminescence (CL) and can in principle provide similar insights. Various groups have investigated the usefulness of IL for materials analysis on different sample classes [45–48]. All implementations had in common that they utilized a retractable parabolic mirror above the sample with an aperture for the primary beam to extract the signal onto a suitable spectrometer.

However, despite some interesting results obtained on the defect formation and diffusion in ionic crystals [49] and high sensitivity toward defects in these materials [50], it has been found that unavoidable ion beam damage hinders an effective application of this method to real-world problems [48].

3.1.3.6 Scanning transmission ion microscopy in HIM

For STIM very thin samples ($\lesssim$50 nm depending on elemental composition) can be investigated and various detectors have been developed in the past.

The simplest implementation utilizes a sleeve on top of the sample so only SE emitted from the bottom surface can be collected. This approach has a surprisingly high efficiency and has been one of the first approaches toward STIM [27]. To detect the transmitted ions instead of the backside emitted SEs, an ion-to-electron conversion mirror can be employed. This approach has been used to obtain mass thickness contrast in bright field (BF) [51, 52] as well as annular dark field (ADF) [27, 53] mode.

The combination of bright- and dark-field imaging has been used to demonstrate the presence of thickness fringes in thin MgO samples [54].

The advantage of active detectors has quickly been realized and the first invocation based on active detectors utilized an annular MCP to obtain mass contrast on gold-silica core-shell nanoparticles [55]. This detector could only perform ADF as no means is provided to analyze the unscattered beam. A different approach enabling both BF and ADF is based on a silicon diode array and has been used to obtain mass thickness contrast and observe channeling [56, 57]. However, the setup is beam sensitive and the detector degrades quickly under the intense primary ion beam irradiation leading to dead pixels [58].

To overcome the beam-induced degradation of the expensive sensing layers, a different approach has been suggested by Serralta et al. [59]. They use an MCP in combination with a delay line to build a position and time-sensitive detector. Count rates of the delay line are limited to 5 Mcps, and the active area is 5 cm × 5 cm. Also, this detector has demonstrated the capability to investigate channeling and mass thickness contrast. The latter imaging modality has been particularly useful for the visualization of biological structures in a nanotoxicological context [44] while the former has been used to identify grain structures in polycrystalline samples [60], the higher SE yield in the case of ions (2–8) as compared to electrons (≈1).

3.1.4 Other Instrumentation and Accessories

3.1.4.1 Vacuum system

The operation of the GFIS source requires UHV conditions to ensure no unwanted impurities are present in the source volume impeding the lifetime of the trimer. Consequently, ultra-pure source gases are to be used and an additional chemical getter is used in the gas delivery system. The column itself is made from stainless steel to allow the use of CF technology for all feed-throughs and viewports. The middle section of the column is pumped by an ion-getter pump. These pumps are usually not the best option when pumping noble gases at relatively high pressures. This is a compromise considering the possible negative influence on vibrations caused by an additional turbo molecular pump as is found on earlier versions of the instrument.

The main chamber of the microscope is made from mild steel and can only reach a pressure of around 5×10^{-7} mbar to 5×10^{-8} mbar. This is enabled by the use of a dedicated sample load lock and regular cleaning of the sample interior using a downstream plasma cleaning process [61]. While better than many other scanning charged particle microscopes, it has been shown that a lower pressure can improve the imaging and milling performance [44, 62, 63].

3.1.4.2 Cryosystem

The cooling of the source to near liquid nitrogen temperatures is achieved by using a solid nitrogen Dewar. In comparison to earlier implementations of HIM using a pulse-tube cooler, this solution is vibration-free and provides temperature stability better than 0.5 K over the cooling cycle. The necessary refills with liquid nitrogen can be scheduled in the early morning and evening allowing for approximately 20 h of operation.

3.1.4.3 Vibration isolation

Special emphasis has to be given to the vibration isolation of the system. The magnification of roughly 1 makes the system very susceptible to vibrations as now demagnification of any movement of the source relative to the sample occurs. Mechanical decoupling from floor vibrations is achieved by the use of a two-stage passive air damper system in combination with a heavy granite block to reduce the system's eigenfrequencies. Additional acoustic shielding is applied by some users but is not strictly necessary if operated in a quiet environment. The use of Helmholtz coils to reduce the influence of electromagnetic AC fields on the other hand is standard and added to all machines.

3.1.5 Other Addons in HIM

In the following subsections, a non-exhaustive list of addons sometimes found on the HIM, which are not used for charged particle image creation is given.

3.1.5.1 Charge compensation

Different from a sample inside an SEM, the sample in the HIM will always charge positively. This is a result of the single positive charge delivered by the incoming primary ion and the multiple negative charges removed from the sample by the emitted SEs. As a result, charge compensation can be performed with an installed electron flood gun, which delivers electrons with a few hundred electron volts of energy to the sample. The user has to control the energy, direction, on-time, and current of the electron beam as well as the delays before and after applying the electron flood beam.

While also relevant for other fields, it is biology [64] and health research [65] that benefit mostly from the ability to avoid a potentially damaging conductive coating [21, 66]. As a consequence, the HIM outperforms different variants of SEM in terms of minimal FOV and resolved surface detail [64].

3.1.5.2 Gas injection system

The ion beam technology counterpart to large-scale 3D printing and nanoscale-focused electron beam-induced deposition (FEBID) [67, 68] is focused ion beam-induced deposition (FIBID) [69, 70]. A gas injection system (GIS) is needed in this case. A GIS enables the delivery of a precursor gas to a position close to the ion beam impact point using a—usually heated— needle. The most common GIS system on the HIM provides three different precursors, which are applied via a single nozzle. As the precursor gas cannot be stopped from diffusing upwards through the ion optical column and reaching the source volume regular cleaning of the sample chamber using the downstream plasma cleaner (see subsection 3.1.5.1) is mandatory.

It could be shown that the low sputter yield of helium in comparison to Ga-FIBID and the different interaction volume and SE spectrum as compared to FEBID enables the fabrication of deposits with higher metal content and smaller critical dimensions [71, 72].

3.1.5.3 Heating & cooling

Sample heating has been found useful in reducing unwanted specimen amorphization. Besides the use of dedicated heating stages [73] laser-based systems that allow direct heating of the irradiated area have been tested [74].

Sample cooling requires more in-depth modification of the sample chamber to avoid contamination by the cold sample. Recently, an effort has been made to develop a dedicated HIM optimized for the characterization of biological samples [44]. Besides the addition of a SIMS and STIM detector also cryotransfer and cryostage have been added to the system. This included the development of a new stage capable of managing the additional load, due to the heat transport copper braids as well as the lining of the entire sample chamber with cryocooled copper plates.

3.1.5.4 Atomic force microscopy

In an effort to enable correlative imaging attempts have been made to include an atomic force microscope (AFM) into the HIM. While more often performed in standard FIB systems, the large sample stage leaves little room for the inclusion of a standard AFM system utilizing laser readout of the cantilever movement. Andany et al. [75] employed a tip-scanning configuration [76] and a self-sensing readout [77] for measuring the cantilever deflection.

3.1.5.5 Others

In situ electrical probing of the sample has been performed using micromanipulators to directly obtain—among other properties—the critical fluence for milling encapsulated graphene [78] and to monitor the accumulated interface damage in spin-orbit torque devices [79]. Again, the limited space inside the main sample chamber makes the placement and operation of the manipulators challenging and additional infrared cameras are recommended.

Magnets have been used in order to apply a magnetic field during irradiation to study resolution limits and domain patterns during magnetic patterning of a prototypical in-plane exchange-biased layer system [80].

While the analysis of i-SE yields and energy distribution [81] has the potential to obtain quantitative materials contrast [82], only limited efforts have been undertaken to utilize this often-neglected signal. Petrov et al. [83] used a retarding grid and a special sample holder to obtain the i-SE energy distribution. Their results fit well with the predictions made by Ohya et al. [84] and promise besides materials contrast also a pathway to reduce the sensitivity toward carbon contamination [85].

3.2 Graphene Nanostructure Devices with a Prospective Fabrication Tool – Helium-Ion Beams

Graphene is one of the most famous representations of two-dimensional materials with its simplified and ideally intrinsic structure and properties [87]. One layer structure that consists of a single kind of atom builds up its super physical and chemical properties, such as high mechanical strength, high electrical and thermal conductivities, and a high surface-to-volume ratio [88–90]. This kind of simple structure also boots massive simulation studies by modern computational methods, like ab initio calculations and molecular dynamics simulations [91, 92], due to the greatly reduced computation power requirement compared to the bulk materials. However, it is because of the simple one-layer atom structure and nano-size device as well, that device fabrication becomes extremely difficult with the normal silicon-based processes.

In this part, the helium-ion beam milling (HIBM) technique is introduced to help solve this problem by coupling it with advanced graphene nano-electromechanical system (NEMS) technology. Two kinds of typical nanostructure devices, sub-10 nm graphene nanoribbon (GNR) and sub-10 nm graphene nanomesh (GNM) will be discussed followed by their applications.

3.2.1 Sub-10 nm Graphene Nanoribbon

With its great electrical carrier mobility of more than 200,000 $cm^2V^{-1}s^{-1}$, graphene is recommended as a promising candidate for the next generation of logic systems and low-power electronics devices [93]. However, graphene does not have an electronic bandgap, which limits its application in semiconductor logic devices [94]. One of the most effective ways to target this problem is to narrow down the GNR. Due to the quantum confinement from the geometry, the energy band opening was predicted to be observed in sub-10 nm GNR [95, 96]. Then it drives the high motivation to realize the fabrication of the working devices. Even with the standard E-beam lithography (EBL) followed by plasma etching, it is still quite challenging to blow 20 nm with an unavoidable large edge roughness [97]. Therefore, several kinds of sophisticated methods have been developed to fabricate the sub-10 nm GNRs in recent years, such as transmission electron microscope (TEM) carving [98], unzipped carbon nanotubes [99], and chemical synthesis [100, 101]. Here, the HIBM technique is utilized to realize the ultra-narrow GNR devices. Compared to the TEM carving method, it solves the extremely slow patterning speed issue; and compared to the other physical and chemical synthesis methods, the HIBM technique has high compatibility with standard silicon-based processes. In the following parts, the advanced graphene NEMS technology combined with the HIMB technique is shown in detail first. Then the functional devices are demonstrated for the presentational applications.

3.2.1.1 Nano-fabrication processes

3.2.1.1.1 *Graphene NEMS technology*

With the advanced graphene NEMS technology, the suspended graphene nanoribbon can be fabricated in hundreds of nanometers width massively with commercial chemical vapor deposition (CVD) graphene. The fabrication steps are shown below in Figure 3.7.

It starts with a commercial CVD graphene transferred to a silicon substrate covered by a 285 nm thermal SiO_2 layer. For the patterning of the electrode, it is separated into two steps to achieve good adhesion and good electrical conductivity as well. In the first one, the EBL with positive photoresist PMMA/MMA is used to form the patterning followed by the reactive ion etching (RIE) process. In this case, the graphene underneath is removed, and the Cr/Au (5 nm/80 nm) metal layer deposited by E-beam evaporation can achieve a good adhesion with the substrate after the lift-off process (Figure 3.7(a)). In the second one, the patterning is located between the first metal layer and the CVD graphene to make good electrical contact (Figure 3.7(b)). It uses the same processes as the first metal layer without the RIE step. Subsequently, the GNRs are shaped by EBL with the negative photoresist hydrogen silsesquioxane (HSQ). After this step, the designed GNRs are covered with HSQ, as shown in Figure 3.7(c). After the RIE processes, the other uncovered graphene parts are removed. The last suspension step is combined with buffered hydrofluoric acid (BHF) wet etching and CO_2 critical point drying (Figure 3.7(d)). The 1:7 BHF is used, and the HSQ mark and the part of the SiO_2 layer under the GNRs are removed in 45 s. There is about a 100 nm-depth air gap under the suspended

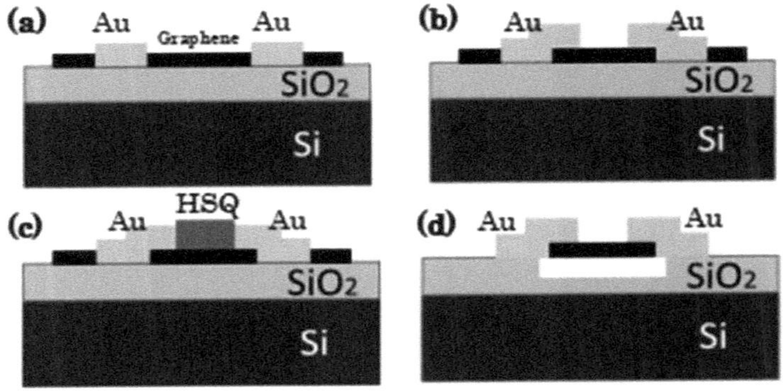

Figure 3.7 The main steps of the graphene NEMS technology. (a) Patterning the first electrode layer after transferring the CVD graphene. (b) Patterning the second electrode layer. (c) GNR formation by using HSQ. (d) Suspending the GNR by HF etching [102]. Reprinted with permission from Liu F. et al., IEEE SNW 2020. Copyright 2020 IEEE Publishing.

Figure 3.8 The scanning electron microscope image of the suspended graphene nanoribbon device [103]. Reprinted with permission from Schmidt M. et al., Small, 15, 1903025. Copyright 2019 Wiley Online Library.

graphene. In order to avoid the surface tension effect, the sample is dried in a CO_2 critical point dryer without being taken from the solution directly. The wide-suspended GNR device is shown in Figure 3.8.

3.2.1.1.2 *Helium-ion beam milling technique*

One of the issues with GNR from this kind of NEMS technology is the large edge roughness including folding and collapsing due to the EBL, RIE, and BHF etching processes as well as the grain boundaries (Figure 3.9a). The HIBM technique can be used to reshape the graphene ribbons into much smoother edges (Figure 3.9(b)). This process is carried out in a high vacuum chamber ($<5 \times 10^{-7}$ mbar) with the ZEISS ORION Nanofab system [104, 105]. Before the HIBM process, a hydrogen atmosphere annealing is needed to remove the photoresist residue, which will be re-sputtered by the helium ions. The configuration of the helium-ion beam is set to 77000 μs dwell time and 1 nm pitch (from beam center to center) with 1 pA beam current [102]. The energetic helium-ion beam is focused to carve the suspended graphene. Compared with the electron beam, the helium-ion beam does not require a large beam flux to break down the bonds of the crystal lattice. One single helium ion under 30 keV acceleration has enough energy to displace the target carbon atom in suspended graphene. Furthermore, compared with the supported graphene device carving, the helium ions' knock-on damage and scattering issues can be avoided in the suspended devices [104]. This is quite important for monolayered and fragile graphene devices.

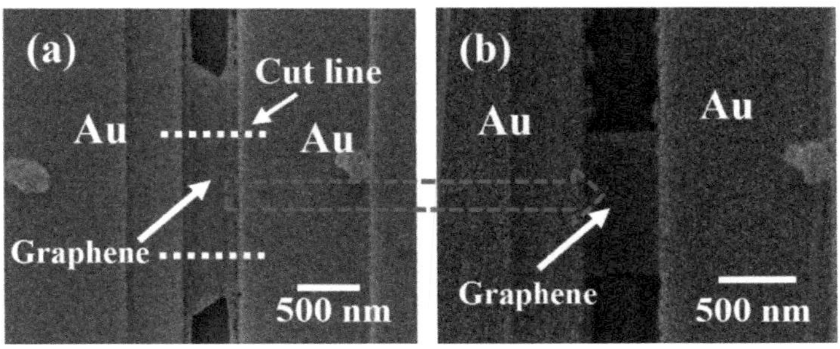

Figure 3.9 (a) The folding collapsing and on the edges of the suspended graphene. (b) The same device as in (a) with flat edges after reshaping by HIBM [16]. Reprinted with permission from Liu F. et al., IEEE SNW 2020. Copyright 2020 IEEE Publishing.

3.2.1.2 Sub-10 nm graphene nanoribbon functional devices

In this part, the techniques above are utilized to build a functional device. The 6 nm wide monolayer and bilayer GNRs are fabricated and characterized electrically for transport gap engineering [103]. The transport gap opening is observed at about 0.16–0.8 eV at room temperature.

In order to achieve the sub-10 nm GNR precisely, it does not directly reshape from the large area of the graphene ribbon. The large area of high-dose helium ion exposure will induce substrate surface swelling and helium bubble formation [104], which causes the SiO_2 substrate to touch the nanoribbon. In this case, the sub-100 nm GNRs are pre-patterned by using the graphene NEMS technology first (Figure 3.10(d), bottom left, 300 nm long, and 40 nm wide). Then two 50 × 60 nm^2 rectangular areas on both sides of the graphene nanoribbon are milled with then helium-ion beam with a sub-10 nm gap in between, as shown in Figure 3.10(a). The final sub-10 nm wide (middle part) device is shown in Figure 3.10(d).

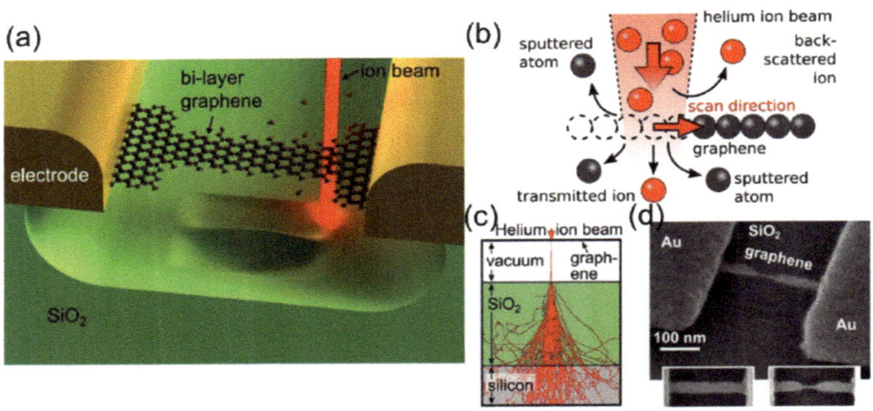

Figure 3.10 HIBM for suspended graphene. (a) Schematic depiction of the milling process to narrow down the ribbon into sub-10. (b) HIB milling mechanism on suspended graphene. (c) The trajectories of the helium ions from the Monte Carlo simulation. (d) Helium-ion microscopic image for the 6 nm wide device after HIBM. Top views before HIBM (bottom left), and after HIBM (bottom right) are shown as insets [103]. Reprinted with permission from Schmidt M. et al., Small, 15, 1903025. Copyright 2019 Wiley Online Library.

For the electrical measurement, the devices are characterized in a cryogenic probe station with a high vacuum chamber ($<1 \times 10^{-6}$ mbar). The stage is cooled down by the open-flow liquid N_2, which has a better vibration control than that done by a compressor. The temperature range is from 150 K to 300 K. The measurement is a typical I_D-V_D characteristic by applying a drain voltage to one side of the electrode and measuring the current through the device. Before the HIBM process, the device in Figure 3.4(d) shows a linear response in I_D-V_D measurement (Figure 3.11(a)). After reducing the width in the middle part of the GNR by HIBM, the I_D-V_D measurement shows a clear nonlinearity (Figure 3.11(b)). To zoom in on the low current region of Figure 3.11(b), a 0.82 eV transport gap is observed in Figure 3.11(c).

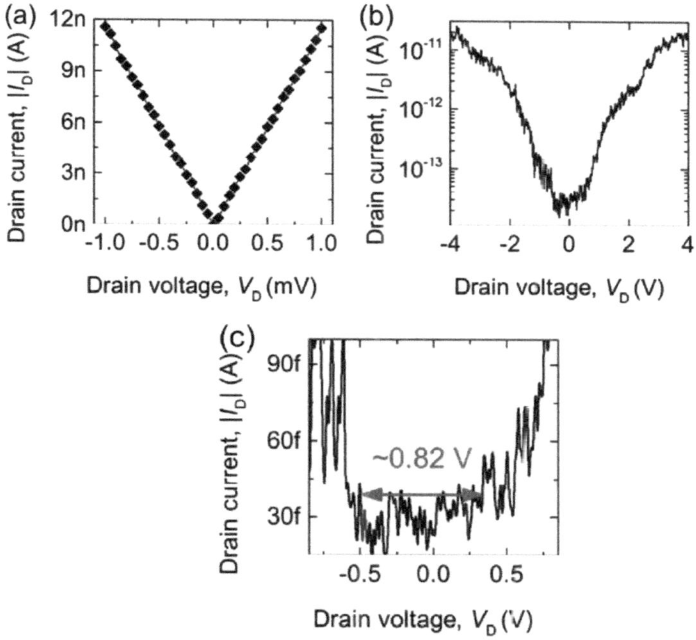

Figure 3.11 Electrical characterization of the sub-10 nm graphene device. (a) I_D–V_D measurement before HIBM. (b) I_D–V_D measurement for the same device in an after HIBM. (c) Zoom in to the low-current region in b to show the transport gap opening of 0.82 ± 0.2 eV [103]. Reprinted with permission from Schmidt M. et al., Small, 15, 1903025. Copyright 2019 Wiley Online Library.

Another four CVD GNRs with the same HIBM process are characterized in the same way from 150 K to 300 K, which also show stable electrical performance [103]. This experiment confirms the reliability and stability of the HIBM technique for sub-10 nm GNR fabrication.

3.2.2 Sub-10 nm Graphene Nanomesh

As discussed in the previous part, the ultra-narrow GNR is a promising candidate for the application of next-generation field-effect transistors (FETs) with an energy gap opening at room temperature. However, its driving current is too low for a functional logic device in an integrated circuit. Moreover, the sub-10 nm wide monolayer bears a fragile structure, which limits its reliability. Here, the graphene nanomesh (GNM) is introduced as another typical artificial graphene nanostructure for the application of both electrical engineering and phonon engineering [106]. By introducing periodical nanopores on a graphene sheet, the GNM is generated with a new crystalline structure based on the original crystal lattice of graphene (Figure 3.12). It has been reported that the transport gap opening also can be observed in GNM devices with larger driving currents and

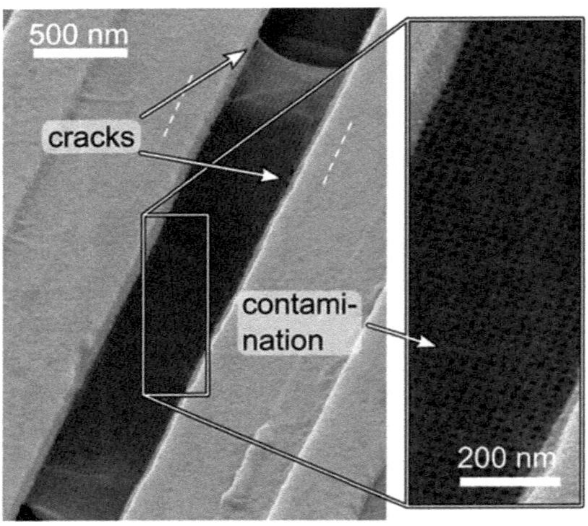

Figure 3.12 A large-area GNM device with 25 nm pitch (nanopore center to center) [112]. Reprinted with permission from Schmidt M. et al., ACS Appl. Mater. Interfaces, 10, 10362–8. Copyright 2018 ACS Publications.

higher on/off ratios than narrow graphene nanoribbons both in theoretical and experimental studies [107–110]. Moreover, this kind of artificial structure especially in the nanoscale has been suggested for phononic crystal applications to control phonon transmission [111]. In the following contents, the sub-10 nm GNM device is discussed including its fabrication by HIBM and the applications in activation energy (E_A) tuning and thermal rectification.

3.2.2.1 Nano-fabrication processes

At first, the suspended GNRs are prepared by using the same graphene NEMS technology, as shown in Figure 3.13. Here the dimension is typically set as 500 nm long and 1200 nm wide. Then the HIBM technique as discussed before is utilized to introduce the sub-10 nm diameter nanopores periodically on the suspended ribbons. The helium-ion beam configuration is set to 1 pA beam current and 45000 μs dwell time [113]. The nanopore diameter can be estimated to be 6 nm approximately in the helium-ion microscope image (Figure 3.13(b)). There are also some missing nanopores observed on the GNM, due to the resist residue on the surface of the GNM even after annealing. Next, how to make a stable suspended GNM device will be discussed in detail.

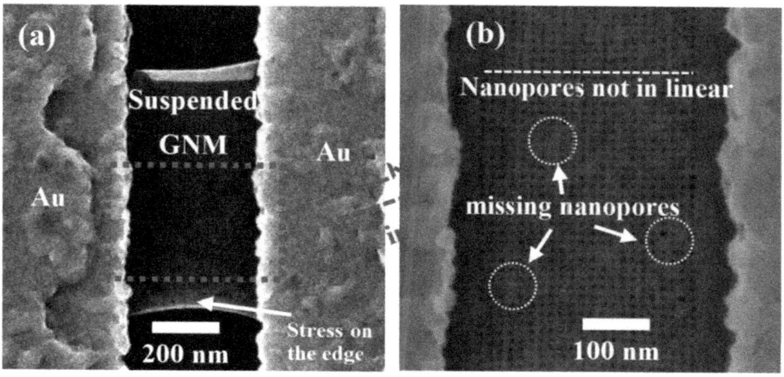

Figure 3.13 (a) Helium-ion microscope image for a typical large-area GNM device. (b) 6 nm diameter periodical nanopores and missing nanopores in the zoom-in area [113]. Reprinted with permission from Liu F. et al., Micromachines, 11, 387. Copyright 2020 MDPI.

Ideally, the meshed area where the helium-ion beam drills the periodical nanopores with a typical pitch should cover the whole area of the suspended graphene ribbons, as shown in Figure 3.14(a). However, the collisions in the connection parts between the graphene and metal were observed after the HIBM (Figure 3.14(b)). It is suggested that the new artificial structures with nanopores rearrange the stress distribution, especially in the side area, which supports the whole weight of the suspended graphene [113]. This can be implied by the middle part sinking slightly according to the non-linear distribution of the periodical nanopores in Figure 3.13(b). That means it should keep a certain unmeshed gap in the connection parts between the metal and graphene to maintain enough supporting strength [113]. For a similar reason, the suspended edges of the graphene also need to clamp the suspended GNM with a stressful condition. In this case, another unmeshed gap also should be reserved at the suspended edge as shown in Figure 3.15(c) to avoid the cracks from edges to center as shown in Figure 3.15(b). Here, based on the experimental experience [113], the meshed area keeps a 50 nm gap from the metal part and a 100 nm gap from the suspended edges (Figure 3.15(c)).

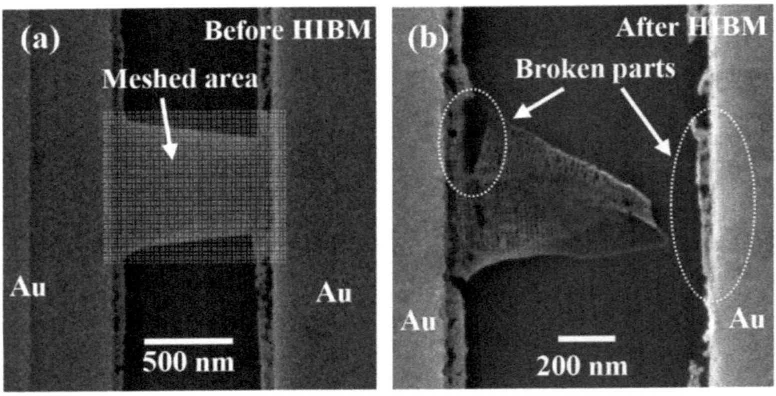

Figure 3.14 (a) The image before the HIBM, the dotted area is to drill nanopores by the HIBM. (b) The suspended graphene in (a) fell down after the HIBM [113]. Reprinted with permission from Liu F. et al., Micromachines, 11, 387. Copyright 2020 MDPI.

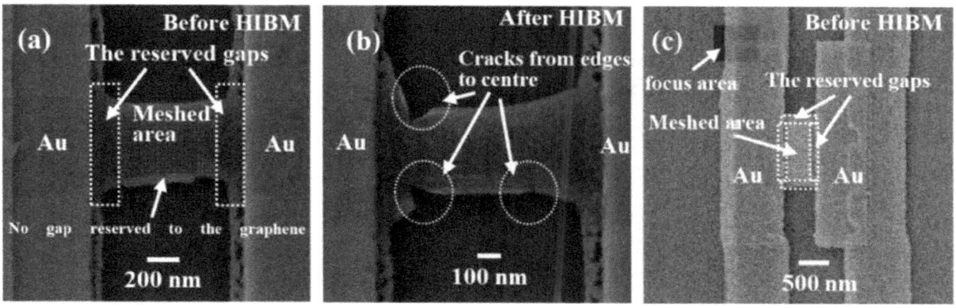

Figure 3.15 (a) No reserved gaps at the suspended graphene edges before the HIBM. (b) Cracks from edges to center after HIBM for the device in (a). (c) Reserved gaps surround the meshed area to enhance the stabilities [113]. Reprinted with permission from Liu F. et al., Micromachines, 11, 387. Copyright 2020 MDPI.

Figure 3.16 (a) The 15 nm-pitch GNM device. (b) The 12 nm-pitch GNM device. (c) Schematic drawing to illustrate the defective region [113]. Reprinted with permission from Liu F. et al., Micromachines, 11, 387. Copyright 2020 MDPI.

With establishing the stable fabrication process for a large area of suspended GNM, the smallest pitch is one of the key parameters to show to what extent this advanced technique can go for graphene nanostructure modification. Then in the next step, the pitches are changed from 50 nm down to 12 nm including 30 nm, 25 nm, 20 nm, 18 nm, and 15 nm in between. It is implemented on seven graphene ribbon devices with the same fabrication processes. The experiment shows that the GNM devices with ≥ 15 nm pitch are patterned successfully (Figure 3.16(a)). After this, the collapses are observed in the

12 nm-pitch GNM device as shown in Figure 3.10(b). It is suggested that although the nanopore diameters are estimated to be 6 nm from the images, there is roughly a 3 nm defective region surrounding the nanopores due to the Gaussian beam tail [113, 114]. The suspended structure in this region is quite fragile and it causes large-area breakdown (Figure 3.16(c)).

3.2.2.2 Applications in electronic engineering: Activation energy tuning

Based on the fabrication above, five graphene GNM devices with different pitches (50 nm, 30 nm, 25 nm, 20 nm, and 15 nm) are measured with a standard transistor-type characterization from 10 K to 300 K in a close-cycle liquid He cryogenic prober station (Figure 3.17). As the GNM devices are much more stable than the ultra-narrow suspended GNR devices, the vibration from the compressor to the station can be borne by the devices. In order to avoid the electrostatic force pulling down the suspended GNM during the I_D-V_{BG} measurement [89], only the I_D-V_D measurements are implemented on these five GNM devices with V_{BG} = 0 V.

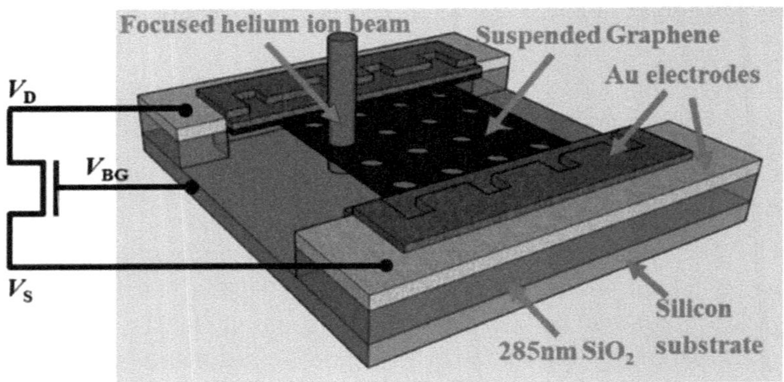

Figure 3.17 Schematic 3D structure for the suspended GNM devices. The back gate voltage (V_{BG}) is applied from the silicon substrate, and the drain voltage (V_D) and source voltage (V_S) are applied from the two sides metal electrodes, respectively [113]. Reprinted with permission from Liu F. et al., Micromachines, 11, 387. Copyright 2020 MDPI.

The example of measurement results at 300 K is shown in Figure 3.18(a). From the linear relationship region of I_D-V_D, the conductance of each GNM device is extracted at the typical temperature between 10 K to 300 K, respectively (Figure 3.18(b)). Before going to the next step of the analysis, the 100 nm reserved gaps on the suspended edges in Figure 3.15(c) should be mentioned about their contribution to the electrical conductance. Based on the experimental results [113], it implies that the conductance contribution from the 100 nm reserved gaps can be counted out, due to the strong tensile stress suppressing the electron transport [115–117]. Then it comes back to discuss the conductance decreasing, due to the different nanopore pitches, as shown in Figure 3.18(b). It is also clearly observed that the conductance decrease becomes larger at the low temperature. Then the Arrhenius plot method is introduced to analyze thermal activation energy in these devices. Before that, the only variable parameter "pitch" is converted to "porosity," which can illustrate the global feature of each device more clearly such as 15 nm to 12.57%, 20 nm to 7.07%, 25 nm to 4.52%, 30 nm to 3.14%, and 50 nm to 1.13%. Subsequently, the original data in Figure 3.18(b) is replotted to porosity versus conductance at different temperatures, shown in Figure 3.19(a), which dedicates the conductance of the higher porosity GNM device has a larger decrease with the temperature change.

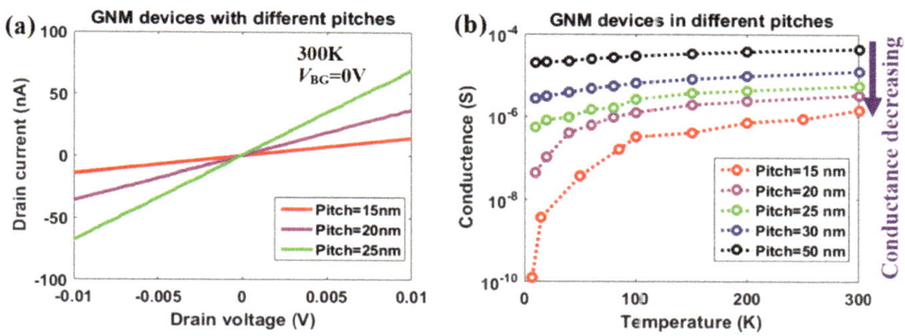

Figure 3.18 (a) The I_D-V_D measurement example for the typical GNM devices. (b) The conductance extracting by linear fitting for each GNM device from 10 K to 300 K, respectively [113]. Reprinted with permission from Liu F. et al., Micromachines, 11, 387. Copyright 2020 MDPI.

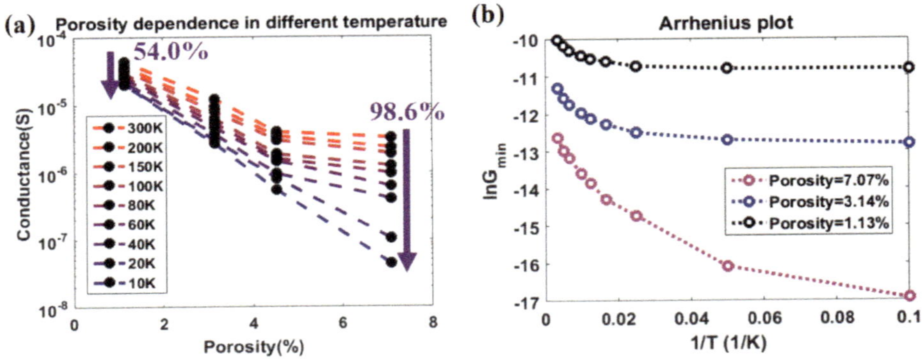

Figure 3.19 (a) The conductance for the GNM devices with different porosity from 300 K to 10 K. (b) The temperature dependence of ln G_{min} for the typical GNM devices [113]. Reprinted with permission from Liu F. et al., Micromachines, 11, 387. Copyright 2020 MDPI.

Figure 3.19(b) shows the Arrhenius plot example for the typical GNM devices. The G_{min} is the conductance minima at each temperature for a certain device. As V_{BG} = 0 V is around the charge neutrality point, and the conductance is a constant in I_D-V_D measurement, then the value of conductance in Figure 3.11(b) can be treated as the G_{min}. The G_{min} is related to the thermal activation energy (E_A) at the high-temperature region as:

$$G_{min} \propto \exp\left(-\frac{EA}{2k_B T}\right). \tag{3.1}$$

Where E_A is suggested to build the transport energy gap, T is the temperature, and k_B is Boltzmann's constant. Based on Eq. (3.1), E_A can be extracted by the linear fitting for the high-temperature region (Figure 3.19(b)). The results of E_A are shown in Figure 3.20 for those GNM devices with different porosities. It demonstrates that E_A can be systematically tuned by controlling the porosity of the GNM. And this "porosity" is restricted to the periodical nanopore patterning. The E_A tunability is suggested to be caused by quantum confinement and strong electron localization [118, 119]. The neck between nanopores can be treated as a sub-10 nm graphene nanoribbon including the defective region, in which the energy gaps are generated as similar as in the small GNRs. As discussed before, ultra-narrow GNRs face a big challenge in fabrication and stability issues, then the GNM devices show a new way to solve these issues for the application of electronic engineering.

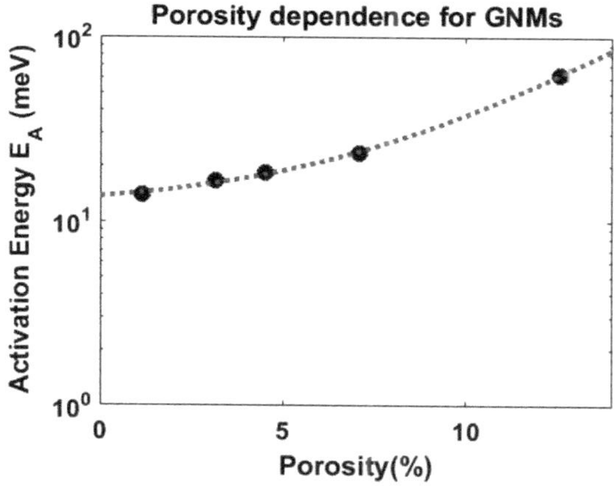

Figure 3.20 The EA turnability with different porosities in GNM devices. Black points are experimental data. The red dot line is curve fitting [113]. Reprinted with permission from Liu F. et al., Micromachines, 11, 387. Copyright 2020 MDPI.

3.2.2.3 Application in phonon engineering: Thermal rectification

In the GNM devices, the original graphene structure is artificially modified with a nanometer periodical configuration. This forms a 2D phononic crystal structure for phonon engineering to control the phonon transmission with artificial nanostructures [111, 120–122]. In the following content, the GNM is used as the phononic crystals to introduce the asymmetry in a suspended graphene ribbon for the application of graphene thermal rectification.

The thermal rectification is that the heat flow in the forward direction is larger than in the backward direction, which is similar to an electrical diode [124–126]. Introducing asymmetry is one of the key points for realizing thermal rectification [127]. For this reason, the GNM structure is patterned on the half area of the suspended graphene ribbon to induce the asymmetry and build the thermal rectifier, as shown in Figure 3.21. The device fabrication is also based on the graphene NEMS technology and the HIBM as discussed before. The different thing is in the first metal layer patterning. Here, the first metal layer needs to work as the heater and heat sink. Without changing the fabrication processes, two gold nanowires are designed as a four-probe measurement setting for precise heating and heat sink, as shown in Figure 3.22(a). Zoom in to the middle part in Figure 3.22(b), the suspended graphene is 1.2 μm wide and 500 nm long. For the meshed area, the pitch is managed to be 20 nm and 30 nm. In order to keep the stability of the GNM devices, 100 nm gaps are still reserved on the suspended edges. The meshed area and unmeshed area can be clearly distinguished in Figure 3.22(c).

After finishing the device fabrication, another big challenge is the precise thermal transport characteristic [128]. A "differential thermal leakage (DTL)" method is developed based on the conventional thermal bridge method to evaluate the thermal transport through the graphene bridge in the forward and backward directions with high accuracy.

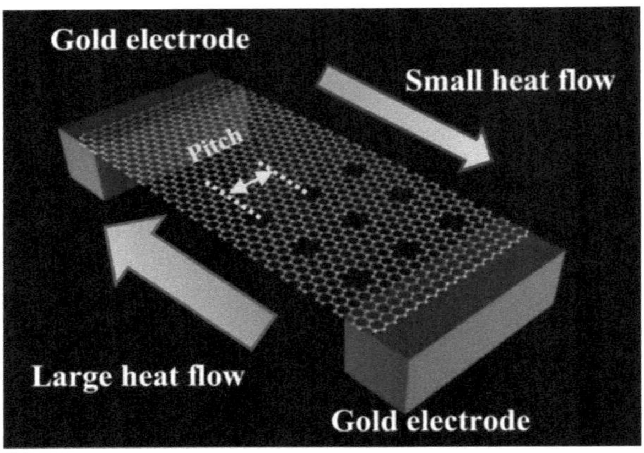

Figure 3.21 Schematic diagram of the graphene-based thermal rectifier with GNM structure [123]. Reprinted with permission from Liu F. et al., Nano Futures, 2021, 5, 045,002. Copyright 2021 IOP Publishing.

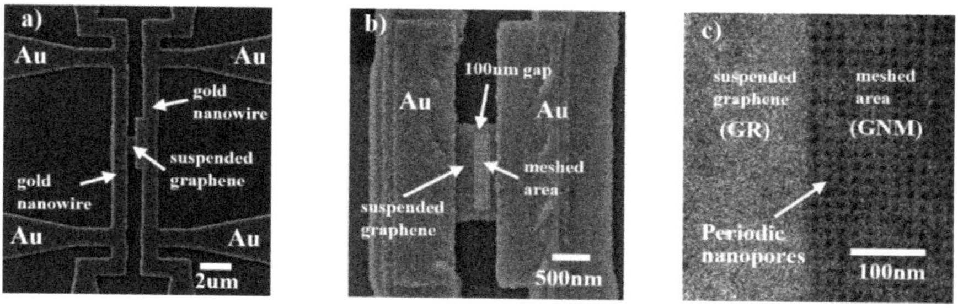

Figure 3.22 (a) The device structure with a four-prob measurement setting in a helium-ion microscope image. (b) The suspended graphene ribbons before HIBM, the white dotted area is the meshed part. (c) The periodic nanopores after HIBM under the helium-ion microscopy [123]. Reprinted with permission from Liu F. et al., Nano Futures, 2021, 5, 045,002. Copyright 2021 IOP Publishing.

The thermal rectification ratio (η_{TR}) is generally used to describe the performance of a thermal rectifier. It is defined as:

$$\eta_{TR}(T_E) = \frac{J_+ - J_-}{J_+}. \tag{3.2}$$

The T_E is the environmental temperature. The J_+ means the forward heat flux which is larger, and the J_- means the backward heat flux which is smaller. In order to converge the η_{TR} in a reasonable regime from 0% to 100%, the J_+ is set as the denominator in Eq. (3.2). In this case, η_{TR} = 0% means that there is no thermal rectification, and η_{TR} = 100% means that it is a perfect thermal rectifier. In the DTL method, these two heat fluxes are converted to the heater temperature change ΔT_R and ΔT_L as shown in Figure 3.23. At one measurement cycle, only the heater is powered, and the heat sink is in a floating condition. For example, in the right part of Figure 3.23(a), a certain amount of heat flow is generated in the right-side heater and passes the graphene bridge to the left-side heat sink When it comes to the steady state, the right-side heater is held at T_{R_G}. Then if the graphene bridge is removed as in the left part of Figure 3.23(a), the right-side heater will increase its temperature to

T_{R_NG} due to the termination of one heat dissipation channel (the suspended graphene channel). Subsequently, the right-side heater temperature change ΔT_R can be calculated as:

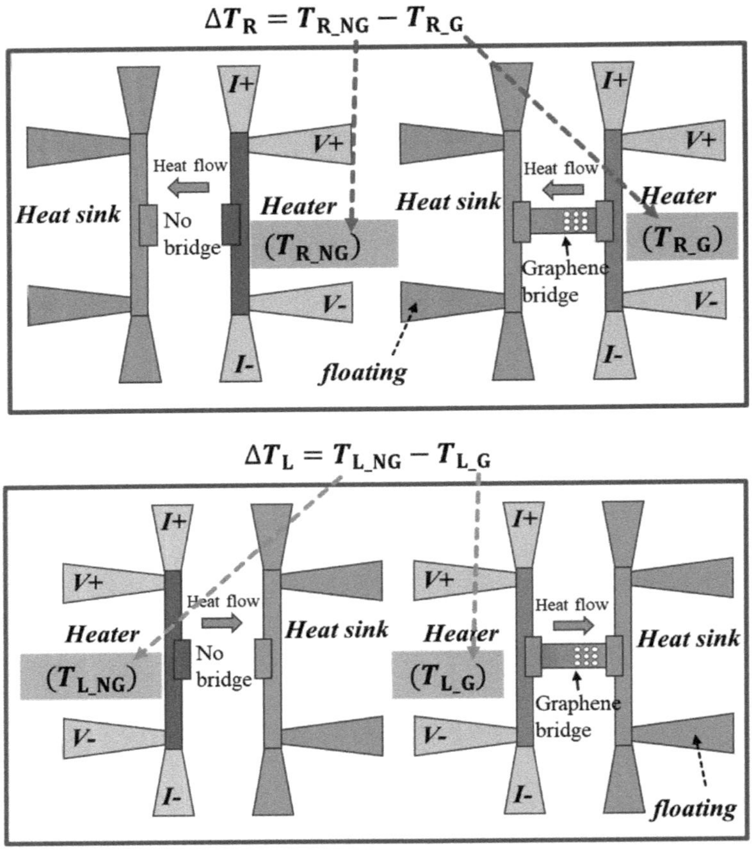

Figure 3.23 The DTL method illustrates how to get the ΔT_R and ΔT_L. ΔT_R evaluates the heat transport from right to left and ΔT_R evaluates that from the opposite direction [123]. Reprinted with permission from Liu F. et al., Nano Futures, 2021, 5, 045,002. Copyright 2021 IOP Publishing.

$$\Delta T_R = T_{R\_NG} - T_{R\_G}. \tag{3.3}$$

Due to the symmetrical device structure, the roles of the heater and heat sink can be swapped, as shown in Figure 3.23(b), the left-side heater temperature change can be achieved in the same way as:

$$\Delta T_L = T_{L\_NG} - T_{L\_G}. \tag{3.4}$$

Then the next question is to solve the relationship between J_R/J_L and $\Delta T_R/\Delta T_L$. J_R means the heat flux from right to left, and J_L means that from left to right. Based on the specific heat capacity calculation equation, J_R and J_L can be written as:

$$J_R = Cm\Delta T_R \tag{3.5}$$

$$J_L = Cm\Delta T_L \tag{3.6}$$

Where C is representative of the specific heat capacity and the m is representative of the heater mass. Corresponding to the later measurement results, J_R is larger than J_L. In this case, η_{TR} can be calculated in Eq. (3.7) without knowing C and m. This is one of the advantages of the DTL method.

$$\eta_{TR}(T_E) = \frac{Cm\Delta T_R - Cm\Delta T_L}{Cm\Delta T_R} = \frac{\Delta T_R - \Delta T_L}{\Delta T_R}. \tag{3.7}$$

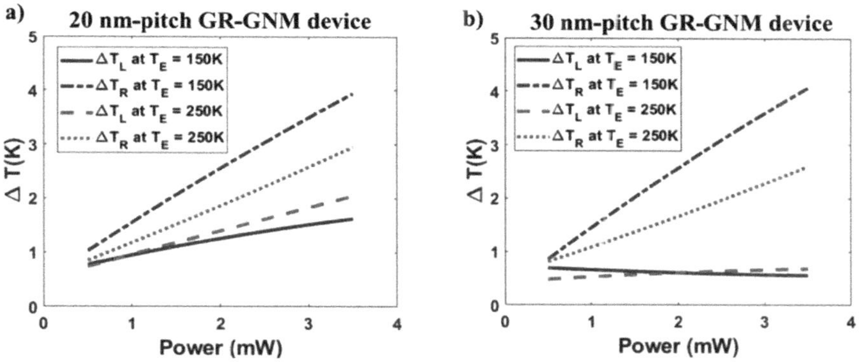

Figure 3.24 DTL measurement results for the two-graphene device at 150 K and 250 K environmental temperatures [123]. Reprinted with permission from Liu F. et al., Nano Futures, 2021, 5, 045,002. Copyright 2021 IOP Publishing.

By using this DTL method, the two devices mentioned above with 20 nm-pitch and 30 nm-pitch phononic crystals are characterized at 150 K and 250 K environmental temperatures. The Keithley 6220/2182A test toolkit is utilized to power the heater and measure the resistance of the heater simultaneously. The power of the heater is set from 0.5 mW to 3.5 mW by considering the accuracy of the measurement and stability of the gold nanowire. The measurement results are shown in Figure 3.24 for the two devices. It clearly demonstrates the thermal rectification phenomenon on both two devices in different environmental temperatures, that the ΔT_{R} values are always larger than the ΔT_{L} values. Based on Eq. (3.7), the η_{TR} in the 20 nm-pitch device with 3.5 mW heating power shows up to 60% value in 150 K. These experiment results are mentioned as a significant achievement for two-dimensional thermal rectification development [129]. It indicates a promising approach to achieving high-performance thermal rectifiers with asymmetrical phononic crystal structures.

3.3 Defect Engineering of Graphene for Carrier Conduction Control

HIM is used for nanofabrication of devices on thin film such as graphene. In this chapter, a new method of defect engineering in graphene using HIM is introduced. Atomic size defects are generated in graphene by helium-ion irradiation with a moderate ion dose, which works as scatterers in the two-dimensional electron system, and such defects enable tuning of electron transport properties. In addition, defective graphene regions can be designed to be arbitrary shape and size, due to the characteristic size of one nanometer by taking advantage of the high spatial resolution of HIM. Here, carrier transport phenomena in defective graphene at cryogenic temperatures and possible applications to field-effect transistors are introduced.

Two-dimensional (2D) materials, such as graphene or transition metal dichalcogenides (TMDCs), have been expected for many kinds of electronic devices, including sensors or transistors. In these devices, the advantages of these 2D materials are in the atomically thin structure that would lead to immunity to short channel effects in scaled transistors and also to the ultra-high sensitivity in molecular sensors. However, in order to take advantage of graphene, electric charge carrier transport properties should be modified by some external treatments. For example, the fabrication of nanoribbons is one of the methods to generate band gaps in graphene. In this case, the edge roughness of the ribbons usually causes an unintended scattering of carriers that would induce carrier conduction properties far from the theoretical expectations. Another method to generate a band gap in graphene is to apply a high electric field perpendicular to the bilayer graphene plane. In this case, a dual-gate structure should be placed on both sides of the bilayer graphene to tune the density of electrons and holes in the graphene channel. Also, the required electric field is very close to the breakdown field of silicon dioxide, but still, the band gap is not enough for transistor operation. Defective graphene was introduced to graphene to control the electronic charrier conduction without generating a band gap in graphene. Here, point defects are generated in graphene sheets, which cause the scattering of electrons and holes in graphene, which induces a transport gap around the Dirac point. By utilizing the transport gap, the defective graphene was applied for a graphene transistor, which will be introduced in this chapter.

From a fundamental point of view, defective graphene is also interesting in electric charger carrier conduction, because the carrier scattering phenomena in graphene are closely related to the fundamental physics of quantum transport in 2D Dirac electron systems. Carrier scattering in graphene will form a closed trajectory in which electrons are spatially localized due to quantum interference between clockwise and counterclockwise orbits on the same trajectory. This phenomenon is well known as Anderson localization. These two anti-parallel orbits are in relation to time-reversal symmetry, and hence the effect should disappear by breaking the time-reversal symmetry. Applying a magnetic field is a reasonable way to break time-reversal symmetry to investigate the Anderson localization phenomena, which should be detected as a negative magneto transport effect. In this chapter, the details of the defective graphene and its quantum transport will be discussed.

Many methods for making defective graphene have been developed, mostly by applying focused ion beams. For example, the gallium-ion beam is typically used as the focused ion beam (FIB) technique, and it was also applied to introduce defects in graphene. However, the size of ions is much larger than carbon atoms in a graphene sheet, which results in forming large craters in graphene by ion bombardments. Instead, in this chapter, we introduce the method using a helium-ion beam to fabricate defective graphene [105, 130]. Here, the small size of helium ion contributes to forming atomic size defects in graphene, and also the ion beam can be focused strongly down to the diameter of about 1 nm with a nanometer-scale spatial resolution. This way, a defective graphene in which ion-irradiated region can be designed using with arbitral shape, by utilizing the HIM.

In this chapter, we discuss the defect formation process of graphene by helium-ion beam irradiation and its effects on carrier transport properties. Here, the advantages of HIM for defective graphene formation are also introduced. Then, quantum tran sport phenomena in defective graphene are discussed in light of the quantum nature of carrier scattering and interference from the fundamental physics viewpoint. Finally, transistor operation in defective graphene will be demonstrated.

3.3.1 Effect of Helium-Ion Irradiation on Carrier Transport in Graphene

Helium-ion beam irradiation is an ideal method to induce point defects in graphene to modify the carrier transport properties of the 2D Dirac electrons. This is because of the small size of helium ions and the focused ion beam of them are available with the high spatial resolution in HIM. The first attempt of applying a helium-ion beam to graphene was intended to form graphene nanoribbons by curving the graphene sheet with it. However, when graphene is attached to a silicon dioxide layer, the surface of the substrate will be swelled by the damage of the ion beam even with the ion dose lower than required to fully remove carbon atoms in graphene [105, 130]. Therefore, at the earliest steps of this technique, graphene was suspended. Instead, the helium-ion beam was applied to graphene for tuning carrier conduction in graphene just by applying the helium-ion beam with a moderate ion dose [131, 132]. One important point in this technique is that due to the high acceleration voltage of 30 kV, the ions can penetrate deep into the substrate and the number of backscattering ions can be very small. Also, the positive charge of the helium ions remains close to the surface of the sample and can be neutralized by an electron shower that is equipped in the HIM system. These are also advantages of HIM for graphene carrier conduction tuning.

In order to see the effect of helium-ion beam irradiation on carrier conduction in graphene, an in situ measurement of current flowing in graphene during the helium-ion irradiation process [133]. Figure 3.25 shows the results. Figure 3.25(a) is an HIM image of the fabricated graphene transistor on a silicon substrate with its oxidized surface. A single-layer graphene flake is seen at the center between two electrodes, which contacts to probes. This sample is placed on the sample stage of the HIM system, which is connected to a source-measure unit out of the system. The current flowing between electrodes is monitored during the helium-ion irradiation process. Before starting the ion irradiation, we conducted some HIM observations of the device with a low ion dose of 5.0×10^{13} ions/cm^2, and then started the main process of ion beam irradiation. Figure 3.25(b) shows the device and current configurations. In order to define the "channel" at the center of the device, Regions 1–4 were irradiated with a heavy dose of 2.0×10^{16} ions/cm^2 to form high-resistance regions. The channel is defined as the unirradiated (pristine) region of 1 μm square. Then, the current flows along the red arrow shown in Figure 3.25(b) when a bias is applied to the drain electrode. Figure 3.25(c) shows the in situ monitoring of the current flowing in the device during the irradiation process at the drain bias of 1 mV. Evan an HIM observation causes a sharp drop of the current, a drastic increase in resistance, which is much stronger than that caused by forming high resistance regions. After that, the channel region was irradiated with a helium-ion beam at moderate ion doses in rectangular regions across the channel with a width of 100 nm as depicted in Figure 3.25(d).

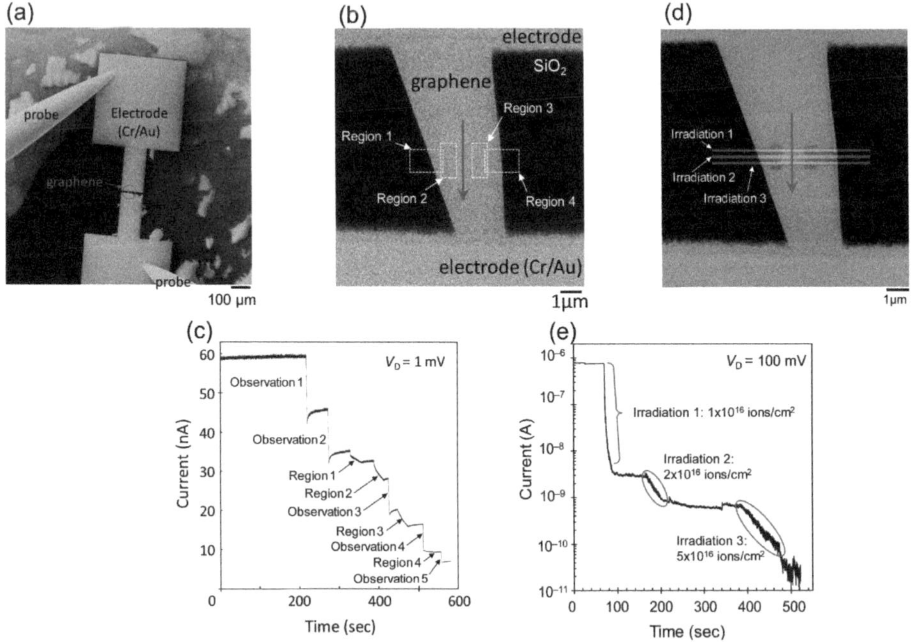

Figure 3.25 Helium-ion irradiation process in a single-layer graphene. (a) A graphene device with Cr/Au electrodes is connected to a measurement system via probes. This device is placed on the sample stage of HIM. (b) A single-layer graphene flake is seen in the gap of electrodes. Highly resistive regions 1–4 are irradiated with a high dose of 2.0×10^{16} ions/cm^2, then the current flows along the red arrow shown at the center of the HIM image. (c) Monitored current during the helium-ion irradiation process shown in (b). The drain bias was fixed at 1 mV. HIM observations and high-dose irradiations were done alternately. (d) Scans of helium-ion irradiation across the channel. Three regions (Irradiation 1–3) were irradiated with different ion doses. The width of each line is 100 nm. A red arrow indicates the current flow. (e) Monitored current during the scans shown in (d). The ion dose of each scan is indicated in the figure.

Each scan line had a 1 nm width from left to right across the channel with a constant spacing of 1 nm between each scan line, and the current flowing across the scanning lines was monitored at the same time of ion irradiation (Figure 3.25(e)). The drain bias in this measurement was 100 mV. After a sharp drop of the current by Irradiation 1, an exponential drop of current was observed in Irradiation 2 and Irradiation 3. Note that the width of the irradiated region is proportional to the time of the irradiation process, implying that the current drop is exponential, not linear, to the width of the irradiated area. This is the hallmark of the localization of electrons.

3.3.2 Carrier Localization in Helium-Ion-Irradiated Graphene

Carrier conduction control by defect engineering is based on carrier scattering by induced defects, which induces localization phenomena by quantum interference. In order to investigate systematically the basic physics of localization, some devices with the helium-ion-irradiated graphene channel were fabricated as shown in Figure 3.26(a) [134]. Here, single-layer graphene was exfoliated from an HOPG with adhesive tape and deposited on a silicon chip with the oxidized surface. The source and drain electrodes of Cr/Au were attached with a common lift-off process. The channel region is defined by the gap between these two electrodes and the gap between heavily dosed insulating regions as colored red in Figure 3.26(a). The channel region was irradiated by helium ions with moderate doses as indicated by the light blue region.

Figure 3.26(b) shows the magnetoresistance, $\Delta R/R_0$, versus the magnetic field perpendicular to the graphene plane, in a defective graphene sample of ion dose of 5×10^{14} ions/cm^2 at various temperatures. Here, ΔR is the resistance deviation from the zero-field value, and R_0 is the absolute value of resistance. As seen in the figure, the magnetoresistance is negative. Negative magnetoresistance implies that the closed orbit induced by the multiple scattering will be broken up by a magnetic field, which breaks the time-reversal symmetry. As a result, the localized state is solved to reduce the resistance in the electron system. One striking result here is that the negative magnetoresistance signal remains up to room temperature of 300 K (Figure 3.26(b)). The negative magnetoresistance is the largest at the charge neutral condition where the Fermi level is at around the Dirac point, as shown in Figure 3.26(c).

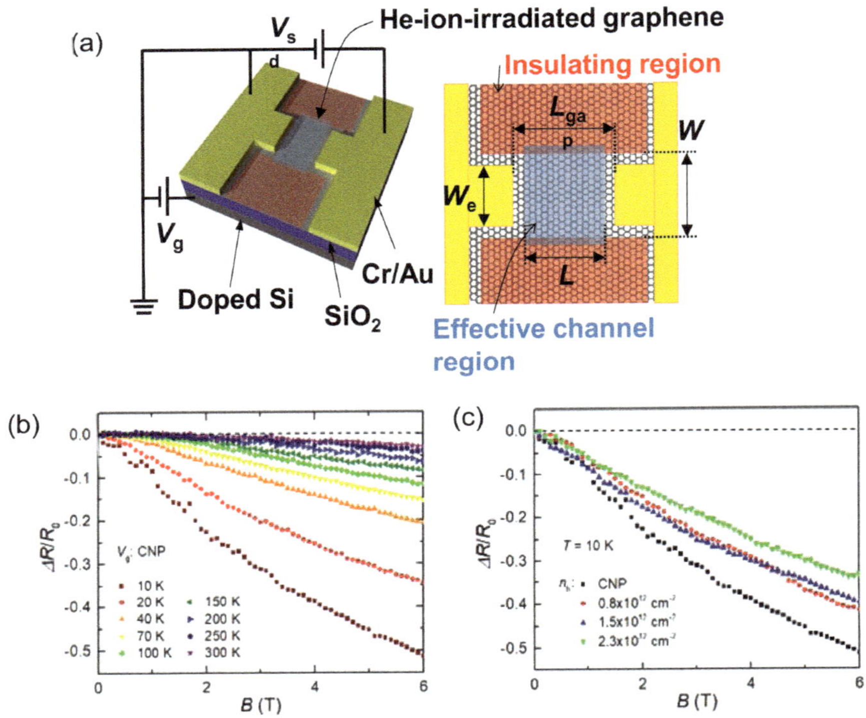

Figure 3.26 (a) Design of fabricated graphene device. Here, $L=W=1$ um, where L and W are channel length and width, respectively. Red regions were irradiated with a high ion dose to turn them into insulating regions, and the light blue region was irradiated with a moderate ion dose for channels. (b) Magnetoresistance under the charge neutral condition at various temperatures. Negative magnetoresistance remained up to 300 K. (c) Magnetoresistance at $T = 10$ K at different carrier densities.

In the strong localization regime, carrier conduction is based on variable range hopping (VRH). Localization length, which is the characteristic length of localization, is estimated from the temperature dependence of resistance. In our defective graphene with a defect density of 5 × 10^{14} ions/cm^2, the localization length is about 20 nm. In this case, the average distance between point defects is evaluated to be 7 nm, which is a reasonable value compared to the localization length. As the ion dose increases, the localization length is lowered, and the localization scheme will be unstable when the ion dose exceeds 2 × 10^{15} ions/cm^2.

3.3.3 Application of Defective Graphene to Transistors

The atomically thin body of the graphene channel, as well as the extremely high mobility of electrons and holes, have been expected to overcome the short-channel effects in the scaled transistors, while the lack of band gap in graphene has been the bottleneck for application. Defective graphene was investigated for application to graphene transistors because the transport gap in defective graphene was expected for the key for transistor operations. In addition to the thickness, it is also the advantage that the ion-irradiated area can be designed arbitrarily, so we elaborated a new concept of transistor, as shown in Figure 3.27. Here, one of the drawbacks of defective graphene is that the carrier scattering causes degraded carrier mobility, so the channel length should be designed as short as possible.

This transistor is attached with two independent top gates that can be biased independently between source and drain contacts. The gap of the top gates was designed to be 20 nm, and the graphene region here will be irradiated with a helium-ion beam at the ion dose of 8.7 × 10^{15} ions/cm^2. The channel width of 30 nm was determined by the gap length between the high-resistance regions where a heavy dose of helium-ion irradiation was applied (Figure 3.27(b and c)). One of the top gates works to fix the transistor type (n or p). When the gate bias of one of the top gates is positive (negative), the transistor works as a n-type (p-type) transistor as the gate biasing operation of the other top gate (Figure 3.27(d and e)). This way, the transistor can work as a polarity-reversible transistor [131]. It has been discussed that this kind of polarity-reversible transistor could reduce power consumption by up to 59% by renewing the circuit architecture [135].

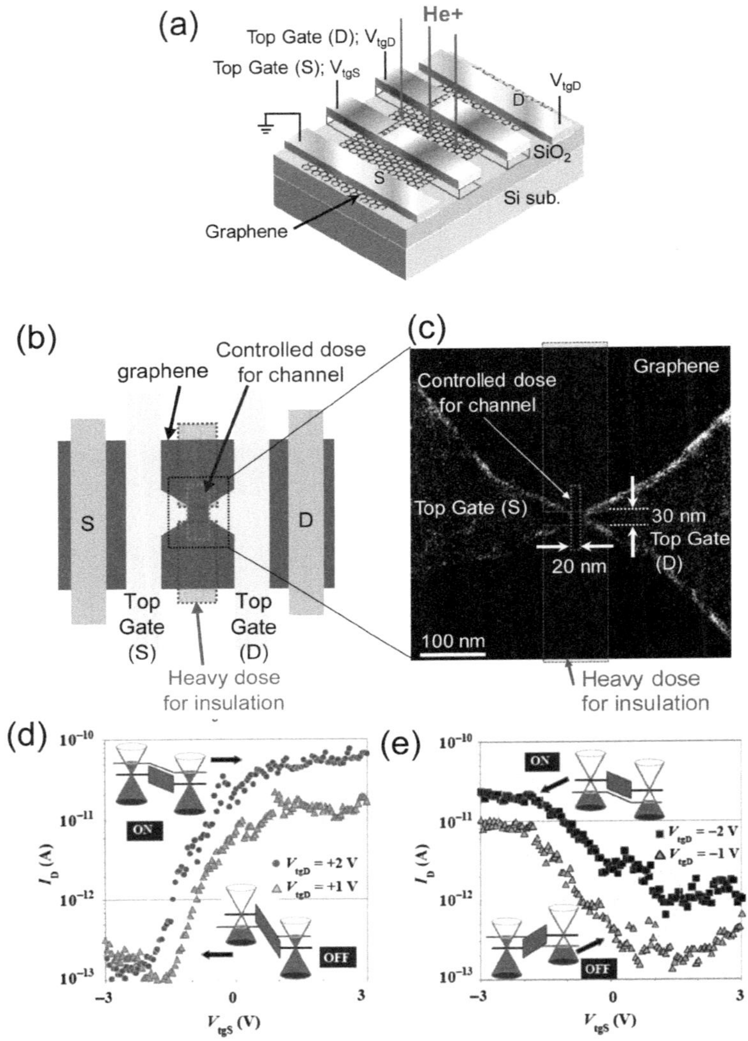

Figure 3.27 (a) Schematic of the fabricated device. (b) A schematic illustration of the ion-irradiation regions and the top gate geometry. Graphene between the two top gates is irradiated by a controlled dose (blue rectangle), while outside of this region (red rectangles), it is irradiated by a heavy dose to turn it into an insulator. (c) A HIM image of the device. Graphene between the two top gates is irradiated with a controlled dose (blue rectangle), while outside of this region (red rectangles) is irradiated by a heavy dose to turn it into an insulator. (d) N-type transistor operation with fixed $V_{tgD} > 0$. (e) P-type transistor operation with fixed $V_{tgD} < 0$.

Another demonstration of the polarity-reversible transistor is shown in Figure 3.28 [136]. A CVD-grown single-layer graphene sheet was transferred onto a silicon substrate with a surface oxide layer, and then, the device was fabricated by a wafer-scale top-down process (Figure 3.28(a)). As shown in Figure 3.28(b), there are two polarity-reversible transistors are combined into one device, applying opposite polarity gate biases to make it work as an inverter (Figure 3.28(c)). For example, V_1, and V_2 are connected to V_{DD} and ground, respectively. With $V_{tg1} < 0$ and $V_{tg2} > 0$, the device works as an inverter. As seen in Figure 3.28(d), a positive (negative) V_{in} generates a negative (positive) V_{out}. This inverter device is a reconfigurable one because the configurations of V_1 and V_2 can be swapped by adjusting the polarity of V_{tg1} and V_{tg2}.

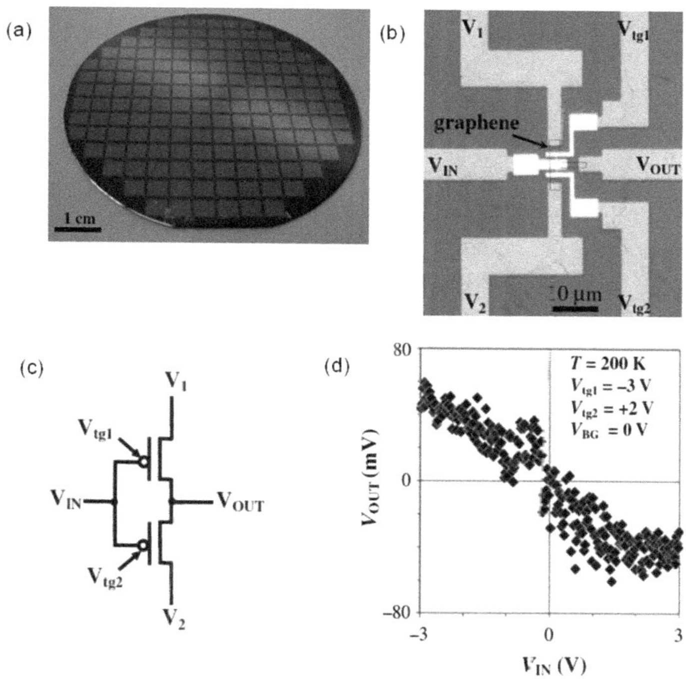

Figure 3.28 (a) Graphene devices fabricated on a silicon wafer. (b) An optical image of a fabricated inverter device. Bias configurations are also indicated. (c) The diagram of a six-terminal inverter device in which two polarity-reversible transistors are combined. (d) Inverter operation of the device.

In this chapter, the carrier transport modification in graphene by helium-ion irradiation is reviewed. This technique is mainly based on enhanced carrier scattering, which will lead to carrier localization and transport gap formation. Defective graphene with a high dose of ion irradiation exhibits negative magnetoresistance even at room temperature, implying strong localization at high temperatures. The generation of transport gap in graphene around the Dirac point enabled the application of defective graphene to transistor devices. In this chapter, as an example of the transistor with a defective graphene channel, low-temperature operations of the polarity-reversible transistor, as well as the inverter operation, were demonstrated. Although a more detailed understanding of carrier scattering in defective graphene is still required, modification of carrier conduction in graphene may open a new field of device application of graphene and other 2D materials.

3.4 Superconducting Oxide Nano Electronics Directly Written with a Focused Helium-Ion Beam

The small 0.5 nanometer chemically inert beam of the helium-ion microscope enables the creation of superconducting electronics directly written into high-transition temperature oxide superconductor material at the nanoscale. In this section, we review the physics of this process and highlight some of the experimental devices created in this manner.

3.4.1 Introduction to Superconductive Electronics

Superconducting electronics (SCE), first predicted [137] and demonstrated [138] in the early sixties, were once thought to be the replacement for semiconductors. The very high clock frequency (100s of GHZ) [139] of superconductor logic motivated large research programs by several corporate research labs; however, innovation in silicon CMOS performance, driven predominantly by scaling, soon out-paced SEC. There is a renewed interest in SEC driven by research in quantum computing [140] and ultra-low power high-performance computing [141]. Unlike CMOS devices, which operate by the movement of charge, superconducting electronics are quantum mechanical in nature and operate by the movement of tiny bits of quantized magnetic flux.

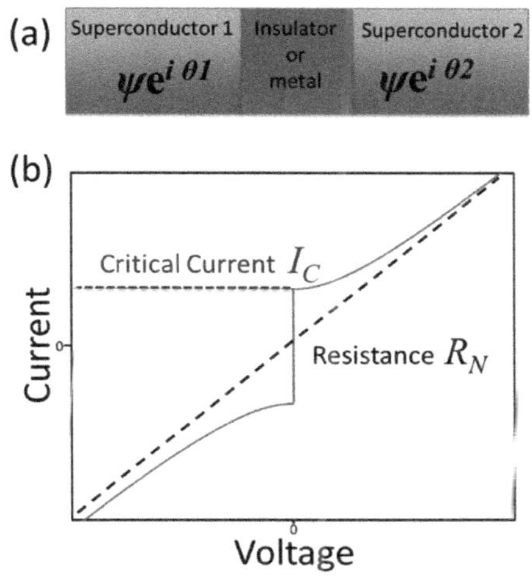

Figure 3.29 (a) Schematic representation of a Josephson junction. (b) Current-voltage characteristics of a typical Josephson junction used for electronics, depicting I_C and R_N.

The basic building block of SEC is the Josephson junction (JJ) [137]. It is analogous to superconductors as a transistor is to semiconductors. A JJ consists of two superconducting electrodes coupled through a weak barrier such as an insulator or normal metal. In a JJ (Figure 3.29) a "supercurrent" will flow in the absence of a potential difference (voltage) if there is a phase difference between the order parameters of the two superconductors. The difference between the phases of the two superconductors is defined as the Josephson phase, $\varphi \equiv (\theta_1 - \theta_2)$. The supercurrent in the simplest case is described by the first Josephson equation [137], $I = I_C \sin \varphi$, where the critical current, I_C is the maximum supercurrent that the junction will sustain and is determined by junction geometry, temperature, and the materials of the electrodes, amongst other factors. Furthermore, if there is a potential difference V, across the junction, the phase difference changes with time and is described by the second Josephson equation [137], $\partial\varphi/\partial t = 2eV/\hbar$, where e is the charge of an electron and $\hbar$ is the reduced Planck's constant. The second Josephson equation is exact and only depends on fundamental constants. If a DC voltage is applied across a JJ, the phase evolves with time at a frequency, $f = 2eV/\hbar$. This results in an AC supercurrent oscillating between the two superconductors at frequency f. The proportionality factor between voltage and frequency is referred to as the Josephson constant, $K_J = 2e/\hbar = 483.597879$ THz/V. This forms the basis for the definition of the volt and the JJ voltage standard [142].

3.4.2 High-Transition Temperature Oxide Superconductors

The vast majority of superconducting ceramics (discovered in the late 1980s [143, 144]) have a high critical temperature T_C ("high-T_C") relative to their elemental metallic counterparts such as Nb and Al, raising the prospect for liquid nitrogen operation of oxide superconducting interconnects, wires, and quantum devices. However, oxide superconductors, regardless of their generally high T_C values, are notoriously challenging to process into nanoscale devices or epitaxially combine with other materials in hybrid structures due to their highly anisotropic properties. Oxide superconductivity tends to vary along specific crystallographic directions [145]. For example, Figure 3.30 shows the unit cell structure of the prototypical oxide superconductor $YBa_2Cu_3O_{7-\Delta}$ (YBCO), which has a T_C of 90 K and an orthorhombic unit cell with a distorted perovskite structure: in the center is a single yttrium atom flanked by two copper oxygen planes. It is widely accepted that strong interaction between electrons in these planes is key to the superconducting state [146]. Electrons are strongly coupled for k-space directions parallel to the Cu–O–Cu bonds but weakly coupled in k-space directions along the diagonals. Hence, the superconductivity is anisotropic and believed to have a $dx^2–y^2$ symmetry ("d-wave" symmetry) [147]. The superconducting coherence lengths in YBCO are also highly anisotropic, being less than 0.2 nm long in the c-axis and 2–3 nm in the *a*-*b* plane [145]. The copper oxide chains at the top and bottom of the unit cells along the *b* direction contribute to electrical transport: the electrical conductivity in the *b*-direction σ_b is the highest, but parallel to the *a*- and *c*-directions σ is 1/3 σ_b and 1/10 σ_b, respectively [145, 148].

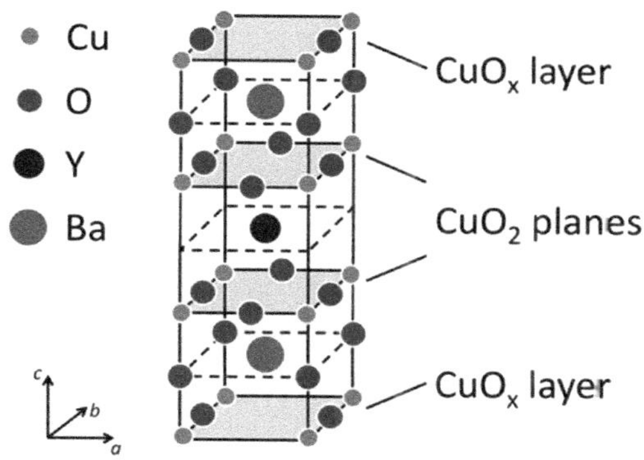

Figure 3.30 The orthorhombic unit cell of the cuprite superconductor YBCO, illustrating the copper oxygen planes and chains. The unit cell directions are *a*, *b*, and *c*.

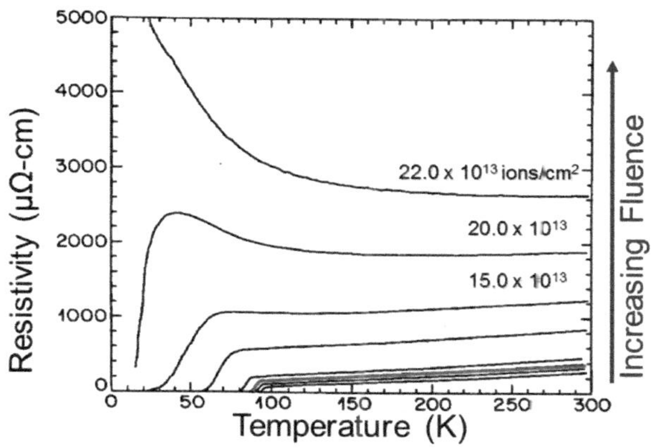

Figure 3.31 Resistivity as a function of temperature for ion irradiated films of YBCO. Figure adapted from [149].

The crystal structure of oxide superconductors is sensitive to point defects caused by ion irradiation. In Figure 3.31, we have plotted the electrical resistivity of a YBCO thin-film versus temperature and neon ion irradiation dose (1 MeV beam voltage) [149] with increasing irradiation, the resistivity rises and T_C decreases. For irradiation doses exceeding 2.2×10^{14} ions/cm^2, the material becomes a disordered insulator. The irradiation-induced metal-to-insulator transition in the normal state comes about because the electron mean free path is sensitive to the disorder of the oxygen lattice. Disorder shortens the mean free path and increases resistivity.

3.4.3 Focused Helium-Ion Beam Direct-Write Patterning of Superconductors

Our Lab at UC Riverside has utilized the disorder-driven superconductor-insulator-transition to establish a protocol for engineering nanoscale oxide devices including JJs with high-T_C oxides [150]. A focused gas field helium ion source beam (FHIB) [151] directly irradiates the material with doses to "write" an insulating state. This process does not involve removing materials. Conducting pathways are patterned into the plane of a thin film. The ion source can be focused as small as 0.5 nm in diameter and can write features with nanometer control [152]. The key to this technique is that the superconducting film must be thin (e.g., up to 50 nm) to prevent the beam from spreading out into the material, ensuring the ion beams pass through the film.

3.4.4 Focused Helium-Ion Beam Josephson Junctions

To use this technique to create JJs, 4 μm wide strips of YBCO were patterned with photolithography into a YBCO thin film, which had an in situ deposited Au contact layer on top (Figure 3.32(a)). The starting YBCO film thickness was 150 nm but the Au was removed and the YBCO was etched to a thickness of ~30 nm in the area (Figure 3.32(b)) intended for junctions to ensure that a 30 kV He^+ beam could fully penetrate the film. Samples were then loaded into a helium-ion microscope and the 30 kV beam was scanned across the 4 μm wide superconducting bridge to create the Josephson tunnel barrier (Figure 3.32(b)).

Figure 3.32 Focused helium-ion beam Josephson junction fabrication. (a) Optical micrograph of a patterned Au/YBCO film on a 5 mm × 5 mm sapphire substrate. (b) The central substrate region showing the gold contact region and YBCO film were etched prior to junction fabrication. This pattern contained four, 4-µm wide contacts for four-point resistance measurements. The red lines represent where the helium-ion beam was scanned to create the junctions. The current (*I*) and voltage (*V*) leads for one of these bridges is labeled. (c) A schematic diagram of the focused helium-ion beam creating a Josephson junction in a YBCO thin film. Figure adapted from [150].

Current as a function of voltage (*I-V*) is shown in Figure 3.33(a) for a typical superconductor-normal metal-superconductor (SNS) JJ for six temperatures. The resistance ~1 Ω is 10 times larger than masked ion irradiated weak links [150]. In contrast to the SNS junction, Figure 3.33(b) shows *I-V* behavior for several temperatures of a YBCO-based superconductor-insulator-superconductor (SIS) JJ fabricated using ion-irradiation with a higher dose (6×10^{16} He^+/cm^2). These JJs do not exhibit hysteresis in the *I-V* characteristic like conventional metal SIS junctions because the Stewart-McCumber parameter [148, 149] $\beta_C = 2\pi R_N{}^2 C/\Phi_0$, where R_N is the normal state resistance, C is the capacitance and Φ_0 is the flux quantum, is very small ($\beta_C \sim 0.001$). This comes about from a small capacitance because of the small area of the planar electrodes normal to the transport direction. Values of the product $I_C R_N$ range between 50 μV and 200 μV and are approximately an order of magnitude higher than masked ion irradiated junctions and comparable with other high-T_C JJs.

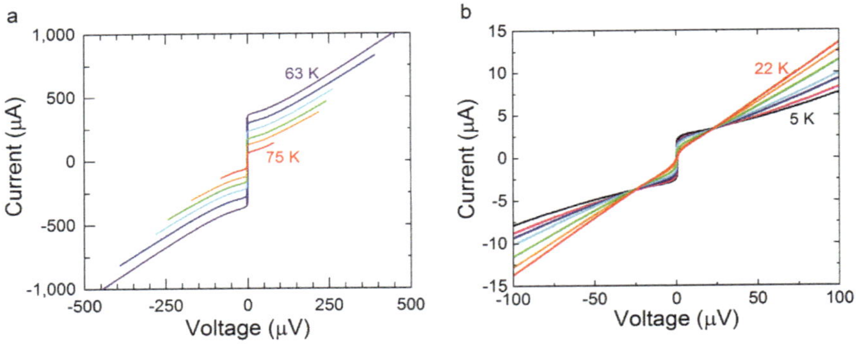

Figure 3.33 (a) *I-V* characteristics measured for temperatures 63 K, 65 K, 67 K, 69 K, and 71 K for a SNS Josephson junction fabricated using a dose of 2×10^{16} He^+/cm^2. (b) *I-V* characteristics measured at 5 K, 10 K, 12 K, 14 K, 16 K, 18 K, and 22 K for a SIS Josephson junction fabricated using a dose of 6×10^{16} He^+/cm^2. Figure adapted from [14].

3.4.4.1 Nanowire Josephson junctions

Historically, the width of YBCO JJs was limited to about 1 micron because etching the film with isotropic Ar^+ milling overheats and degrades fine features. By utilizing the FHIB, we showed [152] that one can aggressively scale the width to nanometer dimensions by irradiating insulating barriers to constrict the current path (Figure 3.34(a)).

In this work, we defined five YBCO bridges with FHIB to have widths ranging from 50 nm to 200 nm, and a 4 μm unpatterned control device. An FHIB JJ was placed in the center of the constriction so that we could accurately determine the width from the electrical properties of the JJ by the measurement of current-voltage characteristics. From these measurements, we extracted I_C and normal state conductance ($1/R_N$) and plotted it in Figure 3.34(b) as a function of width. The data show that

$R_N \propto 1/w$ and $I_C \propto w$ indicating the restricted channel as the only current pathway and the material was not thermally damage. This important demonstration provides confidence that nanoscale insulating and conducting pathways can be directly written into the material with high resolution using the FHIB.

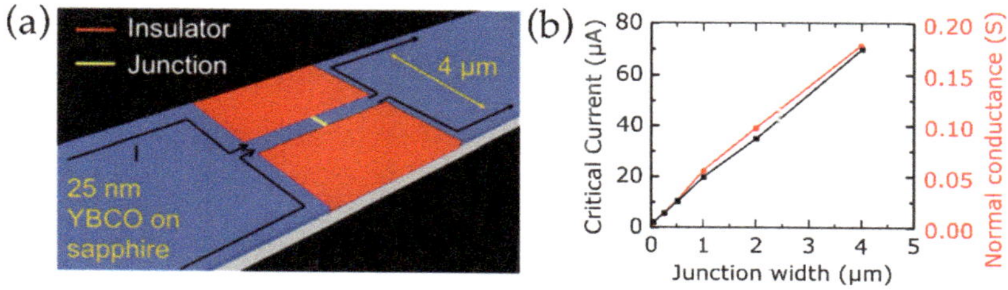

Figure 3.34 Nanowire Josephson Junction. (a) The width is narrowed down by using helium-ion irradiation as opposed to ion milling, which allows for nanoscale features. (b) The critical current and conductance scaled linearly with the junction width. Figure adapted from [152].

3.4.4.2 Helium-ion beam fabrication of passive components

Direct-write FHIB capabilities can also be used to integrate JJs with passive components like inductors, capacitors, and resistors on single-layer YBCO that are scalable and easy to fabricate in the same process steps. Ion irradiation can finely tune the resistivity as a function of dose like that shown in Figure 3.31. At intermediate doses, the resistivity stays constant over a wide range of temperatures, which is a very attractive feature for device stability in an environment with large thermal fluctuations. With this approach, junction shunt resistors as small as ~0.8 Ω can be utilized for low $I_C R$ applications like voltage standards [142].

We have found that for inductors and capacitors, FHIB-generated components typically have an inductance of ~4 pH/ψ and a capacitance of 0.675 fF/μm. With extended FHIB-generated structures using nanostrip meandered inductors and interdigital capacitors, we estimate this technology can achieve practical values of 4.8 nH and 27 fF for inductors and capacitors, respectively, which would be several orders of magnitude higher than the ranges one could obtain from conventional patterning with Ar^+ milling.

3.4.5 Applications

In this subsection, we highlight some of the device applications that have been realized by utilizing FHIB nanoscale patterning of high-T_C superconductors.

3.4.5.1 Superconducting quantum interference devices (SQUIDS)

One of the simplest Josephson junction circuits is a highly sensitive magnetic flux detector called a superconducting quantum interference device (SQUID) [153]. It consists of two JJs connected in parallel within a superconducting loop. In practice, a SQUID is typically current-biased above the I_C and the voltage oscillates as a function of the magnetic flux contained within the loop with a period of the magnetic flux quantum (Φ_0). Using FHIB direct-write junctions and nanoscale insulating lines, we have developed and tested many FHIB DC-SQUIDs in several configurations for magnetic sensing. Here, we describe two structures that utilize nanoscale patterning.

Experimental SQUIDs at nanoscale dimensions are a viable approach to reduce flux noise below 1 $\mu\Phi_0/\mathrm{Hz}^{1/2}$ but as the size decreases so does the sensitivity to magnetic field. This can be overcome by concentrating more flux into the loop of the SQUID. In conventional metal superconductor SQUIDS, this is achieved by creating a superconducting multi-turn transformer. However, in ceramic superconductors making wires and multiple layers is challenging. FHIB provides a potential solution through nanoscale patterning of concentration structures using nanoscale patterning.

3.4.5.1.1 *Nano-slit SQUID*

In work by Cho et al. [154], we created a SQUID consisting of a long but narrow, 10 nm by 20 μm slit, which was directly coupled to a superconducting washer pick-up loop of 100 by 100 μm^2 to increase flux concentration and magnetic field sensitivity (Figure 3.35). The narrow-slit design maximizes the coupling to the field concentrator. The field sensitivity of the magnetometer was measured to be 4 pT/$Hz^{1/2}$ and white flux noise was found to be 1 $\mu\Phi_0/Hz^{1/2}$. This is an excellent value for a high-T_C SQUID of this size and has the potential to serve as a sensitive magnetometer.

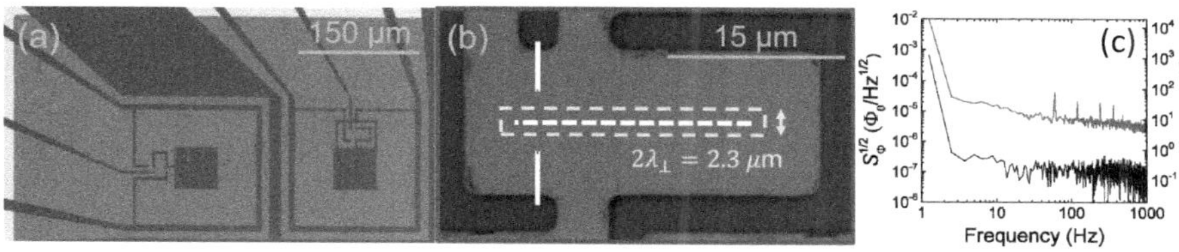

Figure 3.35 (a) Two YBCO nano-slit SQUID directly coupled magnetometers with single-turn control lines. The square washer-like loop is connected in parallel with the nano-slit SQUID for flux concentration. The nano-slit SQUID of the right-most device is enlarged in **(**b**)**. A narrow slit (white dashed line) is irradiated with FHIB into insulating. Two 2-μm wide Josephson junctions are subsequently irradiated (solid red lines) along with two insulators (white rectangles). (c) Noise spectrum for the directly coupled nano-slit SQUID. Figure adapted from [154].

3.4.5.1.2 *Nano-SQUIDs*

To showcase another breakthrough in the complex fabrication of high-T_C devices, a nano-SQUID-based flux-to-voltage trans-impedance amplifier (TIA) was demonstrated [155] with trans-impedance of 0.1 Ω and low 9 nW power dissipation compared to 3.1 mW power dissipation of cryogenic semiconductor TIA. In practice, the nano loop area requires a magnetic field hundreds of times stronger than the Earth's field to inductively couple magnetic flux. However, FHIB is used to write a control line a few nm away from the SQUID as shown in Figure 3.36 for stronger coupling with no power dissipation. FHIB nano-SQUIDs loops of 10 nm and 50 nm have exhibited voltage modulation amplitude of 0.5 mV and 0.8 mV, respectively. Remarkably, the junction I_C ranged from 50 μA to 5 μA with a nearly temperature-independent R_N of 30 Ω. Such tunable features allow SQUID operations over a wide temperature range with a stable output impedance.

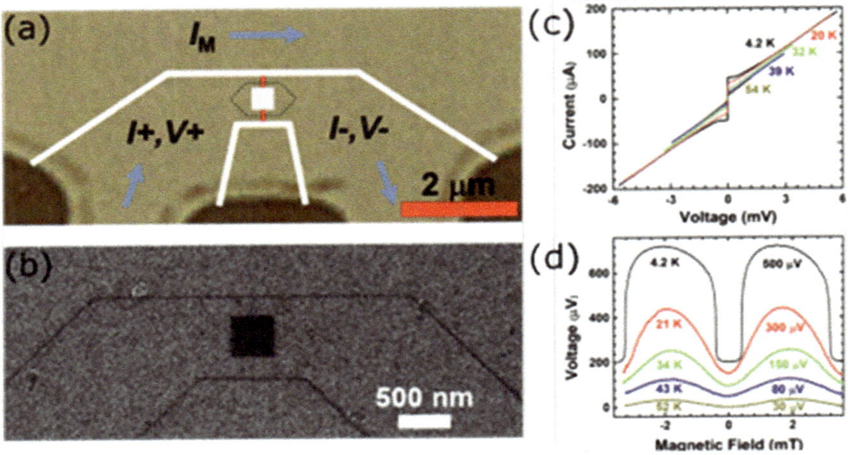

Figure 3.36 (a) Optical image of a nano-SQUID where the red line represents the barrier of the junction, and the white lines and square represent the insulating YBCO. (b) Helium-ion microscopic image of a YBCO nano-SQUID with a 400-nm square loop and 200 nm JJ width. (c) Current-voltage (*I-V*) characteristics of (b) at five different temperatures: 4.2 K, 20 K, 32 K, 39 K, and 54 K. I_C decreased from 50 µA to 5 µA, while the R_N remained at approximately kept at 30 Ω. (d) Voltage-magnetic field (*V-B*) curves from 4.2 K to 52 K where the modulation voltage decreased from 0.5 mV to 0.03 mV. Figure adapted from [155].

3.4.5.2 Digital electronics

Digital electronics can be achieved in Josephson circuits using magnetic flux quanta as a digital storage element referred to as single flux quantum circuits (SFQ) with low power consumption [156]. Narrow SFQ pulses form the basis of rapid SFQ logic (RSFQ) with a possible operating frequency higher than 770 GHz [139]. Using FHIB we created a reset-set flip-flop SFQ circuit in YBCO [157]. The circuit, shown in Figure 3.37, consists of a DC to SFQ converter that launches an SFQ through a Josephson transmission line (JTL), which is stored in a loop when a current pulse is applied to the "Set" electrode. A pulse on the "Reset" circuit on the right creates an anti-aligned SFQ which annihilates the SFQ in storage to reset the logic device.

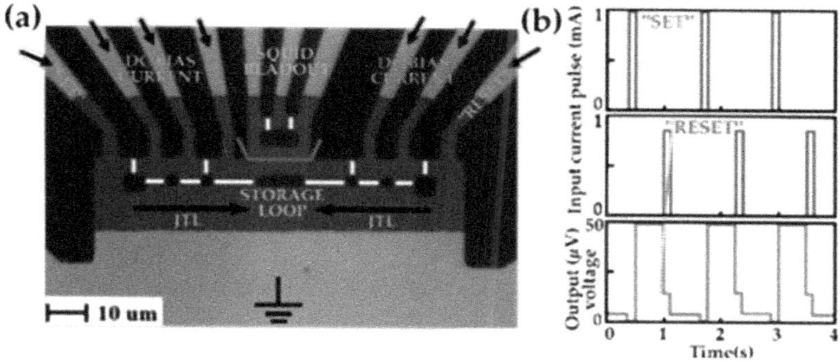

Figure 3.37 (a) The optical photograph of the RSFF device. The center orange area is the exposed YBCO pattern, where the gold electrode layer was chemically etched away. The solid white lines represent the Josephson Junctions, which are directly written on the exposed top YBCO layer by the helium-ion irradiation. The red line is the insulating line written by a much higher irradiation dose, which divides the SQUID from the storage loop. (b) The operation of the RS flip-flop device, presented in the time domain with input DC "Set" and "Reset" pulse along with output voltage pulse, detected by SQUID. Figure adapted from [157].

3.4.5.3 Large-scale Josephson junction array

For some applications, it is desirable to have several JJ spaced out over a relatively large area. One example is a long junction array used for high dynamic range detection [158]. In this approach, the voltage of N JJs is added linearly while the noise power scales like $\sqrt{N}$. Thus, the signal-to-noise ratio increases as $\sqrt{N}$. For this approach to be practical, N needs to be of the order of thousands. The circuits presented thus far in this book chapter have had a small spatial extent of ~100 μm, due to the FHIB's small field of view. To achieve larger circuits, LeFebvre et al. [159] showed that the FHIB JJ process can be automated to create large-scale circuits.

In this work, an array of about 2500 JJs (shown in Figure 3.38) was created using over 200 individual writing fields stitched together. This example shows that it is practical to scale up the FHIB JJ process for the nanomanufacturing of high-temperature superconductor circuits.

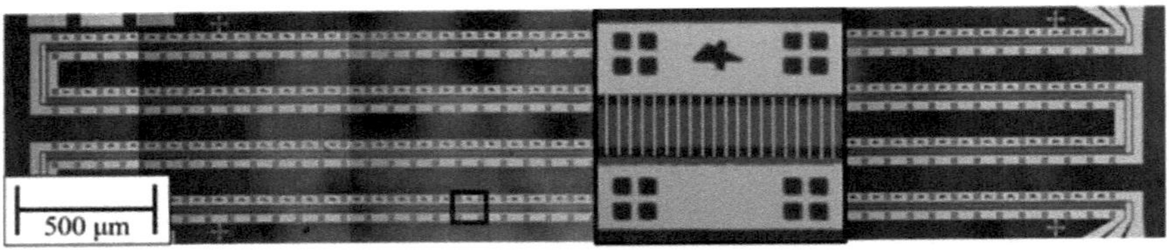

Figure 3.38 Long JJ array for high dynamic range magnetic sensing, which demonstrates process automation and scalability. The inset is a zoom-in of the alignment field that corresponds to the field of view of the FHIB. The vertical lines represent where JJs are written with He-FIB. Figure adapted from [158].

Acknowledgments

G. Hlawacek thanks N. Klingner (Helmholtz Zentrum Dresden-Rossendorf) for help with the graphics and insightful discussions. Support through the COST Action 19140 FIT4NANO is acknowledged. S. Nakaharai acknowledges T. Iwasaki, S. Moriyama, Y. Morita, Y. Wakayama (NIMS), S. Ogawa, T. Iijima (AIST), H. Mizuta, M. Muruganathan, M. Akabori, O. G. Agbonlahor (JAIST), S. Nakamura, Y. Okawa, M. Aono, C. Joachim, E. Verveniotis, S.-L. Li, K. Tsukagoshi (NIMS), S. Sato, N. Yokoyama (Fujitsu Labo.), and more collaborators.

References

1. Ogawa, S., Thompson, W., Stern, L., Scipioni, L., Notte, J., Farkas, L. and Barriss, L. (2010). Helium ion secondary electron mode microscopy for interconnect material imaging, *Japanese Journal of Applied Physics* 49, 4S, p. 04DB12, doi:10.1143/jjap.49.04db12.
2. Koch, J. (1936). Die Herstellung und die nähere Untersuchung einer neuen Alkaliionenquelle, *Zeitschrift für Physik* 100, 1112, pp. 669–684, doi:10.1007/bf01336749.
3. Seliger, R. L., Ward, J. W., Wang, V. and Kubena, R. L. (1979). A high-intensity scanning ion probe with submicrometer spot size, *Applied Physics Letters* 34, 5, p. 310.
4. Levi-Setti, R. (1974). Proton scanning microscopy: feasibility and promise, *Scanning Electron Microscopy* 7, pp. 125–134.
5. Escovitz, W. H., Fox, T. R. and Levi-Setti, R. (1975). Metaphase chromosomes in the scanning proton microscope, *Proceedings,*

annual meeting, Electron Microscopy Society of America 33, pp. 304–305, doi:10.1017/s0424820100115146.

6. Escovitz, W. H., Fox, T. R. and Levi-Setti, R. (1975). Scanning transmission ion microscope with a field ion source, *Proc. Natl. Acad. Sci. U. S. A.* 72, 5, pp. 1826–8.
7. Ward, B. W., Notte, J. A. and Economou, N. P. (2006). Helium ion microscope: a new tool for nanoscale microscopy and metrology, *J. Vac. Sci. Technol. B* 24, 6, p. 2871, doi:10.1116/1.2357967.
8. Miller, M., Cerezo, A., Hetherington, M. and Smith, G. (2006). *Atom Probe Field Ion Microscopy* (Oxford Science Publications).
9. Hren, J. J. and Ranganathan, S. (eds.) (1968). *Field-Ion Microscopy* (Springer US), ISBN 978-1-4899-6241-6, doi:10. 1007/978-1-4899-6513-4.
10. Muller, E. W. (1951). Das Feldionenmikroskop, *Zeitschrift für Phys.* 131, 1, pp. 136–142, doi:10.1007/BF01329651.
11. Tsong, T. T. (2003). Toward single atom chemical analysis with STM, *e-Journal Surf. Sci. Nanotechnol.* 1, pp. 102–105, doi:10.1380/ejssnt.2003.102.
12. Muller, E. W. and Tsong, T. T. (1969). *Field Ion Microscopy, Principles and Applications* (Elsevier Science Ltd, New York), ISBN 9780444000620.
13. Mason, E. A. and McDaniel, E. W. (2005). Appendix III: Tables of properties useful in the estimation of ionneutral interaction energies, in *Transp. Prop. Ions Gases* (WileyVCH Verlag GmbH & Co. KGaA, Weinheim, FRG), pp. 531–539, doi:10.1002/3527602852.app3.
14. Urban, R., Wolkow, R. A. and Pitters, J. L. (2016). *Single Atom Gas Field Ion Sources for Scanning Ion Microscopy*, in [86], chap. 2, pp. 31–61, doi:10.1007/978-3-319-41990-9 2.
15. Notte, J. A. (2019). ZEISS ORION NanoFab – the GFIS ion optical column, *techreport JN 2019.05.02*, Carl Zeiss Microscopy GmbH, URL https://goo.gl/ARiRVk.
16. Notte, J. A. and Huang, J. (2016). *The Helium Ion Microscope*, in [86], chap. 1, pp. 3–30, doi:10.1007/978-3-319-41990-9 1.
17. Hill, R. and Rahman, F. H. M. (2011). Advances in helium ion microscopy, *Nucl. Instruments Methods Phys. Res. Sect. A* 645, 1, pp. 96–101, doi:10.1016/j.nima.2010.12.123.
18. Ernst, N., Bozdech, G., Schmidt, H., Schmidt, W. A. and Larkins, G. L. (1993). On the full-width-at-half-maximum of field ion energy distributions, *Applied Surface Science* 67, 1–4, pp. 111–117, doi:10.1016/0169-4332(93)90301-Q.

19. Notte, J. A., Ward, B. W., Economou, N. P., Hill, R., Percival, R., Farkas, L., McVey, S., Seiler, D. G., Diebold, A. C., McDonald, R., Garner, C. M., Herr, D., Khosla, R. P. and Secula, E. M. (2007). An introduction to the helium ion microscope, in D. G. Seiler, A. C. Diebold, R. McDonald, M. Garner, D. Herr, R. P. Khosla and E. M. Secula(eds.), *AIP Conf. Proc.* 931 (AIP), pp. 489–496, doi:10.1063/1.2799423.

20. Hlawacek, G., Veligura, V., van Gastel, R. and Poelsema, B. (2014). Helium ion microscopy, *Journal of Vacuum Science and Technology B* 32, 2, p. 020801, doi:10.1116/1.4863676.

21. Hill, R., Notte, J. A. and Scipioni, L. (2012). *Scanning Helium Ion Microscopy*, (Elsevier Science & Technology Books), ISBN 9780123943965, pp. 65–148, doi:10.1016/b978-0-12-394396-5.00002-6.

22. Barth, J. E. and Kruit, P. (1996). Addition of different contributions to the charged particle probe size, *Optik (Stuttg)* 101, pp. 101–109.

23. Notte, J. A. (2012). Charged particle microscopy: why mass matters, *Micros. Today* 20, 05, pp. 16–22, doi:10.1017/ S1551929512000715.

24. Joy, D. C., Prasad, M. S. and Meyer, H. M. (2004). Experimental secondary electron spectra under SEM conditions, *Journal of Microscopy* 215, Pt 1, pp. 77–85, doi:10.1111/j.0022-2720. 2004.01345.x.

25. Ramachandra, R., Griffin, B. J. and Joy, D. C. (2009). A model of secondary electron imaging in the helium ion scanning microscope, *Ultra-microscopy* 109, 6, pp. 748–757, doi:10.1016/j. ultramic.2009.01.013.

26. Everhart, T. E. and Thornley, R. F. M. (1960). Wide-band detector for micro-microampere low-energy electron currents, *Journal of Scientific Instruments* 37, 7, pp. 246–248, doi:10. 1088/0950-7671/37/7/307.

27. Scipioni, L., Sanford, C. A., Notte, J. A., Thompson, B. and McVey, S. (2009). Understanding imaging modes in the helium ion microscope, *J. Vac. Sci. Technol. B* 27, 6, p. 3250, doi:10.1116/1.3258634.

28. van Gastel, R., Hlawacek, G., Zandvliet, H. J. W. and Poelsema, B. (2012). Subsurface analysis of semiconductor structures with helium ion microscopy, *Microelectronics Reliability* 52, 9–10, pp. 2104–2109, doi:10.1016/j.microrel. 2012.06.130.

29. Veligura, V., Hlawacek, G., van Gastel, R., Zandvliet, H. J. W. and Poelsema, B. (2012). Channeling in helium ion microscopy: mapping of crystal orientation, *Beilstein Journal of Nanotechnology* 3, pp. 501–506, doi:10.3762/bjnano.3.57.

30. Kimura, K., Joumori, S., Oota, Y., Nakajima, K. and Suzuki, M. (2004). High-resolution RBS: a powerful tool for atomic level characterization,

Nuclear Instruments and Methods in Physics Research Section B: Beam Interactions with Materials and Atoms 219–220, pp. 351–357, doi:10.1016/j.nimb.2004. 01.081.

31. Brongersma, H. H., Grehl, T., van Hal, P. A., Kuijpers, N. C. W., Mathijssen, S. G. J., Schofield, E. R., Smith, R. A. P. and ter Veen, H. R. J. (2010). High-sensitivity and highresolution low-energy ion scattering, *Vacuum* 84, 8, pp. 1005–1007, doi:10.1016/j.vacuum.2009.11.016.

32. ter Veen, H. R. J., Kim, T., Wachs, I. E. and Brongersma, H. H. (2009). Applications of high sensitivity-low energy ion scattering (HS-LEIS) in heterogeneous catalysis, *Catalysis Today* 140, 3–4, pp. 197–201, doi:10.1016/j.cattod.2008.10.012.

33. Heller, R., Klingner, N. and Hlawacek, G. (2016). *Backscattering Spectrometry in the Helium Ion Microscope: Imaging Elemental Compositions on the nm Scale*, in [86], chap. 12, pp. 265–295, doi:10.1007/978-3-319-41990-912.

34. Wollman, D. A., Irwin, K. D., Hilton, G. C., Dulcie L. L., Newbury, D. E. and Martinis, J. M. (1997). High-resolution, energy-dispersive microcalorimeter spectrometer for X-ray microanalysis, *Journal of Microscopy* 188, 3, pp. 196–223, doi:10.1046/j.1365-2818.1997.2670824.x.

35. Sijbrandij, S., Notte, J. A., Scipioni, L., Huynh, C. and Sanford, C. A. (2010). Analysis and metrology with a focused helium ion beam, *J. Vac. Sci. Technol. B* 28, 1, p. 73, doi:10.1116/1.3271254.

36. van Gastel, R., Hlawacek, G., Dutta, S. and Poelsema, B. (2015). Backscattered helium spectroscopy in the helium ion microscope: principles, resolution and applications, *Nuclear Instruments and Methods in Physics Research Section B: Beam Interactions with Materials and Atoms* 344, pp. 44–49, doi:10.1016/j.nimb.2014.11.073.

37. Klingner, N., Heller, R., Hlawacek, G., von Borany, J., Notte, J. A., Huang, J. and Facsko, S. (2016). Nanometer scale elemental analysis in the helium ion microscope using time of flight spectrometry, *Ultramicroscopy* 162, pp. 91–97, doi:10.1016/j.ultramic.2015.12.005.

38. Klingner, N., Heller, R., Hlawacek, G., Facsko, S. and von Borany, J. (2019). Time-of-flight secondary ion mass spectrometry in the helium ion microscope, *Ultramicroscopy* 198, pp. 10–17, doi:10.1016/j.ultramic.2018.12.014.

39. Wheatcroft, L., Klingner, N., Heller, R., Hlawacek, G., Ozkaya, D., Cookson, J. and Inkson, B. J. (2020). Visualization and chemical characterization of the cathode electrolyte interphase using He-ion microscopy and in situ time-of-flight secondary ion mass spectrometry,

ACS Applied Energy Materials 3, 9, pp. 8822–8832, doi:10.1021/acsaem.0c01333.

40. Gaderbauer, W., Arndt, M., Truglas, T., Steck, T., Klingner, N., Stifter, D., Faderl, J. and Groiss, H. (2020). Effects of alloying elements on surface oxides of hot-dip galvanized press hardened steel, *Surface and Coatings Technology* 404, p. 126466, doi:10.1016/j.surfcoat.2020.126466.
41. Wirtz, T., Castro, O. D., Audinot, J.-N. and Philipp, P. (2019). Imaging and analytics on the helium ion microscope, *Annual Review of Analytical Chemistry* 12, 1, pp. 523–543, doi:10.1146/annurev-anchem-061318-115457.
42. Wirtz, T., Dowsett, D. and Philipp, P. (2016). *SIMS on the Helium Ion Microscope: A Powerful Tool for High-Resolution High-Sensitivity Nano-Analytics*, in [86], chap. 13, pp. 297–323, doi:10.1007/978-3-319-41990-9 13.
43. Wirtz, T., Castro, O. D., Biesemeier, A., Hoang, H. Q. and Audinot, J.-N. (2020). Magnetic sector SIMS system with continuous focal plane detector for advanced analytical capabilities on FIB instruments, *Microscopy and Microanalysis* 26, S2, pp. 1972–1974, doi:10.1017/s1431927620019984.
44. Castro, O. D., Biesemeier, A., Serralta, E., Bouton, O., Barrahma, R., Hoang, Q. H., Cambier, S., Taubitz, T., Klingner, N., Hlawacek, G., Pinto, S. D., Gnauck, P., Lucas, F., Bebeacua, C. and Wirtz, T. (2021). npSCOPE: a new multimodal instrument for in situ correlative analysis of nanoparticles, *Analytical Chemistry* 93, 43, pp. 14417–14424, doi:10.1021/acs.analchem.1c02337.
45. Boden, S. A., Franklin, T. M. W., Scipioni, L., Bagnall, D. M. and Rutt, H. N. (2012). Ionoluminescence in the helium ion microscope, *Microscopy and Microanalysis* 18, 6, pp. 1253–1262, doi:10.1017/S1431927612013463.
46. Veligura, V. and Hlawacek, G. (2016). *Ionoluminescence*, in [86], chap. 14, pp. 325–351, doi:10.1007/978-3-319-41990-9 14.
47. Ogawa, S., Iijima, T., Awata, S., Sugie, R., Kawasaki, N. and Otsuka, Y. (2012). Characterization of damage in SiO_2 during helium ion microscope observation by luminescence and TEM-EELS, *Microscopy and Microanalysis* 18, S2, pp. 814–815, doi:10.1017/S1431927612005922.
48. Veligura, V., Hlawacek, G., van Gastel, R., Zandvliet, H. J. W. and Poelsema, B. (2015). Investigation of ionoluminescence of

semiconductor materials using helium ion microscopy, *Journal of Luminescence* 157, pp. 321–326, doi:10.1016/j.jlumin.2014. 09.016.

49. Veligura, V., Hlawacek, G., van Gastel, R., Zandvliet, H. J. W. and Poelsema, B. (2014). A high resolution ionoluminescence study of defect creation and interaction, *Journal of Physics: Condensed Matter* 26, 16, p. 165401, doi:10.1088/0953-8984/ 26/16/165401.

50. Veligura, V., Hlawacek, G., Jahn, U., van Gastel, R., Zandvliet, H. J. W. and Poelsema, B. (2014). Creation and physical aspects of luminescent patterns using helium ion microscopy, *Journal of Applied Physics* 115, 18, p. 183502, doi:10.1063/1.4875480.

51. Scipioni, L., Ferranti, D. C., Smentkowski, V. S. and Potyrailo, R. A. (2010). Fabrication and initial characterization of ultrahigh aspect ratio vias in gold using the helium ion microscope, *J. Vac. Sci. Technol. B* 28, 6, p. C6P18, doi:10.1116/1.3517514.

52. Hall, A. R. (2013). In situ thickness assessment during ion milling of a free-standing membrane using transmission helium ion microscopy, *Microscopy and Microanalysis* 19, 3, pp. 740–744, doi:10.1017/ s1431927613000500.

53. Emmrich, D. (2019). Mikroskopie und Modifikation von Nanostrukturen mit Helium-Ionen, phdthesis, University of Bielefeld.

54. Notte, J. A., Hill, R., McVey, S. M., Ramachandra, R., Griffin, B. J. and Joy, D. C. (2010). Diffraction imaging in a He^+ ion beam scanning transmission microscope, *Microscopy and Microanalysis* 16, 05, pp. 599–603, doi:10. 1017/S1431927610093682.

55. Woehl, T. J., White, R. M. and Keller, R. R. (2016). Darkfield scanning transmission ion microscopy via detection of forward-scattered helium ions with a microchannel plate, *Microscopy and Microanalysis* 22, 2009, pp. 544–550, doi:10.1017/s1431927616000775.

56. Kavanagh, K. L., Herrmann, C. and Notte, J. A. (2017). Camera for transmission He^+ ion microscopy, *J. Vac. Sci. Technol. B, Nanotechnol. Microelectron. Mater. Process. Meas. Phenom.* 35, 6, p. 06G902.

57. Wang, J., Huang, S. H. Y., Herrmann, C., Scott, S. A., Schiettekatte, F. and Kavanagh, K. L. (2018). Focussed helium ion channeling through si nanomembranes, *Journal of Vacuum Science & Technology B, Nanotechnology and Microelectronics: Materials, Processing, Measurement, and Phenomena* 36, 2, p. 021203, doi:10.1116/1.5020667.

58. Kavanagh, K. L., Bunevich, A. and Motapothula, M. R. (2020). Transmission helium ion microscopy of graphene, doi:10.48550/ ARXIV.2004.01682.

59. Serralta, E., Klingner, N., Castro, O. D., Mousley, M., Eswara, S., Pinto, S. D., Wirtz, T. and Hlawacek, G. (2020). Scanning transmission imaging in the helium ion microscope using a microchannel plate with a delay line detector, *Beilstein Journal of Nanotechnology* 11, pp. 1854–1864, doi:10.3762/ bjnano.11.167.
60. Tabean, S., Mousley, M., Pauly, C., Castro, O. D., Serralta, E., Klingner, N., Mücklich, F., Hlawacek, G., Wirtz, T. and Eswara, S. (2022). Quantitative nanoscale imaging using transmission he ion channelling contrast: proof-of-concept and application to study isolated crystalline defects, *Ultramicroscopy* 233, p. 113439, doi:10.1016/j.ultramic.2021. 113439.
61. Bell, D. C., Lemme, M. C., Stern, L. A. and Marcus, C. M. (2009). Precision material modification and patterning with He ions, *J. Vac. Sci. Technol. B* 27, 6, p. 2755, doi:10.1116/ 1.3237113.
62. van Gastel, R., Barriss, L., Sanford, C. A., Hlawacek, G., Scipioni, L., Merkle, A., Voci, D., Fenner, C., Zandvliet, H. J. W. and Poelsema, B. (2011). Design and performance of a near ultra high vacuum helium ion microscope, *Microscopy and Microanalysis* 17, S2, pp. 928–929, doi:10. 1017/S1431927611005514.
63. Veligura, V., Hlawacek, G., Berkelaar, R. P., van Gastel, R., Zandvliet, H. J. W. and Poelsema, B. (2013). Digging gold: keV He^+ ion interaction with Au, *Beilstein Journal of Nanotechnology* 4, pp. 453–460, doi:10.3762/bjnano.4.53.
64. Joens, M. S., Huynh, C., Kasuboski, J. M., Ferranti, C., Sigal, Y. J., Zeitvogel, F., Obst, M., Burkhardt, C. J., Curran, K. P., Chalasani, S. H., Stern, L. A., Goetze, B. and Fitzpatrick, J. A. J. (2013). Helium ion microscopy (HIM) for the imaging of biological samples at sub-nanometer resolution, *Scientific Reports* 3, p. 3514, doi:10.1038/srep03514.
65. Frese, N., Schmerer, P., Wortmann, M., Schürmann, M., König, M., Westphal, M., Weber, F., Sudhoff, H. and Gölzhäuser, A. (2021). Imaging of sars-cov-2 infected vero e6 cells by helium ion microscopy, *Beilstein Journal of Nanotechnology* 12, 1, pp. 172–179, doi:10.3762/ bjnano.12.13.
66. Schürmann, M., Frese, N., Beyer, A., Heimann, P., Widera, D., Mönkemöller, V., Huser, T., Kaltschmidt, B., Kaltschmidt, C. and Gölzhäuser, A. (2015). Helium ion microscopy visualizes lipid nanodomains in mammalian cells, *Small* 11, 43, pp. 5781–5789, doi:10.1002/smll.201501540.
67. Utke, I., Hoffmann, P. and Melngailis, J. (2008). Gas-assisted focused electron beam and ion beam processing and fabrication, *J. Vac. Sci. Technol. B* 26, 2008, p. 1197, doi:10.1116/1.2955728.

68. Utke, I., Swiderek, P., Höflich, K., Madajska, K., Jurczyk, J., Martinović, P. and Szymańska, I. B. (2022). Coordination and organometallic precursors of group 10 and 11: focused electron beam induced deposition of metals and insight gained from chemical vapour deposition, atomic layer deposition, and fundamental surface and gas phase studies, *Coordination Chemistry Reviews* 458, p. 213851, doi:10.1016/j.ccr.2021. 213851.

69. Alkemade, P. F. A. and Miro, H. (2014). Focused helium-ion-beam-induced deposition, *Applied Physics A* 117, 4, pp. 1727–1747, doi:10.1007/s00339-014-8763-y.

70. Shorubalko, I., Pillatsch, L. and Utke, I. (2016). *Direct–Write Milling and Deposition with Noble Gases*, in [86], chap. 15, pp. 355–393, doi:10.1007/978-3-319-41990-9 15.

71. Chen, F.-C., Chen, Y.-P., Huang, Y.-J. and Chien, S.C. (2010). Morphological study on pentacene thin-film transistors: the influence of grain boundary on the electrical properties, *J. Phys. D. Appl. Phys.* 43, 40, p. 405103, doi:10.1088/0022-3727/43/40/405103.

72. Córdoba, R., Mailly, D., Rezaev, R. O., Smirnova, E. I., Schmidt, O. G., Fomin, V. M., Zeitler, U., Guillamón, I., Suderow, H. and Teresa, J. M. D. (2019). Three-dimensional superconducting nanohelices grown by he^{+}-focused-ion-beam direct writing, *Nano Letters* 19, 12, pp. 8597–8604, doi:10.1021/acs.nanolett.9b03153.

73. Xu, X., Heinig, K.-H., Möller, W., Engelmann, H.-J., Klingner, N., Gharbi, A., Tiron, R., von Borany, J. and Hlawacek, G. (2019). Morphology modification of Si nanopillars under ion irradiation at elevated temperatures: plastic deformation and controlled thinning to 10 nm, *Semiconductor Science and Technology* 35, 1, p. 015021, doi:10.1088/1361-6641/ab57ba.

74. Stanford, M. G., Lewis, B. B., Iberi, V., Fowlkes, J. D., Tan, S., Livengood, R. H. and Rack, P. D. (2016). In situ mitigation of subsurface and peripheral focused ion beam damage via simultaneous pulsed laser heating, *Small* 12, 13, pp. 1779–1787, doi:10.1002/smll.201503680.

75. Andany, S. H., Hlawacek, G., Hummel, S., Brillard, C., Kangül, M. and Fantner, G. E. (2020). An atomic force microscope integrated with a helium ion microscope for correlative nanoscale characterization, *Beilstein Journal of Nanotechnology* 11, pp. 1272–1279, doi:10.3762/bjnano.11. 111.

76. Kreith, J., Strunz, T., Fantner, E. J., Fantner, G. E. and Cordill, M. J. (2017). A versatile atomic force microscope integrated with a scanning electron microscope, *Review of Scientific Instruments* 88, 5, p. 053704, doi:10.1063/1. 4983317.

77. Tortonese, M., Barrett, R. C. and Quate, C. F. (1993). Atomic resolution with an atomic force microscope using piezoresistive detection, *Applied Physics Letters* 62, 8, pp. 834–836, doi:10.1063/1.108593.

78. Nanda, G., Hlawacek, G., Goswami, S., Watanabe, K., Taniguchi, T. and Alkemade, P. F. A. (2017). Electronic transport in helium-ion-beam etched encapsulated graphene nanoribbons, *Carbon* 119, pp. 419–425, doi:10.1016/j.carbon. 2017.04.062.

79. Dunne, P., Fowley, C., Hlawacek, G., Kurian, J., Atcheson, G., Colis, S., Teichert, N., Kundys, B., Venkatesan, M., Lindner, J., Deac, A. M., Hermans, T. M., Coey, J. M. D. and Doudin, B. (2020). Helium ion microscopy for reduced spin orbit torque switching currents, *Nano Letters* 20, 10, pp. 7036–7042, doi:10.1021/acs.nanolett.0c02060.

80. Gaul, A., Emmrich, D., Ueltzhöffer, T., Huckfeldt, H., Dŏganay, H., Hackl, J., Khan, M. I., Gottlob, D. M., Hartmann, G., Beyer, A., Holzinger, D., Nemšák, S., Schneider, C. M., Gölzhäuser, A., Reiss, G. and Ehresmann, A. (2018). Size limits of magnetic-domain engineering in continuous in-plane exchange-bias prototype films, *Beilstein Journal of Nanotechnology* 9, pp. 2968–2979, doi:10.3762/bjnano.9.276.

81. Petrov, Y. V. and Vyvenko, O. F. (2016). *Secondary Electron Generation in the Helium Ion Microscope: Basics and Imaging*, in [86], chap. 5, pp. 119–146, doi:10.1007/978-3-319-41990-95.

82. Rodenburg, C., Jepson, M. A. E., Boden, S. A. and Bagnall, D. M. (2014). Helium ion microscopy and energy selective scanning electron microscopy – two advanced microscopy techniques with complementary applications, *Journal of Physics: Conference Series* 522, p. 012049, doi:10.1088/ 1742-6596/522/1/012049.

83. Petrov, Y. V. and Vyvenko, O. F. (2011). Secondary electron emission spectra and energy selective imaging in helium ion microscope, in J. T. Thomas, D. D. Desjardins, J. J. Güell and K. L. Bernier (eds.), *Proc. SPIE*, 8036 (SPIE, Orlando, Florida), pp. 80360O–1–80360O–10, doi:10.1117/12.886347.

84. Ohya, K., Yamanaka, T., Inai, K. and Ishitani, T. (2009). Comparison of secondary electron emission in helium ion microscope with gallium ion and electron microscopes, *Nucl. Instruments Methods Phys. Res. Sect. B* 267, 4, pp. 584–589, doi:10.1016/j.nimb.2008.11.003.

85. Rodenburg, C., Jepson, M. A. E., Bosch, E. G. T. and Dapor, M. (2010). Energy selective scanning electron microscopy to reduce the effect of contamination layers on scanning electron microscope dopant

mapping, *Ultramicroscopy* 110, 9, pp. 1185–1191, doi:10.1016/j.ultramic.2010.04.008.

86. Hlawacek, G. and Gölzhäuser, A. (eds.) (2016). *Helium Ion Microscopy, NanoScience and Technology* (Springer International Publishing, Switzerland), ISBN 978-3-31941988-6, doi:10.1007/978-3-319-41990-9.

87. Novoselov, K. S. and Geim, A. K. (2007). The rise of graphene, *Nat. Mater.* 6, pp. 183–191.

88. Meyer, J. C., Geim, A. K., Katsnelson, M. I., Novoselov, K. S., Booth, T. J. and Roth, S. (2007). The structure of suspended graphene sheets, *Nature* 446, pp. 60–63.

89. Sun, J., Wang, W., Muruganathan, M. and Mizuta, H. (2014). Low pull-in voltage graphene electromechanical switch fabricated with a polymer sacrificial spacer, *Appl. Phys. Lett.* 105, pp. 033103.1 033103.4.

90. Nika, D. L. and Balandin, A. A. (2017). Phonons and thermal transport in graphene and graphene-based materials, *Rep. Prog. Phys.* 80, pp. 036502.1–036502.2.

91. Mohr, M., Papagelis, K., Maultzsch, J. and Thomsen, C. (2009). Two-dimensional electronic and vibrational band structure of uniaxially strained graphene from ab initio calculations, *Phys. Rev. B* 80, pp. 2665–2668.

92. Seifoori, S. and Hajabdollahi, H. (2015). Impact behavior of single-layered graphene sheets based on analytical model and molecular dynamics simulation, *Appl. Surf. Sci.* 351, pp. 565–572.

93. Bolotin, K. I., Sikes, K. J., Jiang, Z., Klima, M., Fudenberg, G., Hone, J., Kim, P. and Stormer, H. L. (2008). Ultrahigh electron mobility in suspended graphene, *Solid State Commun.* 146, pp. 351–355.

94. Cresti, A., Nemec, N., Biel, B., Niebler, G., Triozon, F., Cuniberti, G., et al. (2008). Charge transport in disordered graphene-based low dimensional materials, *Nano Res.* 1, pp. 361–394.

95. Yang, L., Park, C. H., Son, Y. W., Cohen, M. L. and Louie, S. G. (2007). Quasiparticle energies and band gaps in graphene nanoribbons, *Phys. Rev. Lett.* 99, pp. 186801.1–186801.4.

96. Barone, V., Hod, O. and Scuseria, G. E. (2006). Electronic structure and stability of semiconducting graphene nanoribbons, *Nano Lett.* 6, pp. 2748–2754.

97. Han, M. Y., Ozyilmaz, B., Zhang, Y. B. and Kim, P. (2007). Energy band-gap engineering of graphene nanoribbons, *Phys. Rev. Lett.* 98, pp. 206805.1–206805.4.

98. Wang, Q., Kitaura, R., Suzuki, S., Miyauchi, Y. and Shinohara, H. (2016). Fabrication and in situ transmission electron microscope characterization of free-standing graphene nanoribbon devices, *ACS Nano* 10, pp. 1475–1480.

99. Jiao, L., Wang, X., Diankov, G, Wang, H. and Dai, H. (2010). Facile synthesis of high-quality graphene nanoribbons, *Nat. Nanotechnol.* 5, pp. 321–325.

100. Li, X., Wang, X., Zhang, L., Lee, S. and Dai, H. (2018). Chemically derived, ultrasmooth graphene nanoribbon semiconductors, *Science* 319, pp. 1229–1232.

101. Stankovich, S., Piner, R. D., Nguyen, S. T. and Ruoff, R. S. (2006). Synthesis and exfoliation of isocyanate-treated graphene oxide nanoplatelets, *Carbon* 44, pp. 3342–3347.

102. Liu, F., Muruganathan, M., Ogawa, S., Morita, Y., Schmidt, M. and Mizuta, H. (2020). Half-meshed and fully-meshed suspended graphene for transport gap engineering, *Proc. 2020 IEEE SIlicon Nanoelectronics Workshop, SNW*, pp. 69–70.

103. Schmidt, M. E., Muruganathan, M., Kanzaki, T., Iwasaki, T., Hammam, A. M. M., Suzuki, S., Ogawa, S. and Mizuta, H. (2019). Dielectric-screening reduction induced large transport gap in suspended sub-10 nm graphene nanoribbon functional devices, *Small* 15, pp. 1903025.1–1903025.7.

104. Pickard, D. and Scipioni, L. (2009). Graphene nano-ribbon patterning in the orion plus, *Zeiss Application Note*, Carls Zeiss, Peabody, MA 2009.

105. Lemme, M. C., Bell, D. C., Stern, L. A., Baugher, B. W. H., Jarillo-Herrero, P. and Marcus, C. M. (2009). Etching of graphene devices with a helium ion beam, *ACS Nano* 3, pp. 2674–2676.

106. Bai, J., Zhong, X., Jiang, S., Huang, Y. and Duan, X. (2010). Graphene nanomesh, *Nat. Nanotechnol.* 5, pp. 190–194.

107. Berrada, S., Nguyen, V. H., Querlioz, D., Saint-Martin, J., Alarcón, A., Chassat, C., Bournel, A. and Dollfus, P. (2013). Graphene nanomesh transistor with high on/off ratio and good saturation behavior, *Appl. Phys. Lett.* 103, pp. 183509.1–183509.5.

108. Ouyang, F., Peng, S., Liu, Z. and Liu, Z. (2011). Bandgap opening in graphene antidot lattices: the missing half, *ACS Nano* 5, pp. 4023–4030.

109. Nguyen, V. H., Nguyen, M. C., Nguyen, H. V. and Dollfus, P. (2013). Disorder effects on electronic bandgap and transport in graphene-

nanomesh-based structures, *J. Appl. Phys.* 113, pp. 013702.1–013702.9.

110. Oswald, W. and Wu, Z. (2012). Energy gaps in graphene nanomeshes, *Phys. Rev. B* 85, pp. 115431.1–115431.5.

111. Maldovan, M. (2013). Sound and heat revolutions in phononics, *Nature* 503, pp. 209–217.

112. Schmidt, M. E., Iwasaki, T., Muruganathan, M., Haque, M., Van, N. H., Ogawa, S. and Mizuta, H. (2018). Structurally controlled large-area 10 nm pitch graphene nanomesh by focused helium ion beam milling, *ACS Appl. Mater. Interfaces* 10, pp. 10362–10368.

113. Liu, F., Wang, Z., Nakanao, S., Ogawa, S., Morita, Y., Schmidt, M., Haque, M., Muruganathan, M. and Mizuta, H. (2020). Conductance tunable suspended graphene nanomesh by helium ion beam milling, *Micromachines* 11, pp. 387.1–387.13.

114. Emmrich, D., Beyer, A., Nadzeyka, A., Bauerdick, S., Meyer, J. C., Kotakoski, J. and Gölzhäuser, A. (2016). Nanopore fabrication and characterization by helium ion microscopy, *Appl. Phys. Lett.* 108, pp. 163103.1–163103.4.

115. Lu, Y. and Guo, J. (2010). Band gap of strained graphene nanoribbons, *Nano Res.* 3, pp. 189–199.

116. Zhang, Y. J., Heiranian, M., Janicek, B., Budrikis, Z., Zapperi, S., Huang, P. Y., Johnson, H.T., Aluru, N.R., Lyding, J.W. and Mason, N. (2018). Strain modulation of graphene by nanoscale substrate curvatures: a molecular view, *Nano Lett.* 18, pp. 2098–2104.

117. Reserbat-Plantey, A., Kalita, D., Han, Z., Ferlazzo, L., Autier-Laurent, S., Komatsu, K., Li, C., Weil, R., Ralko, A., Marty, L., et al. (2014). Strain superlattices and macroscale suspension of graphene induced by corrugated substrates, *Nano Lett.* 14, pp. 5044–5051.

118. Nakaharai, S., Iijima, T., Ogawa, S., Suzuki, S., Li, S.-L., Tsukagoshi, K., Sato, S. and Yokoyama, N. (2013). Conduction tuning of graphene based on defect-induced localization, *ACS Nano* 7, pp. 5694–5700.

119. Han, M.Y., Brant, J.C. and Kim, P. (2010). Electron transport in disordered graphene nanoribbons, *Phys. Rev. Lett.* 104 pp. 056801.1–056801.4.

120. Volz, S., Ordonez-Miranda, J., Shchepetov, A., Prunnila, M, Ahopelto, J., Pezeril, T., Vaudel, G., Gusev, V., Ruello, P., Weig, E. M., Schubert, M., Hettich, M., Grossman, M., Dekorsy, T., Alzina, F., Graczykowski, B., Chavez-Angel, E., Reparaz, J. S., Wagner, M. R., Sotomayor-Torres, C. M., Xiong, S., Neogi, S. and Donadio, D. (2016). Nanophononics: state of the art and perspectives, *Eur. Phys. J. B* 89, pp. 15.1–15.20.

121. Hopkins, P. E., Reinke, C. M., Su, M. F., Olsson, R. H., Shaner, E. A., Leseman, Z. C., Serrano, J. R., Phinney, L. M. and El-Kady, I. (2011). Reduction in the thermal conductivity of single crystalline silicon by phononic crystal patterning, *Nano Lett.* 11, pp. 107–112.

122. Zen, N., Puurtinen, T. A., Isotalo, T. J., Chaudhuri, S. and Maasilta, I. J. (2014). Engineering thermal conductance using a two-dimensional phononic crystal, *Nat. Commun.* 5, pp. 3435.1–3435.9.

123. Liu, F., Muruganathan, M., Feng, Y., Ogawa, S., Morita, Y., Liu, C., Guo, J., Schmidt, M. and Mizuta, H. (2021). Thermal rectification on asymmetric suspended graphene nanomesh devices, *Nano Futures* 5, pp. 045002.1–045002.12.

124. Li, B., Wang, L. and Casati, G. (2004). Thermal diode: rectification of heat flux, *Phys. Rev. Lett.* 93, pp. 184301.1–184301.4.

125. Chang, C., Okawa, D., Majumdar, A. and Zettl, A. (2006). Solid-state thermal rectifier, *Science* 314, pp. 1121–1124.

126. Dames, C. (2009). Solid-state thermal rectification with existing bulk materials, *J. Heat Transfer* 131, pp. 061301.1–061301.7.

127. Zeng, N. (2008). Mechanisms causing thermal rectification: the influence of phonon frequency, asymmetry, and nonlinear interactions, *Phys. Rev. B* 78. 024305.1–024305.5.

128. Xu, X., Li, B. and Chen, J. (2016). Phonon thermal conduction in novel 2D materials, *J. Phys.: Condens. Matter* 28, pp. 483001.1–483001.20.

129. Zhao, S., Zhou, Y. and Wang, H. (2022). Review of thermal rectification experiments and theoretical calculations in 2D materials, *Int. J. Heat Mass Transf.* 195, pp. 123218.1–123218.31.

130. Bell, D. C., Lemme, M. C., Stern, L. A., Williams, J. R. and Marcus, C. M. (2009). Precision cutting and patterning of graphene with helium ions nanotechnology, 20, 455301.

131. Nakaharai, S., Iijima, T., Ogawa, S., Suzuki, S., Tsukagoshi, K., Sato, S. and Yokoyama, N. (2012). Electrostatically-reversible polarity of dual-gated graphene transistors with He ion irradiated channel: toward reconfigurable CMOS applications, *Technical Digest - International Electron Devices Meeting, IEDM2012*, pp. 4.2.1–4.2.4.

132. Nakaharai, S., Iijima, T., Ogawa, S., Suzuki, S., Li, S.-L., Tsukagoshi, K., Sato, S. and Yokoyama, N. (2013). Conduction tuning of graphene based on defect-induced localization, *ACS Nano* 7, pp. 5694–5670.

133. Nakaharai, S., Iwasaki, T., Morita, Y., Moriyama, S. and Ogawa, S. (2022). Electron transport tuning of graphene by helium ion irradiation nano express, 3, p. 024002.

134. Iwasaki, T., Nakamura, S., Agbonlahor, O. G., Muruganathan, M., Akabori, M., Morita, Y., Moriyama, S., Ogawa, S., Wakayama, Y., Mizuta, H. and Nakaharai, S. (2021). Room-temperature negative magnetoresistance of helium-ionirradiated defective graphene in the strong Anderson localization regime, *Carbon* 175, pp. 87–92.

135. Ben-Jamaa, M., H, Mohanran K. and De Micheli, G. (2011). An efficient gate library for ambipolar CNTFET logic, *IEEE Trans. Computer-Aided Des. Integr. Circuits Syst.* 30, pp. 242–255.

136. Nakaharai, S., Iijima, T., Ogawa, S., Yagi, K., Harada, N., Hayashi, K., Kondo, D., Takahashi, M., Li, S.-L., Tsukagoshi, K., Sato, S. and Yokoyama, N. (2015). Wafer-scale fabrication of transistors using CVD-grown graphene and its application to inverter circuit Jpn, *Journ. Appl. Phys.* 50, p. 04DN06.

137. Josephson, B. D. (1962). Possible new effects in superconductive tunnelling, *Physics Letters* 1, 7, pp. 251–253.

138. Anderson, P. W. and Rowell, J. M. (1963). Probable observation of the Josephson superconducting tunneling effect, *Physical Review Letters* 10, 6, p. 230.

139. Rylyakov, W. A. V., Patel, V., Lukens, J. E. and Likharev, K. K. (1999). Rapid single flux quantum T-flip flop operating up to 770 GHz, *IEEE Transactions on Applied Superconductivity* 9, 2, pp. 3212–3215, doi:10.1109/77.783712.

140. Berggren, K. K. (2004). Quantum computing with superconductors, *Proc. of the IEEE* 92, 10, pp. 1630–1638.

141. Holmes, D. S., Kadin, A. M. and Johnson, M. W. (2015). Superconducting computing in large-scale hybrid systems, *Computer* 48, 12, pp. 34–42.

142. Hamilton, C. A. (2000). Josephson voltage standards, *Review of scientific instruments* 71, 10, pp. 3611–3623.

143. Bednorz, J. G. and Müller, K. A. (1986). Possible high Tc superconductivity in the Ba-La-Cu-O system, *Zeitschrift für Physik B Condensed Matter* 64, 2, pp. 189–193.

144. Wu, M.-K. et al. (1987). Superconductivity at 93 K in a new mixed-phase Y-Ba-Cu-O compound system at ambient pressure, *Physical review letters* 58, 9, p. 908.

145. Kim, J.-T., Giapintzakis, J. and Ginsberg, D. (1996). Anisotropy of the resistivity in the a-b plane of a superconducting $YBa_2Cu_3O_7$-δ single crystal, *Journal of superconductivity* 9, 5, pp. 481–484.

146. Dagotto, E. (1994). Correlated electrons in high-temperature superconductors, *Reviews of Modern Physics* 66, 3, p. 763.

147. Müller, K. (2002). On the macroscopic s-and d-wave symmetry in cuprate superconductors, *Philosophical magazine letters* 82, 5, pp. 279–288.

148. Kleiner, D. Ludwig, R. F. Dantsker, E. and Clarke J. (1999). High-transition-temperature superconducting quantum interference devices, *Reviews of Modern Physics* 71, 3, p. 631.

149. White, E. et al. (1988). Controllable reduction of critical currents in $YBa_2Cu_3O_7$-δ films, *Applied Physics Letters* 53, 11, pp. 1010–1012.

150. Cybart, S. A. et al. (2015). Nano Josephson superconducting tunnel junctions in $YBa_2Cu_3O_7$–δ directly patterned with a focused helium ion beam, *Nature Nanotechnology* 10, 7, pp. 598–602, doi:10.1038/nnano.2015.76.

151. Notte, J., Economou, N. and Ward, B. (2007). New microscope technology emerges from research labs, *R&D MAGAZINE* 49, 8, pp. 38–39.

152. Cho, E. Y., Zhou, Y. W., Cho, J. Y. and Cybart, S. A. (2018). Superconducting nano Josephson junctions patterned with a focused helium ion beam, *Applied Physics Letters* 113, 2, p. 022604.

153. Lambe, R. J., Mercereau, J. and Silver, A. (1965). Macroscopic quantum interference in superconductors, *Physical Review* 140, 5A, p. A1628.

154. Cho, E. Y., Li, H., LeFebvre, J. C., Zhou, Y. W., Dynes, R. C. and Cybart, S. A. (2018). Direct-coupled micro-magnetometer with Y-Ba-Cu-O nano-slit SQUID fabricated with a focused helium ion beam, *Appl. Phys. Lett.* 113, 16, p. 162602, doi:10.1063/1.5048776.

155. Li, H. et al. (2020). High-transition-temperature nanoscale superconducting quantum interference devices directly written with a focused helium ion beam, *Applied Physics Letters* 116, 7, p. 070601.

156. Mukhanov, N., Yoshikawa, I., Nevirkovets, P. and Hidaka, M. (2019). Josephson junctions for digital applications, in *Fundamentals and Frontiers of the Josephson Effect*, Springer, 2019, pp. 611–701.

157. Cai, H., Li, H., Cho, E. Y., LeFebvre, J. C. and Cybart, S. A. (2021). $YBa_2Cu_3O_7$-\delta single flux quantum flip flop directly written with a focused helium ion beam, *IEEE Transactions on Applied Superconductivity* 31, 5, pp. 1–5, doi:10.1109/TASC.2021.3069904.

158. LeFebvre, J. C., Cho, E. Y., Li, H. Pratt, K. and Cybart, S. A (2019). Series arrays of planar long Josephson junctions for high dynamic range magnetic flux detection, *AIP Advances* 9, 10, p. 105215.

159. LeFebvre, J. C., Cho, E. Y, Li, H., Cai, H. and Cybart, S. A. (2022). Flux focused series arrays of long Josephson junctions for high-dynamic range magnetic field sensing, *Journal of Applied Physics* 131, 16, p. 163902.

Chapter 4

Diamond Quantum Sensors Based on Spin-Qubit of NV Centers

Mutsuko Hatano

Tokyo Institute of Technology, Meguro, Tokyo, Japan

hatano.m.ab@m.titech.ac.jp

Quantum sensors encompass a diverse class of devices that exploit quantum coherence to detect weak or nanoscale signals. As their behavior is tied to physical constants, quantum sensors can achieve accuracy, repeatability, and precision approaching fundamental limits [1, 2]. As a result, these sensors have shown utility in a wide range of applications in both science and engineering leading to innovations. Among many solid-state spin systems, nitrogen-vacancy (NV) centers in diamonds have attracted considerable attention as a leading platform in many fields of quantum technologies, from quantum sensing to information processing and communication applications. They have superior physical properties, and their quantum coherence is preserved even at room temperature. The energy levels of NV centers are sensitive to magnetic fields, electric fields, strain, and temperature, and the spatial resolution is scalable from nanometers to millimeters. Therefore, diamond quantum sensors are distinctive sensors with multimodality and span scalable application fields. To realize an ideal quantum sensor for applications, the technical

Outlooking beyond Nanoelectronics and Nanosystems: Ultra Scaling, Pervasiveness, Sustainable Integration, and Biotic Cross-Inspiration
Edited by Simon Deleonibus

ISBN 978-981-5129-32-8 (Hardcover), 978-1-003-50990-5 (eBook)
www.jennystanford.com

challenges in the sensor materials and the quantum control techniques, such as coherent microwave spin manipulation, effective optical spin-state initialization, and readout, have been developed. This chapter briefly explains the sensing principle using NV centers and then introduces sensor systems for biomedical applications, and monitoring of batteries for electric vehicles and internal power devices.

4.1 Introduction

Diamond is an excellent host for spin-based qubits, and the spin in diamond has excellent properties. In 1997, an optically detected magnetic resonance (ODMR) of a single NV center was manipulated at room temperature for the first time [3]. This discovery has led to the development of the field of diamond quantum technology. NV centers have proven to be one of the most specific localized electronic structures to serve as spin-based qubits and are promising candidates for quantum sensing [4–8]. Magnetometers have been first demonstrated with single NV [6, 7] and ensemble NV centers [8]. Both single and ensemble NV centers diamond magnetometers [9] have been used in applications spanning pure and applied sciences, including condensed matter physics [10], neuroscience and living systems biology [11, 12], nuclear magnetic resonance [13–16], and earth and planetary science [17].

In this chapter, the sensing principle is briefly explained. Then, technical challenges in realizing applications of linking the material, quantum control, and applications are introduced. For applications, demonstrations of magnetocardiography of living rats [18] and monitoring of batteries for electric vehicles [19–21] and internal power devices [22–24] are explained.

4.2 Physical Properties and Materials

4.2.1 Physical Properties of the NV Center in Diamond and Quantum Sensing Using NV Centers

The NV center is a point defect in a diamond comprising a substitutional nitrogen atom and lattice vacancy, as shown in

Figure 4.1(a). The neutral NV center (NV^0) is a stable charge state; however, a negatively charged NV center (NV^-) is more important for quantum sensing and quantum information applications [9] because optical initialization and readout of the spin state can be performed. In this chapter, the NV^- center is considered and denoted by simply expressing the NV center. The spin state of electrons localized at the NV center exhibits a spin-1 (S = 1) triplet ground state with a long spin lifetime at room temperature, i.e., the spin coherence time T_2. In fact, up to a few milliseconds of T_2 are achievable [25, 26] even at room temperature. The energy levels of the ground state are sensitive to magnetic fields, electric fields, strain, and temperature, and can be initialized and read out optically [9]. The simplified energy level diagram of the NV center is shown in Figure 4.1(b). Diamond is a wide bandgap semiconductor (5.5 eV), and the energy level of the NV center is just the mid-gap of the bandgap. The ground and first excited states are triplet states (S = 1), and the energy difference is 637 nm. The photoluminescence spectrum with a zero-phonon line at this energy is experimentally observed. A metastable singlet state (S = 0) occurs between the ground and first excited states. One of the most important properties of the NV center is the optical spin polarization, which is mediated by this singlet state. Electrons with $|ms = \pm 1\rangle$ states have a certain probability of decaying through this singlet state to the $|0\rangle$ ground state. Although infrared light (1042 nm) is emitted, the energy is significantly different from direct luminescence. Thus, this decay is termed non-radiative decay. Continuous optical pumping results in spin polarization into the $|0\rangle$ state. Furthermore, the spin-state population can be optically read out by observing the photoluminescence intensity. The more the $|0\rangle$ state is populated, the more intense luminescence is observed. The photoluminescence spectrum of the NV^- spins shows a broad range from 637 nm to 830 nm (Figure 4.1(c)). The NV^0 also exhibits a broad photoluminescence spectrum with a zero-phonon line (ZPL) at 575 nm. Notably, the sharp peak at 572 nm is a first-order Raman peak of the diamond.

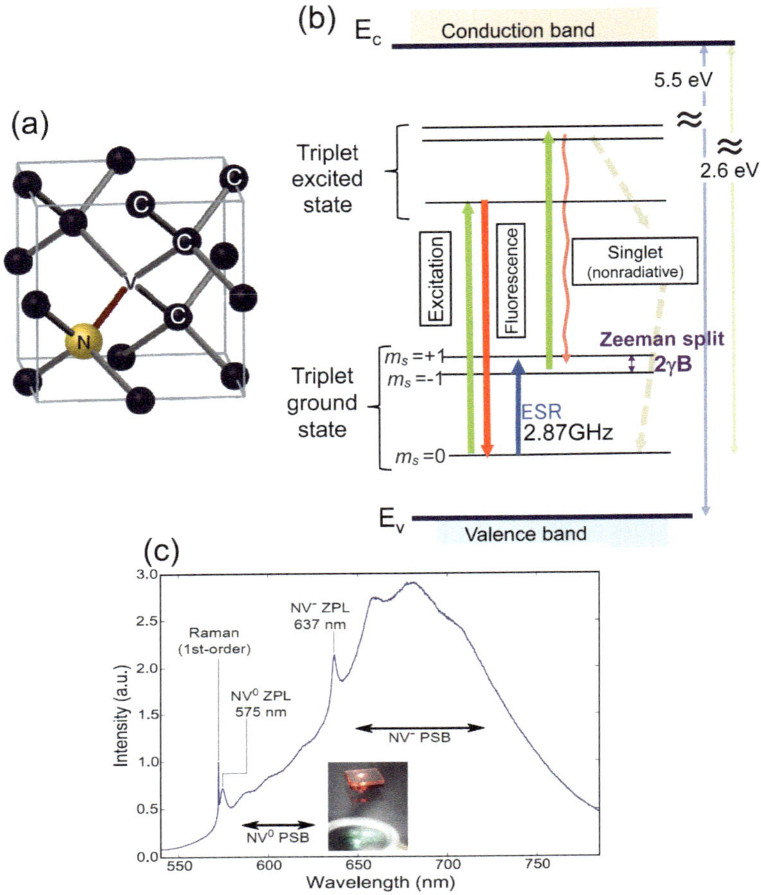

Figure 4.1 Overviews of the NV center quantum system: (a) Structure of diamond crystal with an NV center. (b) Energy level diagram of the NV center and the principle of spin-dependent fluorescence. The energy level of magnetic sublevels ms = ±1 shift proportional to the magnetic field. (c) Typical photoluminescence spectrum. (Courtesy of Tokyo Institute of Technology).

In conjunction with spin-state manipulation using microwave radiation, ODMR can be utilized. As aforementioned, the quantum state of the NV center can be detected by fluorescence because the quantum states 0 and ± 1 differ in the amount of red fluorescence when excited by a green light source. The energy levels of the state ± 1 split depending on the magnetic field due to the Zeeman effect. Therefore, the system is used as a highly sensitive magnetometer. The ground-state spin can be transited between $|0\rangle$ and $|-1\rangle$, or $|0\rangle$ and $|+1\rangle$ by irradiating with microwaves. This transition occurs only when a microwave with a resonant frequency is applied. The method of measuring the fluorescence intensity while changing the frequency of the microwave and searching for the resonance frequency is the basic procedure to observe ODMR. Figure 4.2(a) shows the changes in the ODMR spectrum, due to the application of a magnetic field B_{ext}. When no static magnetic field is applied, the energy of $|-1\rangle$ and $|+1\rangle$ are degenerate, and one resonant peak appears at 2.87 GHz. When a static magnetic field is applied, degeneration is released and the peak splits by the Zeeman effect, as shown in Figure 4.2(b). The split width of this peak is proportional to the magnitude of the magnetic field in the direction of the NV axis. When applied with a tilt of θ to the NV axis, the peak interval of ODMR is $\Delta f = 2\gamma B \cdot \cos\chi$. Because four equivalent NV axis orientations are present in the diamond, the direction and rotation can be detected using the ODMR spectra. Magnetic field imaging can be performed using CCD or CMOS imagers for fluorescence intensity detection; thus, magnetic field vector imaging can be performed. This is a function that other sensors do not possess. In the case of AC magnetic fields, such as the measurement target of NMR, the dynamic decoupling protocol enables high-sensitivity measurements.

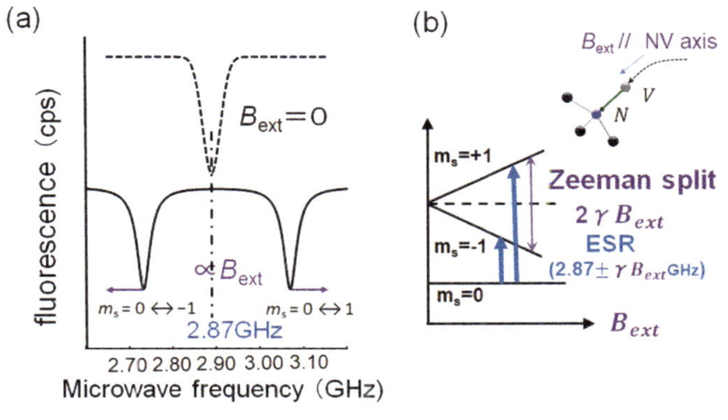

Figure 4.2 Overviews of the NV center quantum system: (a) Schematic of the NV center in diamond. (b) Electronic energy diagram of the NV center. The energy level of magnetic sublevels ms = ±1 shift proportional to the magnetic field. (Courtesy of Tokyo Institute of Technology).

Table 4.1 shows a comparison of the quantum magnetic sensors SQUID and atomic vapor cells. The sensitivity of diamond sensors should be improved, but diamond sensors have various potential. Diamond sensors can operate at a wide temperature range (several milliKelvins to 600 K). Moreover, they are superior in the dynamic range and linearity of the magnetic field. They operate in extreme environments, in liquids and solids. The spatial resolution is scalable from nanometers to millimeters using single to ensemble NV centers. Diamond sensors have the unique functions of vector magnetic field imaging and multimodality such as temperature (which is equivalent to the energy) and magnetic field (which is equivalent to the information). Notably, by using the device technology and integrating the sensor arrays on-chip, the NV center exhibits miniaturization possibility.

Table 4.1 Comparison of quantum magnetic sensors (Courtesy of Tokyo Institute of Technology)

Quantum sensors	SQUID (Super-conductor)	Atomic vapor cells	Diamond (NV centers)
Sensitivity ($\sqrt{\text{Hz}}$)	≈ fT	< fT	~ pT → fT • good dynami range • good linearity
Spatial resolution	> 10 μm	~ mm	Scalability: nano to macro
Temp. Environment	LT	> RT	RT (mK - 600 K) in the atmosphere Robustness in extreme environments: under high temperature and high pressure in liquid and in solid
Vector imaging	–	Δ	○ (the distinctive advantage)
Multimodalities	–	–	◎ Temp. (energy) & Magnetic (information)
Miniturization	–	○	◎ On-Chip integration

Diamond sensors are particularly suitable for sensing biomedical systems and monitoring the social infrastructure. Figure 4.3 shows the potential of the scalable application of diamond magnetic quantum sensors and diamond materials with NV centers. Spatial resolution for scalable applications can be achieved by varying the morphology of the diamond material with NV centers. Sensors in the nanometer range contribute to the structural analysis of proteins and biomolecules, intracellular characterization, elucidation of condensed matter physics, and analysis in semiconductor devices. For their sensing, nano-diamond, single NV center, and nanoscale probe-type sensors [27] are suitable. For sensors from the mesoscopic to the macroscopic range, ensemble NV centers are effective in improving the signal and sensitivity. Ensemble NV centers are formed at a thickness approximately equal to the polarization size of the magnetization vector of the nearby measured object. Therefore, in mesoscopic applications, δ-doping to form shallow NV centers on the diamond surface is efficient. For macroscale applications, the absolute number of NV centers should be increased to the extent possible using the full thickness of the diamond substrate. Furthermore, synthesizing diamond films containing NV centers on heterogeneous substrates such as Si wafers is effective in increasing the area. Heteroepitaxial growth produced by chemical vapor deposition (CVD) has been studied [28, 29].

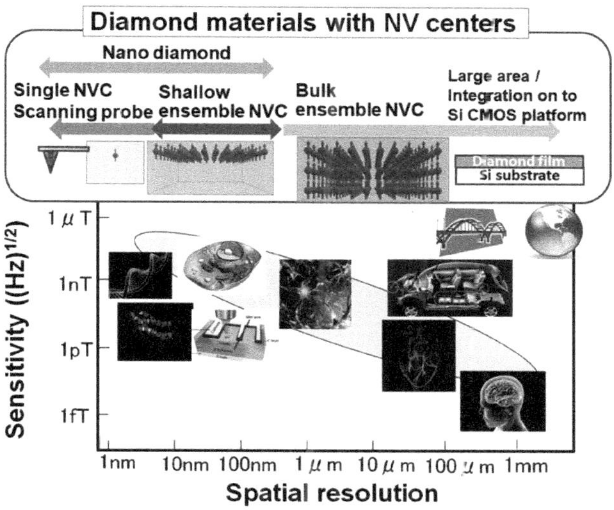

Figure 4.3 Potential on the scalable application of diamond magnetic quantum sensors and materials with NV centers (Courtesy of Tokyo Institute of Technology).

4.2.2 Formation of NV Centers in Diamond for Quantum Sensors

Diamond films or bulk diamonds are commonly synthesized using high-pressure high-temperature (HPHT) or plasma-enhanced chemical vapor deposition (PE-CVD). In HPHT-synthesized diamonds, nitrogen easily incorporates into the diamond lattice forming substitutional nitrogen. Nitrogen concentration can be controlled by carrying the atomic composition of the metal-solvent catalyst to get the nitrogen. The formation of NV centers in a diamond can be classified into two methods: forming NV centers by beam irradiation after synthesis or by PE-CVD during diamond synthesis as explained subsequently. The former irradiation methods include ion implantation and electron-beam irradiation. Because position control is possible, ion implantation has been most widely applied for sensing below the microscopic scale. However, ion implantation usually substantially damages the diamond lattice because of high-energy ion collisions, leading to a short spin coherence time. In the electron-beam irradiation method, a diamond is irradiated with high-energy electron beams, and the electrons kick out carbon atoms in the crystal to form vacancies. Because nitrogen is incorporated into diamond as an impurity during crystal synthesis, heat treatment causes the nitrogen and atomic vacancies to combine to form NV centers. The high AC magnetic sensitivity was obtained with this technique [30].

PE-CVD diamond synthesis can leverage established semiconductor process technology. NV centers are formed during CVD diamond growth, and the thickness of the NV layer can be controlled from nanometers [31] to over a hundred microns [32]. A long spin coherence time T_2 is available in high-purity CVD diamonds by eliminating both environmental paramagnetic and nuclear spins. One of the limits is set by the nuclear spin bath of ^{13}C with natural abundance (1.1%). A further spin coherence time of approximately a few milliseconds is obtained by isotopic engineering [25]. Notably, the high-density NV centers result in high photon emission and high sensitivity because of the reduction of photon-shot noise. Although the spatial resolution is reduced compared with the case of a single NV center, a high-density NV ensemble layer is suitable for imaging applications Diamond is synthesized under highly non-equilibrium conditions where a mixture of methane (CH_4) and hydrogen (H_2) gases exists as plasma, and NV centers are formed during CVD diamond growth with nitrogen (N_2) gas. The plasma environment is developed using a vacuum chamber forming a microwave cavity at 2.45 GHz. We developed high contrast with perfectly aligned NV centers and precise thickness control obtained by CVD for scalable applications [31–35]. Heteroepitaxial diamond on Si substrates is a key technology regarding scalability and Si CMOS hybrid systems. 3C-SiC is used as an intermediate layer between Si and diamond films, and we confirmed the aligned NV centers to the <111> direction in the grains [28, 29].

4.3 Diamond Quantum Sensors and Applications

4.3.1 Millimeter-Scale Magnetocardiography of Living Rats

The most important features of diamond quantum sensors are high sensitivity at room temperature and good biocompatibility. Therefore, the diamond quantum sensor can be placed in close contact with the living body and measure biomagnetic fields with high sensitivity. We demonstrated magnetocardiography (MCG) of living mammalian animals (rats) and the associated electric current estimation with a spatial resolution smaller than the feature size of the heart [18]. The measurement system is shown in Figure 4.4(a). The rats are placed on the stage. The sensor holder and microwave antenna are made thin enough so that the diamond quantum sensor can be close to the heart surface. The diamond quantum sensor is an HPHT crystal with a (111) plane with a size of 5.2 mm × 0.35 mm. The concentrations of the P1 (substitutional nitrogen defect) and NV centers were 15 and 1.8 ppm after electron-beam (EB) irradiation with a total fluence of 5×10^{17} electrons cm^{-2}, followed by post-annealing at 1000 °C for 2 h under a vacuum. It detects magnetic fields perpendicular to the crystal surface (Figure 4.4(c)). This is the magnetic field emitted perpendicularly from the surface of the heart, as shown in Figure 4.4(b). Beyond the sensing using a single NV center, the ensemble NV centers in bulk diamond were employed to improve magnetic sensitivity. The MCG can be acquired by moving the stage in two dimensions relative to the sensor.

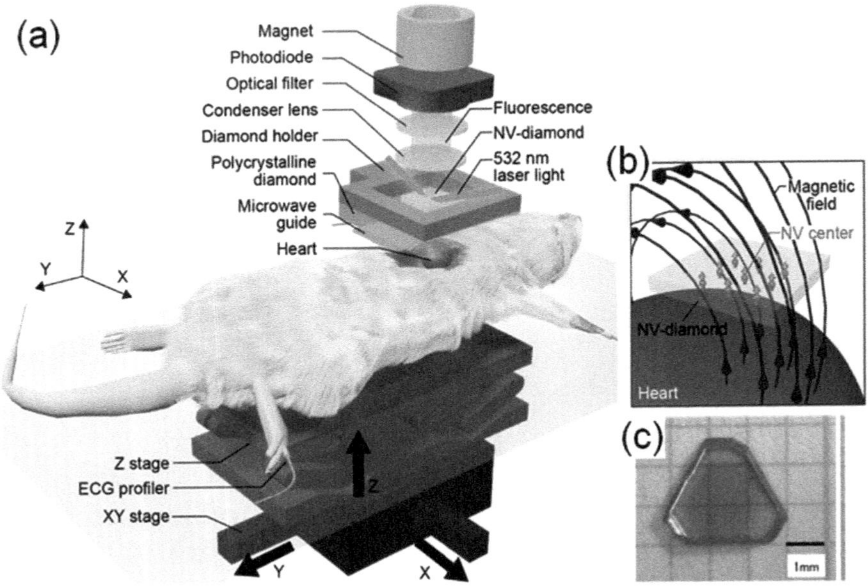

Figure 4.4 Millimeter-scale magnetocardiography of living rats. (a) Setup for rat MCG imaging. (b) Illustration of the magnetic field detected by the NV centers. (c) Optical image of the diamond sample with NV centers. Adapted from Springer Nature, ref [18]. K. Arai et al., millimetre-scale magnetocardiography of living rats with thoracotomy, *Comm. Phys.* 5, 200, 2022.

Figure 4.5(a) shows an optical photograph, and Figure 4.5(b) shows an MRI of a rat's heart. The lower left is the right ventricle and the upper right is the left ventricle. Figures 4.5(c) and (d) present the measured magnetic field map, and Figures 4.5(e) and (f) show vector plots of estimated electric current density reconstructed by the MCG. The upper panels (c), (e), and (g) show images at the R-wave peak of the heartbeat, as shown in Figure 4.5(i). The lower panels (d), (f), and (h) are images 20 ms

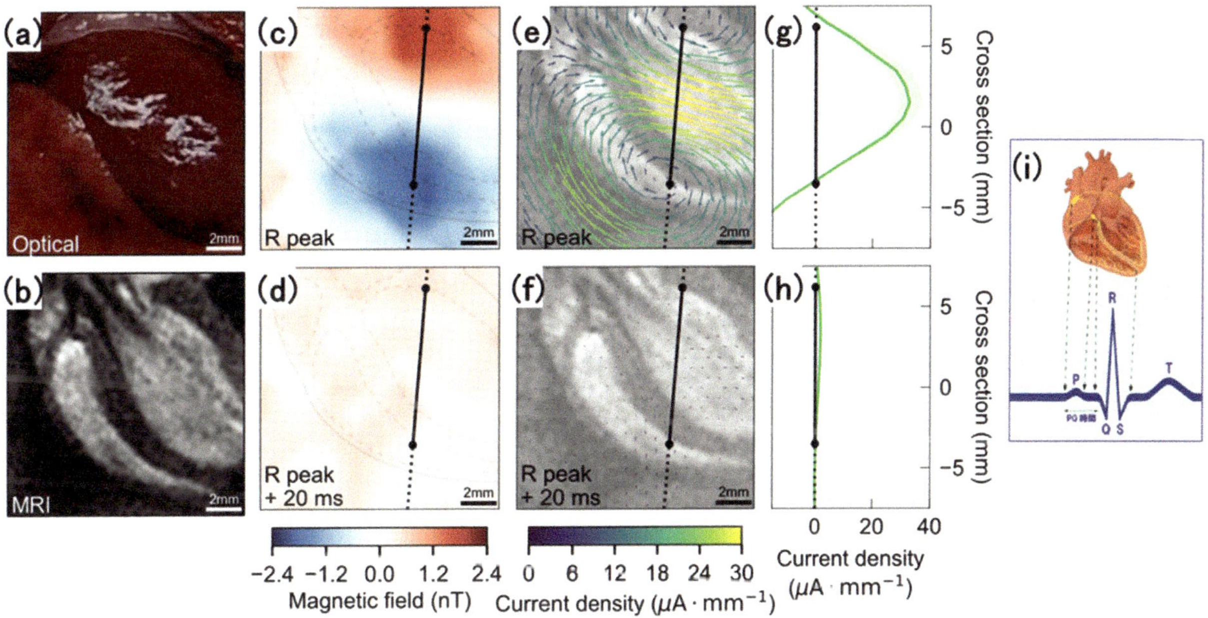

Figure 4.5 Rat MCG imaging results. (a) Optical image and (b) magnetic resonance image of the rat's heart. (c) Measured magnetic field map at the R-wave peak. (d) 20 ms after the R-wave peak. (e), (f) Vector plots of estimated electric current density (c), (d). (g), (h) Normal component of the electric current density vector with respect to the linocuts. (i) Illustration of the timing of the R-wave peak of the heartbeat. Adapted from Springer Nature, ref [18]. K. Arai et al., millimetre-scale magnetocardiography of living rats with thoracotomy, *Comm. Phys.* 5, 200, 2022.

after the R-wave peak. At the timing of the R-wave peak, the current density is localized near the Purkinje fiber bundle. A decrease in current density was observed at 20 ms after the R-wave peak.

Moreover, by approaching the sensor close to the surface of the heart, MCG with millimeter resolution was detected. Conventional sensors typically provide a resolution of a few centimeters, which is currently limited by either the standoff distance or sensor size. High-resolution MCG may reveal the causes of arrhythmia-induced rotational waves and the mechanism of ventricular fibrillation noninvasively. Therefore, advanced contactless diagnostics, which are not possible with ECG, are expected to become possible.

4.3.2 High-Precision Simultaneous Monitoring of Current and Temperature of Electric Vehicle Batteries

Other features of a diamond quantum sensor are the wide dynamic range due to its saturation-free and hysteresis-free properties and the multi-modal capability to detect the magnetic field and temperature simultaneously. A monitoring prototype of an electric vehicle (EV) battery was developed based on these features [19, 20]. The state of charge (SOC) of the EV battery needs to be monitored by integrating the charge/discharge current. The remaining cruising mileage can be estimated by measuring the current with an accuracy of approximately 10 mA while accepting the maximum amplitude of several hundred amperes. Furthermore, EV batteries require temperature control in an environment from −40 °C to + 85 °C.

The prototype is shown in Figure 4.6. The diamond quantum sensor is a 2 × 2 × 1 mm^3 ensemble NV centers in bulk diamond adhered on the top of a multi-mode fiber (Figure 4.6(a)). It is a type-Ib crystal with P1 and NV center concentrations of approximately 100 and 5–6 ppm after 3×10^{18} cm^{-2} EB irradiation. Two sensors are attached on both sides of the busbar connecting the EV battery module and motor engine. The technical issue of the battery monitor is 10 mA detection in a large noise in an automobile environment and stably tracing high current while maintaining accuracy. The differential detection by the two sensors shown in Figure 4.6(b) eliminated the external noise as a common mode. The diamond quantum sensors were implemented within the battery test system to measure the magnetic field and temperature (Figure 4.6(c)). As shown in Figure 4.6(d), a pair of microwave generators generates a pair of resonance frequencies for each sensor. To match the microwave frequency with the resonance frequency, the fluorescence intensity detected by the photodiode is fed back to the microwave generator through the integral circuit. The magnetic field and temperature can be measured as the difference and average of these two microwave frequencies. The noise spectra of the magnetic field and temperature at one sensor are shown in Figure 4.6(e). Both the magnetic field and temperature have sufficiently flat bandwidth to trace the rapid current and temperature changes during rapid acceleration, deceleration, and fast charging. The stability of the analog feedback by the integral circuit is limited to a few megahertz around the resonance frequency. When the resonance frequency exceeds the limit, the center frequency of the microwave generator is digitally corrected intermittently so that the integral circuit output does not exceed the limit. As shown in Figure 4.6(f), 10 mA is detected and ±1000 A is measured.

Then, the accuracy of the SOC measured by the battery monitoring prototype was verified with the standard driving pattern. With the integral circuit in Figure 4.6(d), the frequency difference between the two microwave generators follows the difference between the two resonance frequencies proportional to the busbar current. Therefore, the integral of the frequency difference between the two microwave generators representing the measured SOC coincides with the integral of the difference between the resonance frequencies representing the theoretical SOC.

The WLTC (Worldwide-harmonized Light Vehicles Test Cycle) standard driving pattern, as shown in Figure 4.7(a), was used to verify the SOC estimation accuracy. The WLTC driving pattern consists of city areas, suburban, and freeway portions. The busbar input current shown in Figure 4.7(b) was obtained assuming the weight and engine performance of a typical automobile.

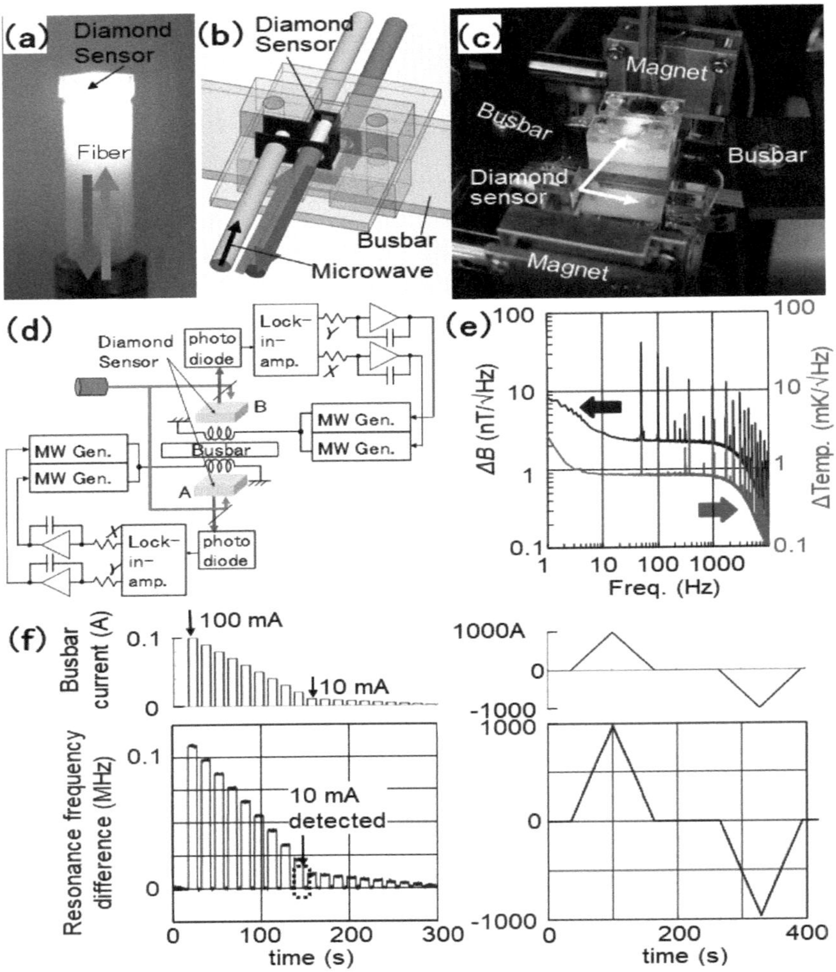

Figure 4.6 Monitoring prototype of electric vehicle battery using diamond sensor. (a) 2 × 2 × 1 mm^3 diamond sensor adhered to the fiber top. (b) Two sensors placed on both sides of the busbar. (c) Sensors and busbar placed between a pair of magnets. (d) Block diagram of the current measurement system. (e) Noise spectrum of the detected magnetic field and the temperature. (f) Detected magnetic field as the resonance frequency difference by the differential detection system when input busbar current is provided as 100 mA ~ 1 mA pulse train (on the left) and ±1000 A triangular waveform (on the right). Adapted from Springer Nature, ref [20]. Y. Hatano et al., high-precision robust monitoring of charge/discharge current over a wide dynamic range for electric vehicle batteries using diamond quantum sensors, *Sci. Rep.* 12, 13991, 2022.

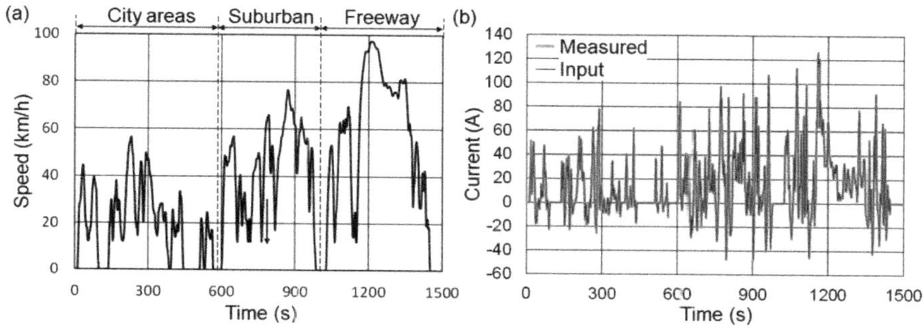

Figure 4.7 Verification of the traceability and accuracy of the busbar current using the standard driving pattern. (a) Driving speed (km/h) in the WLTC mode. (b) Converted busbar current from the driving speed assuming the weight and engine performance of a typical automobile in black line and its measured value by the diamond sensor in green line. The two lines coincide. Reprinted from Springer Nature, ref [20]. Y. Hatano et al., high-precision robust monitoring of charge/discharge current over a wide dynamic range for electric vehicle batteries using diamond quantum sensors, *Sci. Rep*. 12, 13991, 2022.

The current measured by a diamond sensor as the frequency difference between the microwave generators when the input current is provided to the busbar is shown in Figure 4.7(b). The measured current coincides indistinguishably with the input current. Furthermore, as their integral values, the measured SOC, the integral of the frequency difference between the microwave generators, coincide with the theoretical SOC, the integral of the busbar input current, with a cumulative error of 0.1%. This demonstrated the traceability and accuracy of the diamond sensor.

4.3.3 Wide Temperature Operation of Diamond Quantum Sensor

In the monitoring prototype of the EV battery, as shown in Figure 4.6, the sensors and busbar placed between a pair of magnets were installed in a thermostatic bath to confirm the operation at temperatures ranging from –40 °C to 85 °C [19, 20]. However, one of the features of the diamond quantum sensors is their robustness to a wide range of temperature changes. To verify this robustness, the sensors and busbar were mounted in a Dewar and subjected to temperature changes from −150 °C to 150 °C by cooling with liquid nitrogen and heating with a heater [21].

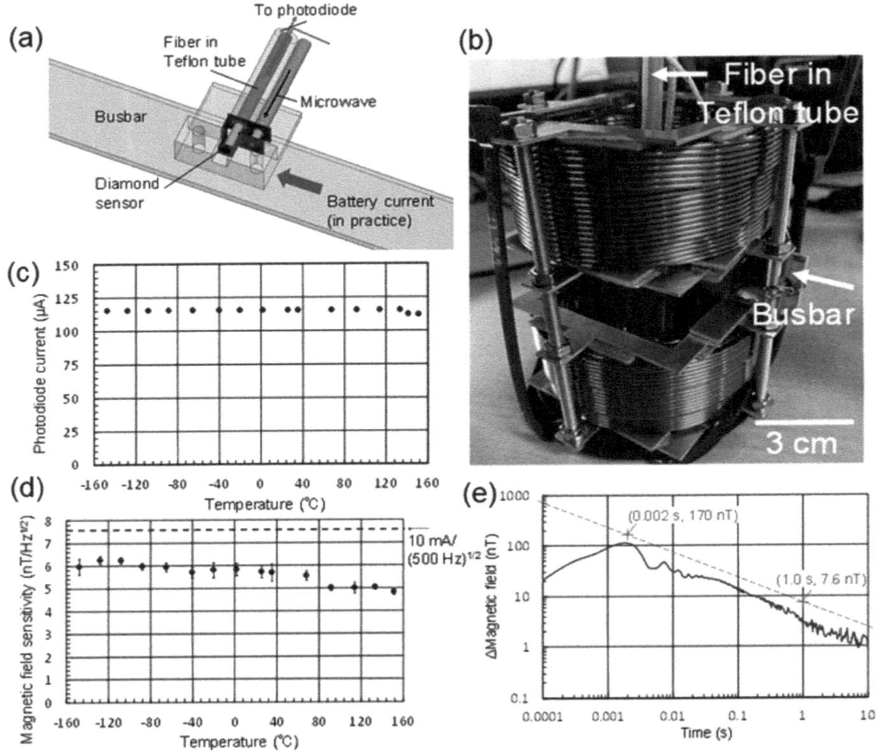

Figure 4.8 Experimental setup for the wide temperature operation of the diamond quantum sensor and the measured results. (a) Diamond sensor attached to the busbar. (b) Magnetic field coils attached on both sides of the busbar to be placed inside the Dewar for temperature control. (c) Temperature dependence of the photodiode current generated by the fluorescence of the diamond sensor. (d) Temperature dependence of the magnetic field sensitivity of the diamond sensor. (e) Allan deviation of the magnetic field noise of the diamond sensor. Adapted from ref [21]. K. Kubota et al., wide temperature operation of a diamond quantum sensor for electric vehicle battery monitoring, *Diam Relat Mater* **124**, 109853.

Figure 4.8(a) and 8(b) show the experimental setups placed directly in the Dewar bottle in an open-air environment. The diamond sensor was an HPHT crystal with P1 and NV center concentrations of 11 and 2.4 ppm after 1×10^{18} cm^{-2} EB irradiation. The static magnetic field was generated by a coil placed on both sides of the busbar. Thermocouples were placed close to the sensor head to monitor the temperature. Control electronics including the microwave generators and lock-in amplifiers and optics including light sources and photodiodes were at room temperature.

To protect the optical fiber from thermal and mechanical stresses, it was covered by a hard Teflon tube. This stabilized the photodiode current detecting the red fluorescence from the diamond sensor to the constant value of approximately 120 μA over a wide temperature range of −150 °C to 150 °C, as shown in Figure 4.8(c). The sensitivity also exhibited a constant value approximately within the range of 5–6 $nT/\sqrt{Hz}$ over the range, as shown in Figure 4.8(d).

In the measurement system shown in Figure 4.8(a), the magnetic field produced in the diamond quantum sensor by the busbar current of 10 mA was 170 nT. In an actual application, the measurement time should be determined to detect 10 mA. As the sensitivity of $nT/\sqrt{Hz}$ is per measurement second, the sensitivity of $170/\sqrt{500} = 7.6$ nT is required in 1 s to measure 170 nT in a bandwidth of 500 Hz (2 ms). Notably, the measured Allan deviation at room temperature was below the straight line connecting (0.002 s, 170 nT) and (1 s, 7.6 nT), as shown in Figure 4.8(e), confirming that a 10-mA current change in a 500-Hz bandwidth could be detected.

The demonstration that the diamond quantum sensor with a sensitivity of 10 mA in a 500-Hz bandwidth can operate over a wide temperature range of −150 °C to +150 °C indicates its usefulness in harsh environmental applications.

4.3.4 Internal Sensing in Power Devices

4.3.4.1 Internal electric-field sensing in diamond power devices

To achieve reliable low-loss and high-breakdown field power devices, the internal electric field should be monitored. However, techniques based on scanning probe microscopy are limited to sensing at the surface only. We demonstrated direct sensing of the internal electric field in diamond power devices using a single NV center [22]. The NV center embedded inside the device acts as a nanoscale electric field sensor. We fabricated vertical diamond pin diodes containing the single NV centers (Figure 4.9(a)).

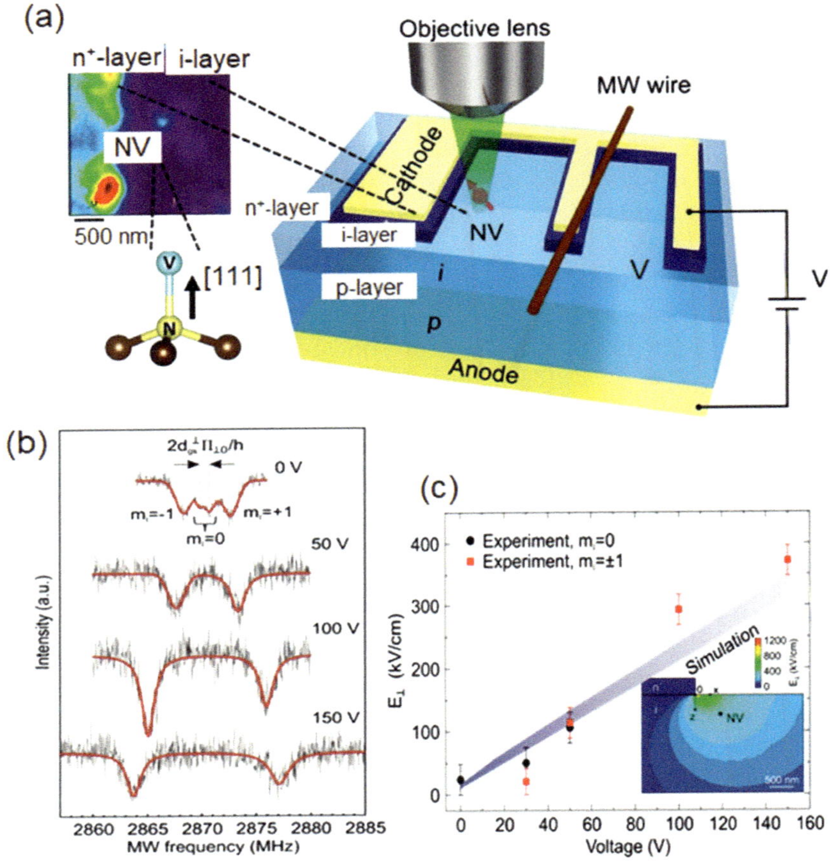

Figure 4.9 Internal electric-field sensing in diamond devices using a single NV center. Diamond device structure (p-i-n diode) and measurement setup. Inset; Top-view confocal fluorescence microscope image of a single NV center. ODMR spectra of the single NV center under reverse-biased conditions up to 150 V. The red lines denote fitting. Transverse electric field measured with the NV center and simulated results. Insert; Simulation of the transverse electric field at 150 V. Reprinted from American Chemical Society, ref [22] T. Iwasaki et al., direct nanoscale sensing of the internal electric field in operating semiconductor devices using single electron spins, *ACS Nano*, 11, 1238, 2017 with the permission of American Chemical Society.

By measuring ODMR under reverse-biased conditions with an applied voltage of approximately 150 V, we observed a large splitting in the magnetic resonance frequencies (Figure 4.9(b)). This indicated that the NV center sensed the transverse electric field in the space-charge region formed in the i-layer. The experimentally obtained electric field values are in good agreement with those calculated by a device simulator ((Figure 4.9(c)). Moreover, we measured the vector electric field in different directions by utilizing NV centers with different NV axes [23]. This quantum sensor using an electron spin in a wide-band-gap material monitors the electric field and local temperature and current in operating semiconductor devices.

4.3.4.2 Internal temperature sensing in 4H-SiC power devices

Hexagonal silicon carbide (4H-SiC) power devices have been used in systems such as trains and vehicles. For safety operation and failure prediction, temperature monitoring is demanded. Similar to NV centers in diamond, negatively charged silicon vacancy, V_{Si}^{-}, centers (Figure 4.10 (a)) in 4H-SiC have been reported to be sensitive to magnetic field [36, 37], electric field [38], and temperature [39, 40]. Thermometric quantum sensors have been reported to measure temperatures inside operating SiC devices [24]. A spatially local pattern of V_{Si} centers in a SiC device (pin diode) has been realized using proton beam writing [41] to measure the temperature change resulting from a forward current injection. Our experiments show that the ODMR signals of V_{Si} centers have unique characteristics caused by the spin–orbit coupling and spin–spin interaction resulting from the crystal structure of 4H-SiC. We characterized the excited state (ES) and the ground state (GS) of V_{Si} centers using ODMR to achieve thermometric quantum sensors.

A confocal microscope image of the diode was obtained to confirm the formation of dots of V_{Si} centers, as shown in Figure 4.10(c). The *I–V* characteristics of the diode are shown in Figure 4.10(d). Typical forward characteristics of a pin diode are observed. Notably, the current increases with increasing voltage, and at a voltage of 30 V, the current is 8.3 mA. This current results in a self-heating effect. The local temperature distribution in the channel was controlled by the current flowing through the pin junction. Figure 4.10(e) shows the excited state

(ES) ODMR signals of V_{Si} centers that were measured at the marked position [light green circle in Figure 4.10(c)]. At 0 mA, we observed the ES ODMR dip at the resonance frequency of

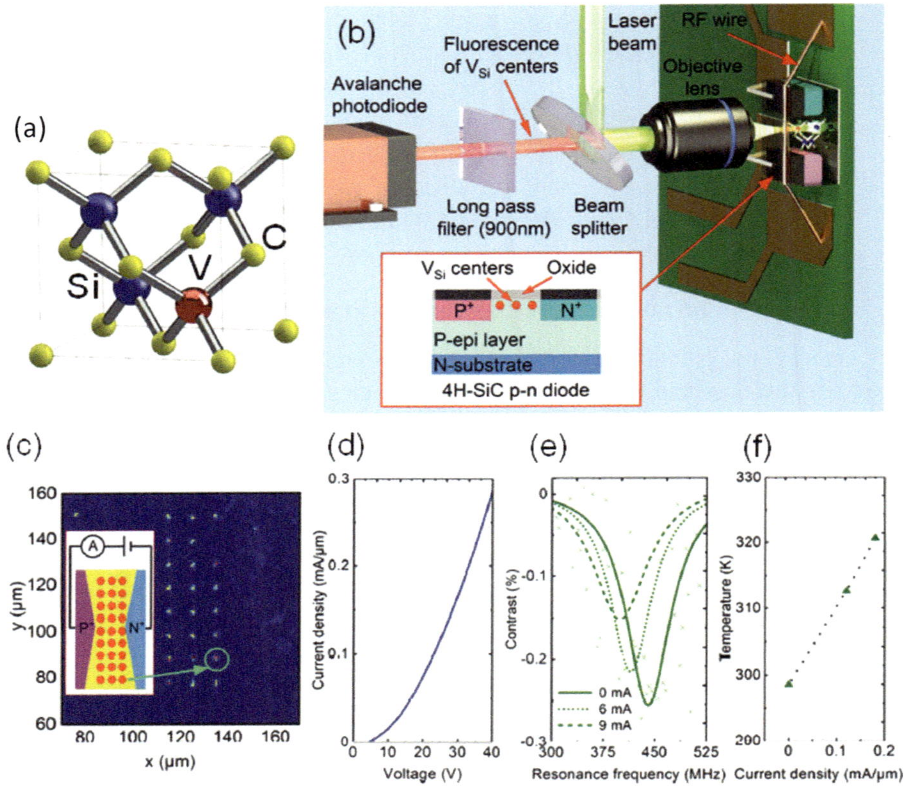

Figure 4.10 Internal temperature sensing in 4H-SiC devices using V_{Si} centers. Schematic of the V_{Si} center in 4H-SiC. Simple diagram of the ODMR experimental setup. Confocal microscope image of the planar 4H-SiC pin diode. Inset: schematic drawing of the equivalent circuit and the diode with dots of V_{Si} centers (red symbols) localized at the pin junction. I–V characteristic of the planar 4H-SiC pin diode. ES ODMR signals at the target position [marked with a light green circle in (c)] in the diode at different currents of 0, 6, and 9 mA. The solid, dotted, and dashed lines are Lorentzian fitting curves. Local temperature of the target position as a function of current. Reprinted from ref [24]. T. M. Hoang et al., Thermometric quantum sensor using excited state of silicon vacancy centers in 4H-SiC devices, *Appl. Phys. Lett.* 118 044001, 2021, with the permission of AIP Publishing.

440 MHz. When the current was increased, this dip shifted to lower resonance frequencies. Based on the relationship between the temperature and positions of the ES ODMR dips, the temperature values of 299, 313, and 321 K at the marked position were monitored, corresponding to currents of 0, 6, and 9 mA, respectively, as shown in Figure 4.10(f). The temperature linearly increases with increasing current. This result indicates that quantum sensing with V_{Si} centers can provide the thermal information of local areas in SiC devices under operation.

4.4 Conclusion

The technical challenges in realizing applications of linking the material and quantum control technology were developed. Millimeter-scale MCG of living rats, high-precision simultaneous monitoring of current and temperature of electric vehicle batteries, and detection of internal sensing in power devices were demonstrated. The diamond quantum sensors have progressed in the last couple of years, from early proof-of-principle experiments to prototype demonstrations. However, many challenges remain, most important is improving sensitivity through quantum protocols and material development. The diamond quantum sensors would contribute to carbon neutrality and the well-being of people near future.

Acknowledgments

This study was supported by MEXT QLEAP Grant Number JPMXS0118067395. The author would like to thank T. Iwasaki, M. Sekino, T. Sakai, A. Kuwahata, K. Arai, Y. Hatano, D. Nishitani, and Lab and Q-LEAP members for their contributions and helpful discussions.

References

1. D. Budker and M. Romalis (2007). Optical magnetometry, *Nat. Phys.*, 3, pp. 227–234.

2. C. L. Degen, F. Reinhard, and P. Cappellaro (2017). Quantum sensing, *Rev. Mod. Phys.*, 89, 035002.
3. A. Gruber, A. Drabenstedt, C. Tietz, L. Fleury, J. Wrachtrup, and C. von Borczyskowski (1997). Scanning confocal optical microscopy and magnetic resonance on single defect centers, *Science*, 276, pp. 2012–2014.
4. F. Jelezko, T. Gaebel, I. Popa, A. Gruber, and J. Wrachtrup (2004). Observation of coherent oscillations in a single electron spin, *Phys. Rev. Lett.*, 92, 076401.
5. J. M. Taylor, P. Cappellaro, L. Childress, L. Jiang, D. Budker, P. R. Hemmer, A. Yacoby, R. Walsworth, and M. D. Lukin (2008). High-sensitivity diamond magnetometer with nanoscale resolution, *Nat. Phys.*, 4, pp. 810–816.
6. J. R. Maze, P. L. Stanwix, J. S. Hodges, S. Hong, J. M. Taylor, P. Cappellaro, L. Jiang, M. V. Gurudev Dutt, E. Togan, A. S. Zibrov, A. Yacoby, R. L. Walsworth, and M. D. Lukin (2008). Nanoscale magnetic sensing with an individual electronic spin in diamond, *Nature*, 455, pp. 644–647.
7. G. Balasubramanian, I. Y. Chan, R. Kolesov, M. Al-Hmoud, J. Tisler, C. Shin, C. Kim, A. Wojcik, P. R. Hemmer, A. Krueger, T. Hanke, A. Leitenstorfer, R. Bratschitsch, F. Jelezko, and J. Wrachtrup (2008). Nanoscale imaging magnetometry with diamond spins under ambient conditions, *Nature*, 455, pp. 648–651.
8. V. M. Acosta, E. Bauch, M. P. Ledbetter, C. Santori, K.-M. C. Fu, P. E. Barclay, R. G. Beausoleil, H. Linget, J. F. Roch, F. Treussart, S. Chemerisov, W. Gawlik, and D. Budker (2009). Diamonds with a high density of nitrogen-vacancy centers for magnetometry applications, *Phys. Rev. B*, 80, 115202.
9. M. W. Doherty, N. B. Manson, P. Delaney, F. Jelezko, J. Wrachtrup, and L. C. L. Hollenberg (2013). The nitrogen-vacancy color center in diamond, *Phys. Rep.*, 528, pp. 1–45.
10. F. Casola, T. van der Sar, and A. Yacoby (2018). Probing condensed matter physics with magnetometry based on nitrogen-vacancy centers in diamond, *Nat. Rev. Mater.*, 3, 17088.
11. J. F. Barry, M. J. Turner, J. M. Schloss, D. R. Glenn, Y. Song, M. D. Lukin, H. Park, and R. L. Walsworth (2016). Optical magnetic detection of single-neuron action potentials using quantum defects in diamond, *Proc. Natl. Acad. Sci. U.S.A.*, 113, pp. 14133–14138.
12. D. L. Sage, K. Arai, D. R. Glenn, S. J. DeVience, L. M. Pham, L. Rahn-Lee, M. D. Lukin, A. Yacoby, A. Komeili, and R. L. Walsworth (2013). Optical magnetic imaging of living cells, *Nature*, 496, pp. 486–489.

13. A. Laraoui, F. Dolde, C. Burk, F. Reinhard, J. Wrachtrup, and C. A. Meriles (2013). High-resolution correlation spectroscopy of 13C spins near a nitrogen-vacancy centre in diamond, *Nat. Commun.*, 4, 1651.

14. T. Staudacher, F. Shi, S. Pezzagna, J. Meijer, J. Du, C. A. Meriles, F. Reinhard, and J. Wrachtrup (2013). Nuclear magnetic resonance spectroscopy on a (5-nanometer)3 sample volume, *Science*, 339, pp. 561–563.

15. H. J. Mamin, M. Kim, M. H. Sherwood, C. T. Rettner, K. Ohno, D. D. Awschalom, and D. Rugar (2013). Nanoscale nuclear magnetic resonance with a nitrogen-vacancy spin sensor, *Science*, 339, pp. 557–560.

16. S. J. DeVience, L. M. Pham, I. Lovchinsky, A. O. Sushkov, N. Bar-Gill, C. Belthangady, F. Casola, M. Corbett, H. Zhang, M. Lukin, H. Park, A. Yacoby, and R. L. Walsworth (2015). Nanoscale NMR spectroscopy and imaging of multiple nuclear species, *Nat. Nanotechnol.*, 10, pp. 129–134.

17. D. R. Glenn, R. R. Fu, P. Kehayias, D. L. Sage, E. A. Lima, B. P. Weiss, and R. L. Walsworth (2017). Micrometer-scale magnetic imaging of geological samples using a quantum diamond microscope, *Geochemistry, Geophys. Geosystems*, 18, pp. 3254–3267.

18. K. Arai, A. Kuwahata, D. Nishitani, I. Fujisaki, R. Matsuki, Y. Nishio, Z. Xin, X. X. Cao, Y. Hatano, S. Onoda. C. Shinei, M. Miyakawa, T. Taniguchi, T. Ohshima, M. Hatano, M. Sekino, and T. Iwasaki (2022). Millimetre-scale magnetocardiography of living rats with thoracotomy, *Commun. Phys.*, 5, 200.

19. Y. Hatano, J. Shin, D. Nishitani, H. Iwatsuka, Y. Masuyama, H. Sugiyama, M. Ishii, S. Onoda, T. Ohshima, K. Arai, T. Iwasaki, and M. Hatano (2021). Simultaneous thermometry and magnetometry using a fiber-coupled quantum diamond sensor, *Appl. Phys. Lett.*, 118, 034001.

20. Y. Hatano, J. Shin, J. Tanigawa, Y. Shigenobu, A. Nakazono, T. Sekiguchi, S. Onoda, T. Ohshima, K. Arai, T. Iwasaki, and M. Hatano (2022). High-precision robust monitoring of charge/discharge current over a wide dynamic range for electric vehicle batteries using diamond quantum sensors, *Sci. Rep.*, 12, 13991.

21. K. Kubota, Y. Hatano, Y. Kainuma, J. Shin, D. Nishitani, C. Shinei, T. Taniguchi, T. Teraji, S. Onoda, T. Ohshima, T. Iwasaki, and M. Hatano (2023). Wide temperature operation of diamond quantum sensor for electric vehicle battery monitoring, *Diam Relat Mater*, 124, 109853.

22. T. Iwasaki, W. Naruki, K. Tahara, T. Makino, H. Kato, M. Ogura, D. Takeuchi, S. Yamasaki, and M. Hatano (2017). Direct nanoscale

sensing of the internal electric field in operating semiconductor devices using single electron spins, *ACS Nano*, 11, pp. 1238–1245.

23. B. Yang, T. Murooka, K. Mizuno, K. Kim, H. Kato, T. Makino, M. Ogura, S. Yamasaki, M. E. Schmidt, H. Mizuta, A. Yacoby, M. Hatano, and T. Iwasaki (2020). Vector electrometry in a wide-gap-semiconductor device using a spin-ensemble quantum sensor, *Phy. Rev. Appl.*, 14, 044049.
24. T. M. Hoang, H. Ishiwata, Y. Masuyama, Y. Yamazaki, K. Kojima, S. Y. Lee, T. Ohshima, T. Iwasaki, D. Hisamoto, and M. Hatano (2021). Thermometric quantum sensor using excited state of silicon vacancy centers in 4H-SiC devices, *Appl. Phys. Lett.*, 118, 044001.
25. G. Balasubramanian, P. Neumann, D. Twitchen, M. Markham, R. Kolesov, N. Mizuochi, J. Isoya, J. Achard, J. Beck, J. Tissler, V. Jacques, P. R. Hemmer, F. Jelezko, and J. Wrachtrup (2009). Ultralong spin coherence time in isotopically engineered diamond, *Nat. Mater.*, 8, pp. 383–387.
26. E. D. Herbschleb, H. Kato, Y. Maruyama, T. Danjo, T. Makino, S. Yamasaki, I. Ohki, K. Hayashi, H. Morishita, M. Fujiwara, and N. Mizuochi (2019). Ultra-long coherence times amongst room-temperature solid-state spins, *Nat. Commun.*, 10, 3766.
27. P. Maletinsky, S. Hong, M. S. Grinolds, B. Hausmann, M. D. Lukin, R. L. Walsworth, M. Loncar, and A. Yacoby (2012). A robust scanning diamond sensor for nanoscale imaging with single nitrogen-vacancy centres, *Nat. Nanotechnol.*, 7, pp. 320–324.
28. J. Yaita, T. Tsuji, M. Hatano, and T. Iwasaki (2018). Preferentially aligned nitrogen-vacancy centers in heteroepitaxial (111) diamonds on Si substrates via 3C-SiC intermediate layers, *Appl. Phys. Express*, 11, 045501.
29. T. Suto, J. Yaita, T. Iwasaki, and M. Hatano (2017). Highly oriented diamond (111) films synthesized by pulse bias-enhanced nucleation and epitaxial grain selection on a 3C-SiC/Si (111) substrate, *Appl. Phys. Lett.*, 110, 062102.
30. T. Wolf, P. Neumann, K. Nakamura, H. Sumiya, T. Ohshima, J. Isoya, and J. Wrachtrup (2015). Subpicotesla diamond magnetometry, *Phys. Rev. X*, 5, 041001.
31. H. Ishiwata, M. Nakajima, K. Tahara, H. Ozawa, T. Iwasaki, and M. Hatano (2017). Perfectly aligned shallow ensemble nitrogen-vacancy centers in (111) diamond, *Appl. Phys. Lett.*, 111, 043103.

32. T. Tsuji, T. Iwasaki, and M. Hatano (2023). Over 100 μm thickness CVD diamond film with perfectly aligned nitrogen-vacancy centers on highly misoriented substrates, *Appl. Phys. Express*, 16, 025503.

33. T. Fukui, Y. Doi, T. Miyazaki, Y. Miyamoto, H. Kato, T. Matsumoto, T. Makino, S. Yamasaki, R. Morimoto, N. Tokuda, M. Hatano, Y. Sakagawa, H. Morishita, T. Tashima, S. Miwa, Y. Suzuki, and N. Mizuochi (2014). Perfect selective alignment of nitrogen-vacancy centers in diamond, *Appl. Phys. Express*, 7, 055201.

34. K. Tahara, H. Ozawa, T. Iwasaki, N. Mizuochi, and M. Hatano (2015). Quantifying selective alignment of ensemble nitrogen-vacancy centers in (111) diamond, *Appli. Phys. Lett.*, 107, 193110.

35. H. Ozawa, K. Tahara, H. Ishiwata, M. Hatano, and T. Iwasaki (2017). Formation of perfectly aligned nitrogen-vacancy-center ensembles in chemical-vapor-deposition-grown diamond (111), *Appl. Phys. Express*, 10, 045501.

36. M. Niethammer, M. Widmann, S. Y. Lee, P. Stenberg, O. Kordina, T. Ohshima, N. T. Son, E. Janzén, and J. Wrachtrup (2016). Vector magnetometry using silicon vacancies in 4H-SiC under ambient conditions, *Phys. Rev. Appl.*, 6, 034001.

37. S. G. Carter, O. O. Soykal, P. Dev, S. E. Economou, and E. R. Glaser (2015). Spin coherence and echo modulation of the silicon vacancy in 4H–SiC at room temperature, *Phys. Rev. B*, 92, 161202(R).

38. G. Wolfowicz, S. J. Whiteley, and D. D. Awschalom (2018). Electrometry by optical charge conversion of deep defects in 4H-SiC, *Proc. Natl. Acad. Sci. U.S.A.*, 115, pp. 7879–7883.

39. D. Simin, V.A. Soltamov, A.V. Poshakinskiy, A.N. Anisimov, R.A. Babunts, D.O. Tolmachev, E.N. Mokhov, M. Trupke, S.A. Tarasenko, A. Sperlich, P.G. Baranov, V. Dyakonov, and G.V. Astakhov (2016). All-optical dc nanotesla magnetometry using silicon vacancy fine structure in isotopically purified silicon carbide, *Phys. Rev. X*, 6, 031014.

40. H. Kraus, V. A. Soltamov, F. Fuchs, D. Simin, A. Sperlich, P. G. Baranov, G. V. Astakhov, and V. Dyakonov (2014). Magnetic field and temperature sensing with atomic-scale spin defects in silicon carbide, *Sci. Rep.*, 4, 5303.

41. Y. Yamazaki, Y. Chiba, T. Makino, S.-I. Sato, N. Yamada, P. G. Baranov, T. Satoh, Y. Hijikata, K. Kojima, S.-Y. Lee, and T. Ohshima (2018). Electrically controllable position-controlled color centers created in SiC pn junction diode by proton beam writing, *J. Mater. Res.*, 33, 3355.

Chapter 5

Thin-Film Batteries and Pervasion to Unconventional Applications

Sami Oukassi

University Grenoble Alpes, CEA, Leti, F-38000 Grenoble, France

Sami.oukassi@cea.fr

This chapter provides an overview of thin-film battery technology. The most important characteristics of this technology are its excellent energy density, high cycle life, safety, and compatibility with highly integrated system designs. Furthermore, the pervasion of this technology is emphasized through a description of an expanding application spectrum, in which henceforth thin-film batteries are considered for different functionalities in combination with electrochemical energy storage. We discuss the importance of materials' ion transport properties and device architecture in the optimization of thin-film battery performance.

5.1 Introduction

The recent technological trends toward small-scale and highly integrated electronic devices have stimulated ever-growing

Outlooking beyond Nanoelectronics and Nanosystems: Ultra Scaling, Pervasiveness, Sustainable Integration, and Biotic Cross-Inspiration
Edited by Simon Deleonibus

ISBN 978-981-5129-32-8 (Hardcover), 978-1-003-50990-5 (eBook)
www.jennystanford.com

demand for innovative energy storage solutions, in particular for the future development of smart medical implants, artificial intelligence robots, microscale wireless sensor networks, and other self-powered microsystems. During the past decade, thin-film batteries (TFBs) have attracted tremendous attention because of their high energy/power densities and great feasibility for integration into miniaturized devices [1–5]. In most applications, a combination of high energy and power densities is usually required, and a compromise is often implemented to meet specifications. Several approaches have been considered in order to optimize the TFB performance, including (i) the device architecture following specific electrode configuration and interfacial engineering, and (ii) the selection of materials combining high ionic conductivity and kinetically fast and reversible electrochemical reaction mechanisms.

In the following, we will present the TFB device, its main performance indicators as well as the optimization approaches implemented. We will also give an overview of the pervasive aspect of TFBs, going beyond energy storage and acquiring multiple additional functionalities.

5.2 Thin-Film Batteries and Key Features

5.2.1 TFB Description

Thin-film batteries (TFBs) are devices whose architecture can be simplified into a stack of electrodes and solid-state electrolytes with a microscale thickness. The important elements of TFBs are almost similar to those of macroscale lithium-ion batteries (LIBs), specifically the positive current collector, cathode, electrolyte, anode materials, and negative current collectors, as illustrated in Figure 5.1. In general, all microfabrication processes are carried out in a clean room environment. The device is made on a silicon, glass, or metal substrate, and different layers are deposited under ultra-high vacuum conditions. The obtained devices exhibit thicknesses of several tens of microns, and dimensions between 1 mm^2 and 100 mm^2. Often one or more encapsulation layers are also integrated in order to ensure a chemical passivation of the device materials with respect to reactions with oxygen and ambient humidity.

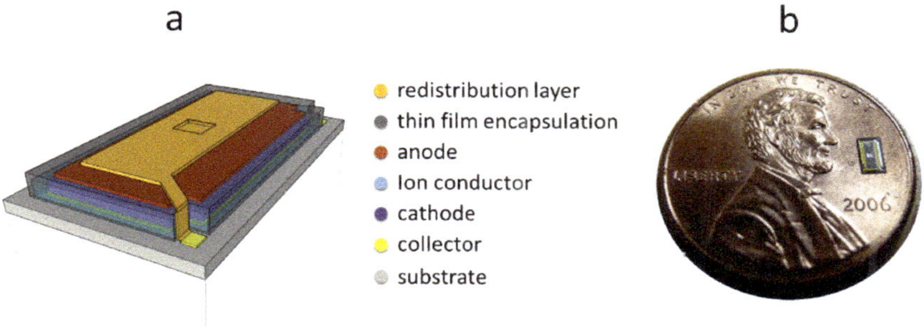

Figure 5.1 Thin-film lithium-ion battery (TFB): (a) Schematic illustration and (b) photograph of a realized 1.7 × 3.1 mm² single device. (Property of "S. Oukassi/CEA-Leti").

In 1992, Bates et al. first prepared the thin-film-ion conductor of lithium phosphorus oxynitride (LiPON), which remains to this day one of the most commonly used ion conductor films for TFBs [6–8]. Subsequently, several achievements have been reported in this field. In 2002, West et al. [9] proposed 50 μm × 50 μm microfabricated TFBs with $LiCoO_2$|LiPON|Li architecture (conventionally mentioned as cathode|ion conductor|anode) for aeronautics applications. In 2007, Notten et al. [10] fabricated a three-dimensional (3D) integrated $LiCoO_2$|LiPON|Si TFB using etched silicon wafers in order to boost areal capacity and decrease internal resistance. In 2017, Bhardwaj et al. [11] presented a TFB with 10 μm cathode in a $LiCoO_2$|LiPON|Li architecture, exhibiting 0.4 $mAh\cdot cm^{-2}$ capacity at 1C discharge rate (C-rate corresponds to the discharge current in amps divided by the TFB capacity in ampere-hours) and greater than 95% capacity retention after 100 cycles at a C/5 discharge rate. In 2019, we reported a high energy density millimeter scale TFB with 20 μm cathode in a $LiCoO_2$|LiPON|Ti architecture, reaching for the first time a discharge capacity of 1 $mAh\cdot cm^{-2}$ with excellent capacity retention over 100 cycles at room temperature and 37 °C [12]. The reported results highlighted the great potential of these TFBs as integrated energy storage solutions in particular for medical implantable, injectable, and wearable solutions. Concomitantly, CYMBET, STMicroelectronics, ILIKA, ENSURGE, and INJECTPOWER have proposed several commercial TFBs, which are expected to be applied in large and growing markets such as digital health, sports, and fitness IoT-connected sensors and wearables.

5.2.2 Key Features

Several parameters can be considered important for measuring the performance of TFBs; the most relevant from an electrical point of view is described in Figure 5.2.

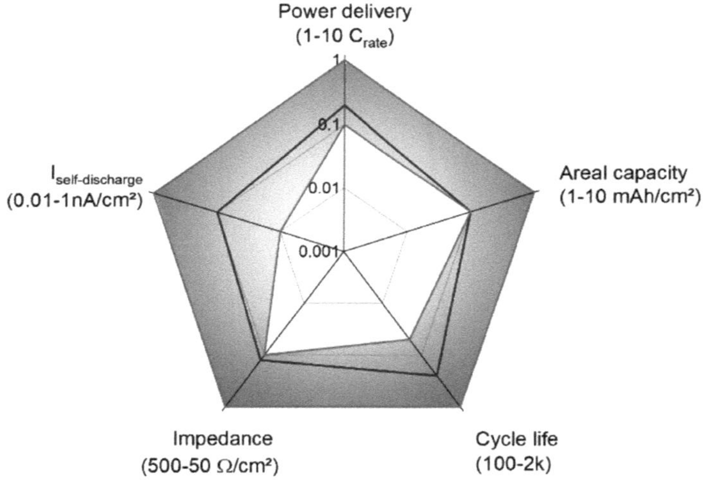

Figure 5.2 TFBs electrochemical performance: Status (blue) in comparison with applications' specifications requests. The value 1 corresponds to the optimal target for each of the specifications: Power delivery is the maximum C-rate at which the TFB might be charged or discharged, areal capacity is the discharge capacity divided by TFB footprint, cycle life is the cycle number a TFB can undergo before a 20% capacity loss, TFB impedance corresponds to the real part extracted at 1 kHz, and self-discharge is the electronic current measured in open circuit voltage conditions. (Property of "S. Oukassi/ CEA-Leti").

The current challenges of TFBs are predominantly the low areal capacity, high internal impedance, and restrained cycle life. Intrinsically, the major obstacles to the optimization of TFBs' electrochemical performance lied in (i) materials development and (ii) mechanical-chemical interface engineering. Cathodes with high capacity and Li-ion diffusivity [13–17], solid electrolytes with high Li-ion conductivity and potential stability window [18–20] in addition to anode-less configuration [21–23] are among the most propitious approaches considered so far for future high energy density and long cycle life TFBs.

This configuration pushes the cathode capacity to its theoretical limits, by elimination of binders and other additives. Moreover, it ensures, in the case of the monolithic cathode, the suppression of grains boundary, which has been shown to be the main limitation in solid-state composite electrodes [24, 25]. Additionally, lithium metal has been considered the best solution for decades, but in the last years, the anode-free solution gained in popularity. In this configuration, the lithium anode is replaced by a simple current collector that acts only as a substrate for the plating and stripping of the lithium upon cycling. By suppressing the excess of lithium, anode-free cells achieve unrivaled energy density [21–23]. In our recent work [12], we propose a model system combining both approaches, using transition metal dioxide, typified by $LiCoO_2$|LiPON|Ti (Figure 5.3) to study the correlation between intrinsic electrode properties and cell electrochemical performance.

Cycling and C-rate capability are among the most important characteristics of TFBs. TFB capacity variation with cycling is shown in Figure 5.4(a). The observed evolution indicates a two-regime pattern. In the first regime (cycles 1–10), the discharge capacity tends to decrease gradually at a low rate of 0.1% per cycle. The discharge capacity decay in the second regime decreases significantly, displaying a very low rate of 0.05% per cycle. Furthermore, the calculated Coulombic efficiency is higher than 99.5%, thus accounting for an excellent lithium-ion insertion/extraction process reversibility. TFB internal impedance variation upon cycling (Figure 5.4(a) inset) exhibits a first stable semi-circle corresponding to LiPON contribution, whereas the low-frequency semi-circle shows a gradual and moderate increase, reaching 10% over 100 cycles, which accounts for a rather stable ion exchange reactions at the bulk and interfaces of the TFB structure. The variation of potential profile and discharge capacity with discharge current density is presented in Figure 5.4(b). The increase of the IR drop (potential difference drop due to current flowing through TFB internal resistance) at the beginning of the discharge operation is correlated to the applied current density and the TFB internal resistance. Moreover, the voltage difference between discharge profile plateaus reflects the increasing TFB polarizing degree with C-rate. The difference between 10C and 0.2C discharge capacities is around 40%, attesting to a very high-rate capability of the considered TFBs, in addition to the very high energy density.

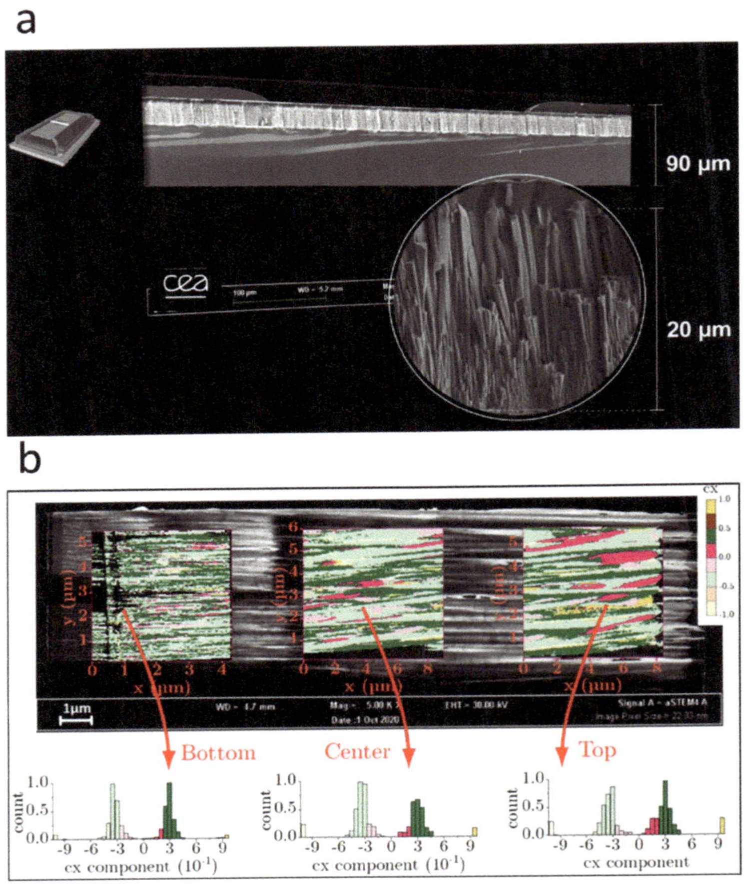

Figure 5.3 Thick cathodes for TFBs: (a) SEM cross-section of TFB, (b) SEM view of $LiCoO_2$ TEM lamella and crystalline orientations mapping obtained by precession electron diffraction. (Histograms of homogeneous distribution with a predominance of the (101) phase in green between the substrate (bottom), bulk (center), and LiPON|$LiCoO_2$ interface (top) acquisition zones). (Property of "S. Oukassi/CEA-Leti").

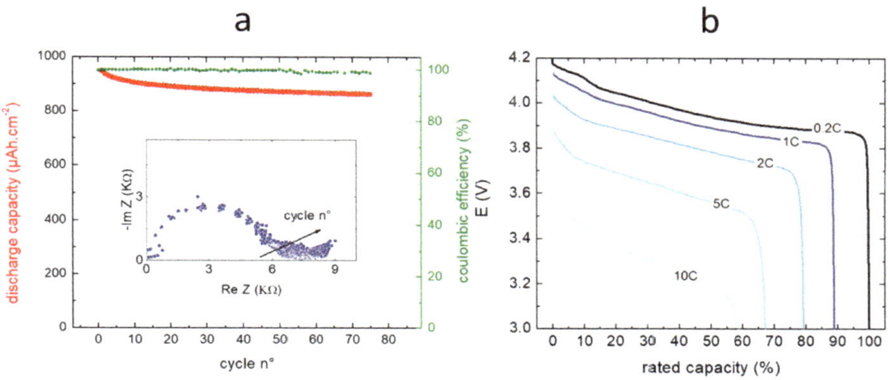

Figure 5.4 TFBs key features [12]: (a) Capacity variation with cycling; Coulombic efficiency (>99.5%) confirms the redox reactions reversibility (inset 4.2 V electrochemical impedance spectroscopy evolution with cycling. Almost no variation is observed (100 cycles), proving the stability of bulk/interfacial reactions). (b) High C-rate TFB performance. 10C corresponding to 6.5 $mA \cdot cm^{-2}$ of current density. (Property of "S. Oukassi/ CEA-Leti").

These results are explained by both a high lithium-ion diffusion coefficient in $LiCoO_2$ measured around 5.10^{-9} $cm^2 \cdot s^{-1}$ and a high LiPON ionic conductivity of 3.10^{-6} $S \cdot cm^{-1}$. It is expected that lowering the solid electrolyte thickness would allow for a further decrease of total internal resistance and consequently a higher rate capability.

Figure 5.5 depicts model prediction results of electrochemical performance variation with $LiCoO_2$ cathode thickness. After validating our model using experimental data up to 30 µm thick cathodes, it has been possible to push the simulation results forward and consider higher thickness, in order to estimate both the maximum expected energy and power densities for an optimal cell design. It is observed that the relation between the cathode thickness and areal capacity/volumetric energy density varies with the applied current (no expected gain above 100 and 50 µm thick $LiCoO_2$ for 0.5 and 2 $mA \cdot cm^{-2}$, respectively). Therefore, it has been proven that increasing the cathode thickness is an option of interest to optimize the storage capacity for applications requiring moderate current densities. Beyond a certain threshold, it turns out that this approach becomes irrelevant as the kinetic limitation of ionic diffusion mechanisms is slowed down for $LiCoO_2$-specific cases.

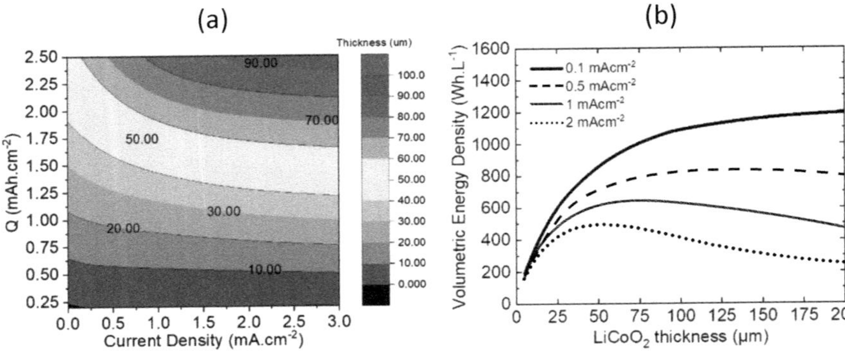

Figure 5.5 Model prediction of TFB electrochemical performance variation with cathode thickness. (a) Areal discharge capacity and (b) volumetric energy density, both at different current densities (the total volume is calculated assuming 50 μm overhead on top of the cathode, and the energy density assumes 4V of average discharge potential). (Property of "S. Oukassi/CEA-Leti").

In order to further enhance the performance, it would be necessary to act at the cathode material level. This additional boost is possible through $LiCoO_2$ doping, which results in a suppression of the semiconductor-to-metal transition and a parallel increase in diffusion coefficient. The different solutions proposed so far seem to be successful, at the expense of a slight decrease in the available capacity [26–29]. While optimization of the electrode design and/or at the material level is limited by the available technologies, an additional degree of freedom is enabled through the charge-discharge protocol optimization, in particular by considering a cascaded discharge approach [30].

TFB electrochemical performance is characterized by an excellent cycle life in comparison with existing energy microstorage solutions (Table 5.1). In addition, TFB integration is more suitable for applications with miniaturization specifications; shape and geometry can be selected with absolute flexibility, providing the thinnest and smallest dimensions reported so far. Unlike other microstorage solutions, TFBs are solder reflow tolerant and can be co-packaged to achieve highly integrated product designs.

Table 5.1 Characteristics of energy microstorage solutions

Battery	Multilayer ceramic	Coin cell	Cylindrical	Thin film
Thickness (mm)	>0.2	>2	>2	<0.1
Footprint (mm^2)	0.5	3	3	<0.1
Capacity ($mAh \cdot cm^{-2}$)	0.75	5	0.5	1.5
Cycle life (cycles)	1000	1000	1000	2000

5.3 Toward More Functions for Thin-Film Batteries

Novel multifunctional electronic devices such as IoT and wearables no longer require a conventional energy storage device, but a power source that can be perfectly integrated with the overall architecture and ensure an additional function [31]. Within this context, a holistic innovation approach covering the material and device, and its integration in a module up to the final system of the application is implemented (Figure 5.6(a)). These functions can be broadly sorted into (but are not limited to) five classes: structural, mechanical, optical, hybrid, and ionotronic TFBs.

Structural TFBs possess mechanical integrity and often exhibit either an ultrathin or ultrathick architecture. They help save weight and are useful in specific applications. We reported the manufacturing and characterization results of a biomimetic-shaped TFB, whose general shape resembled the abdomen of a dragonfly [32]. The final structures presented a total mass of approximately 60 mg per cm^2 of active area, with a mean overall thickness of 100 μm. Experimental cycling results at different C-rates showed almost no capacity fade for current densities between 0.5 and 10 mA, leading to surface energy densities as high as 0.15 $mWh \cdot cm^{-2}$ and power densities up to 5 $mW \cdot cm^{-2}$. The achieved electrochemical results are among the highest values reported in the literature for such miniaturized energy devices and contributed to bringing new solutions for powering MEMS actuation-based micro/nanorobotics.

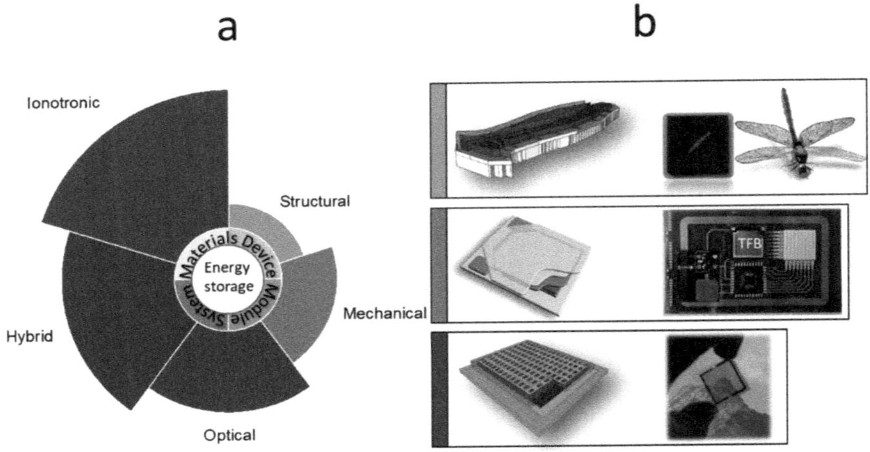

Figure 5.6 Unconventional TFBs: (a) The energy storage is associated with an additional function within the same device, according to an innovation approach from the materials to the system, and (b) an example of TFB realization with (from top to bottom) structural [32] (dragonfly-shaped nanodrone ultra lightweight TFB), mechanical [33] (flexible TFB for system on foil autonomous sensors), and optical [34] (optically semi-transparent TFB) properties. (Property of "S. Oukassi/CEA-Leti")

Mechanical TFBs are endowed with specific mechanical properties such as flexible, stretchable, bendable, rollable, and foldable configurations. In this context, we have developed flexible TFBs using low-temperature sequential physical vapor deposition techniques [33]. A TFB architecture consisting of Ti|V_2O_5|LiPON|Li|encapsulation multilayer was deposited on a flexible polyimide substrate. The obtained TFBs are among the thinnest reported (50 μm) including substrate, active layers, and encapsulation stack. The fabricated TFBs delivered a discharge capacity as high as 0.15 mAh upon galvanostatic cycling at 30 $\mu A \cdot cm^{-2}$ within the 3.8–1.5 V potential range. The capacity decreased during cycling with an average fading rate of 0.25% over 100 cycles. Furthermore, electrochemical characterization showed a clear influence of substrate type on TFB performance between polyimide and silicon. Nevertheless, although there were differences related to the substrate type, properties such as high energy density, lightweight, and flexibility among others that make TFBs attractive for a variety of applications have been maintained, in particular for systems on foil sensors architecture.

Optical TFBs include, among others, transparent and electrochromic devices. In this framework, we have proposed and successfully fabricated a grid-structured all solid inorganic transparent TFBs for the first time [34]. As compared with transparent batteries in the literature, our devices exhibit among the best performances (ranging from discharge capacities of 0.15 to 0.6 mAh at transmittances of 60% and 20%). The fabrication process is viable for industrial large-scale production and allows easy control and fine-tuning of the device's optical and electrochemical properties. In addition, the transparent TFBs show very good cycling behavior over 100 cycles, with an average capacity loss of 0.08% per cycle. Furthermore, the versatility of the architecture opens opportunities for further development such as optimizing the pattern design, stacking multiple devices, varying the balancing of electrode thickness, or exploring the implementation on ultrathin transparent substrates. Based on these superior properties, transparent TFBs have strong potential to be utilized as integrated energy storage units in various emerging applications, including optoelectronics, smart windows, transparent displays, and wearables (Figure 5.6(b), semi-transparent TFBs for smart contact lenses).

Hybrid TFBs consider the integration of energy harvesting and storing operations within the same device framework. Although this concept has been proven functional for bulk lithium-ion batteries, its application to TFBs is limited to the heterogeneous integration of two separate elements at this stage [35, 36]. Ionotronic TFBs are discussed in the next section.

5.3.1 Ionotronics and Energy Storage: A General Trend

A diagnosis of current electronic systems shows that they are almost consisting of functional reconfigurable materials and devices. The devices' properties can be adjusted upon the application of an electric field, yet the change is temporary and the device will retrieve its original state after the stimulation removal. Besides, if the material is also physically reconfigurable at the atomic scale, its properties can be modulated in a nonvolatile yet reversible fashion, thus facilitating a paradigm shift in building highly efficient computing systems. Within this framework, by controlling the internal ion distribution in a solid-state film, a material's chemical composition and physical properties can be reconfigured using an applied electric field. This approach summarizes ionotronics [40, 41]. In this framework, it is of particular interest to combine an ionotronic device approach while concomitantly enabling an additional energy storage function, as illustrated in Figure 5.7. Ameziane et al. [37] used the solid-state TFB technology and demonstrated reversible voltage-controlled switching between perpendicular and in-plane magnetization states in a Co-Pt bilayer (Figure 5.7(a)). In this work, small voltages produce an exceptionally high magnetoelectric coupling efficiency of at least 7700 $fJV^{-1}m^{-1}$ at room temperature, due to the small size and high mobility of lithium ions. Cho et al. reported that via in situ measurements during battery cycling (Figure 5.7(b)), the thermal conductivity of a $LiCoO_2$ cathode reversibly decreases from 5.4 to 3.7 $Wm^{-1}K^{-1}$, and its elastic modulus decreases from 325 to 225 GPa, as it is delithiated from $LiCoO_2$ to $Li_{0.6}CoO_2$. These results provide opportunities for dynamic control of thermal conductivity and thermal management applications. Lee et al. [39] reported on the fabrication of a potentiometric cell, Au|$LiCoO_2$ 5 m| Co_3O_4/$Li_{2.88}PO_{3.73}N_{0.14}$/$Li_2CO_3$/Au, for monitoring CO_2 gas. The sensors showed good reversibility and fast response toward changing CO_2 concentrations from 200 to 3000 ppm.

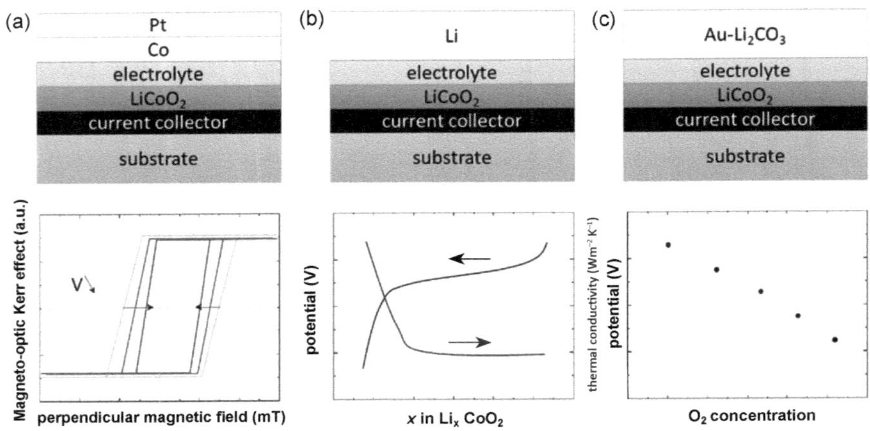

Figure 5.7 Ionotronics and energy storage: (a) TFB for voltage control of perpendicular magnetization [37], (b) electrochemically tunable thermal conductivity of $LiCoO_2$ [38], and (c) potentiometric CO_2 gas sensor with TFB architecture [39]. (Property of "S. Oukassi/CEA-Leti").

5.3.2 In-Memory Energy Concept: Dual Energy/ Information Storage

In this context, two exercises have been implemented: first, we started from a resistive random-access memory and investigated the possibility of storing energy by the nanobattery effect [42]; and secondly, we started from a nanobattery component and investigated the possibility of storing information [43]. The two works are summarized in the following.

Our first work [42] explores the innovative concept of a hybrid dual-behavior device, based on emerging nonvolatile memory technology, for both data retention and energy storage. RRAM (resistive random-access memory) is considered a major candidate as next-generation memory, thanks to its promising performances in terms of scalability and CMOS process compatibility. Its working mechanisms, based on faradaic processes, motivate the study of the feasibility of operating RRAM as an energy storage element.

Our farsighted concept of a memory-battery dual-behavior device is presented in Figure 5.8. The elements are arranged in a crossbar array to maximize integration density. In the envisioned design, subarrays could be dynamically assigned to perform as either memory or battery, depending on real-time needs. In a practical implementation, each element of the array shown in Figure 5.8 comprises a RRAM cell in series with a selection device, in order to avoid leakage and disturbance between cells. When operated in battery mode, the charging operation follows a similar procedure to the writing of a memory cell, while the discharging should be operated by some dedicated control circuitry. This block might be able to address multiple cells in parallel, possibly adjusting the number of selected elements in accordance with the demanded energy, in order for them to supply the energy to the load.

The major challenge, in this case, was to consider a state-of-the-art memory component, in this case, a $CuTe_x|Ta_2O_5|W$ stack optimized for information storage, and to keep the selected materials and the architecture unchanged. Only the electrical protocol was modified in this case to evaluate the energy storage capacity.

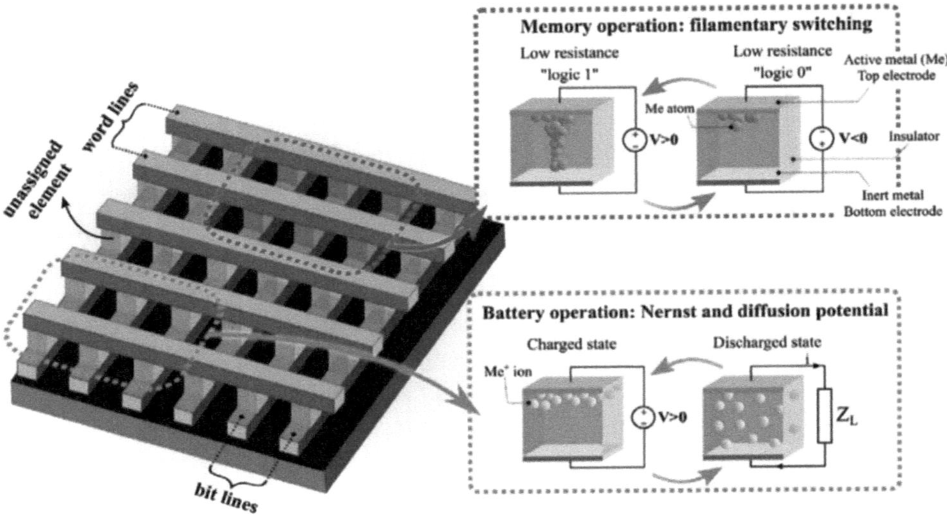

Figure 5.8 Innovative concept of a hybrid dual-behavior RRAM device [42]. In the envisioned implementation, elements are placed at the intersection of word and bit lines, in a crossbar array configuration, which allows maximizing energy and memory density. In red and green, respectively, cells store logic values and energy. (From [42], property of "S. Oukassi/CEA-Leti").

The characterization of RRAM in-memory energy devices conducted in this work allows us to draw some preliminary considerations. It has been proved that the proposed technology allows operating a device simultaneously both as a memory and energy storage element, whenever the cell is storing a logic "0." In fact, electrochemical reactions are expected to take place either before the creation or after the dissolution of the conducting filament. Figure 5.9(a) illustrates the operating regions, in terms of qualitative voltage and current domains, of such dual-behavior RRAM cells. Furthermore, our study put into evidence that the faradaic processes happening inside our samples seem not to involve the whole volume, as a linear trend of the current peak with the area could not be satisfied when enlarging the device diameter, but was instead verified when connecting multiple devices in parallel. Consequently, the energy storage mechanism appears to be a "local" phenomenon, possibly limited to the switchable oxide region of the solid electrolyte. The estimated energy density achievable with our device's peaks around 3.5 pJ μm^{-2} (at ν = 160 mV s^{-1}), while the maximum power density amounts to 80 pW μm^{-2} (for ν = 5 V s^{-1}). The Ragone plot of Figure 5.9(b) benchmarks and compares our devices to other state-of-the-art integrated energy storage elements. Remarkably, the energy and power densities extrapolated appear to rival those of integrated planar supercapacitors. Figure 5.9(c) reports stored energy versus sample area, highlighting that the energy could theoretically be linearly increased with the number of devices connected in parallel, up to values comparable to those of integrated planar supercapacitors.

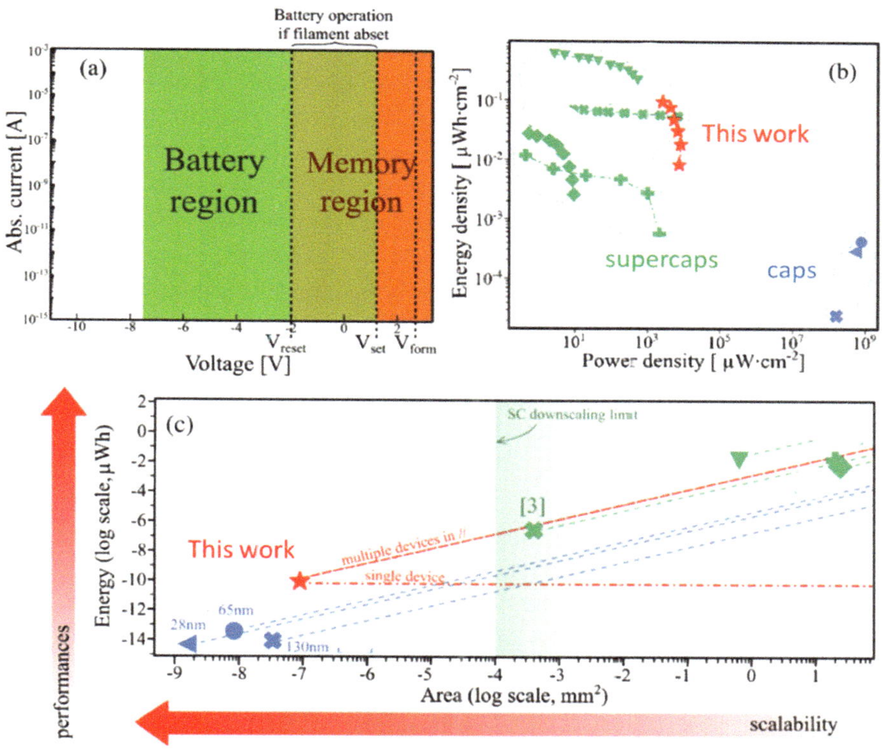

Figure 5.9 Specifications of RRAM-based in-memory energy devices. (a) Reports qualitative operating regions of a device. (b) Ragone plot. In green, some planar supercapacitors are reported; in blue, on-chip capacitors for different technological nodes (130 nm, 65 nm, and 28 nm FDSOI (fully depleted silicon on insulator)); and in red, the value extrapolated for our work. (c) Device area versus energy. (From [42], property of "S. Oukassi/CEA-Leti").

The proposed technology would well adapt to the in-memory computing field and has the potential to broaden its advantages by adding the energy storage feature inside the memory. Further evaluation is demanded, in particular, aiming to quantify the output voltage, Coulombic efficiency and experimental energy, and power density capability. As a next step, new design solutions where the memory and energy performances are optimized could be explored, in order to bring the technology to the application level.

Our second work [43] explores the innovative concept of a hybrid dual-behavior device, based on emerging nonvolatile memory technology, for both data retention and energy storage. Electrochemical random-access memory (ECRAM) is considered a major candidate as next-generation memory, thanks to its promising performances in terms of scalability and CMOS process compatibility. In particular, three-terminal synaptic transistors (SynT) offer an ion conductor-gated control from which ion doping content can be monitored with high energy efficiency via redox reactions, thus decoupling write-read actions and improving the linearity of programming states for neuromorphic computing. Its working mechanisms, based on faradaic processes, motivate the study of the feasibility of operating ECRAM as an energy storage element (Figure 5.10).

Cyclic voltammetry tests confirmed that the cells behave as standard electrochemical storage elements when investigating the impact of the scan rate, maximum positive voltage, and area on the reduction peak. The storage capability of the $LixTiO_2$|LiPON|Ti gate stack amounts up to 1.5 $\mu Ah \cdot cm^{-2}$ have been obtained, two decades higher than deep trench capacitor existing solutions. We demonstrated a high-rate capability of TiO_2 electrode with a capacity fading less than 50% for a 100 times current rate increase. As the current density was switched back to the low current rate, the capacity recovered its initial value. This recovery of the total capacity indicated that the capacity fading with incremental current rate was only related to kinetics limitation, and not material degradation or parasitic reactions.

These characteristics of TiO_2 confirmed a high-quality channel material for fast intercalation operations and high endurance (Figure 5.11(a)).

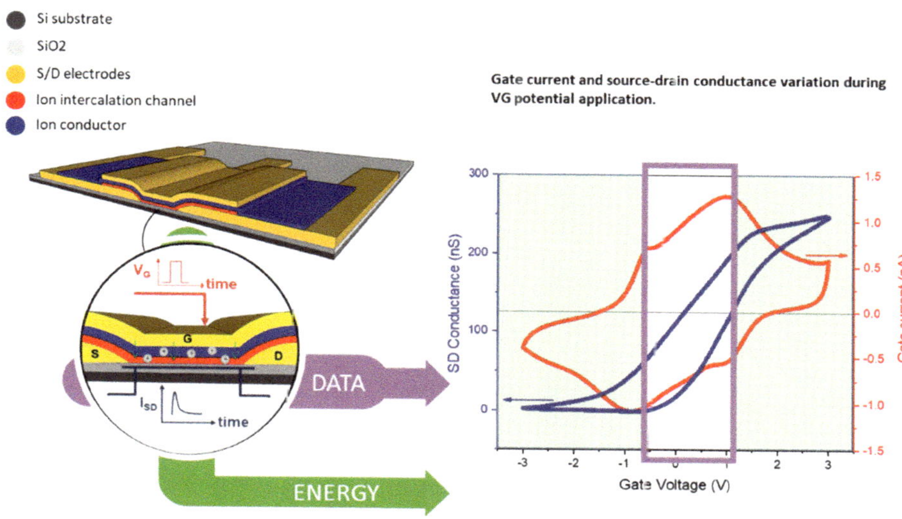

Figure 5.10 Synaptic transistor. Schematic view and corresponding dual energy and data storage properties. Vertical gate stack is a TFB energy storage solution while lateral architecture allows for conductance modulation and data storage. (Property of “S. Oukassi/CEA-Leti”).

For the same device architecture, a nonvolatile conductance modulation (<75 nS) is achieved through reversible lithium intercalation into the channel, and synaptic functions, such as long-term potentiation/depression involve ultralow switching energy of 2 fJ μm^{-2}, among the lowest reported so far. Moreover, this SynT shows excellent endurance (>10^5 weight updates, Figure 5.11(b)) and recognition accuracy (>95% on the modified National Institute of Standards and Technology data test using crossbar simulations). Finally, ECRAM “in-memory energy” technology would be a newfangled approach to meet the needs of various emerging and standard applications, in particular as synaptic elements in an artificial neural network for neuromorphic computing with high learning precision.

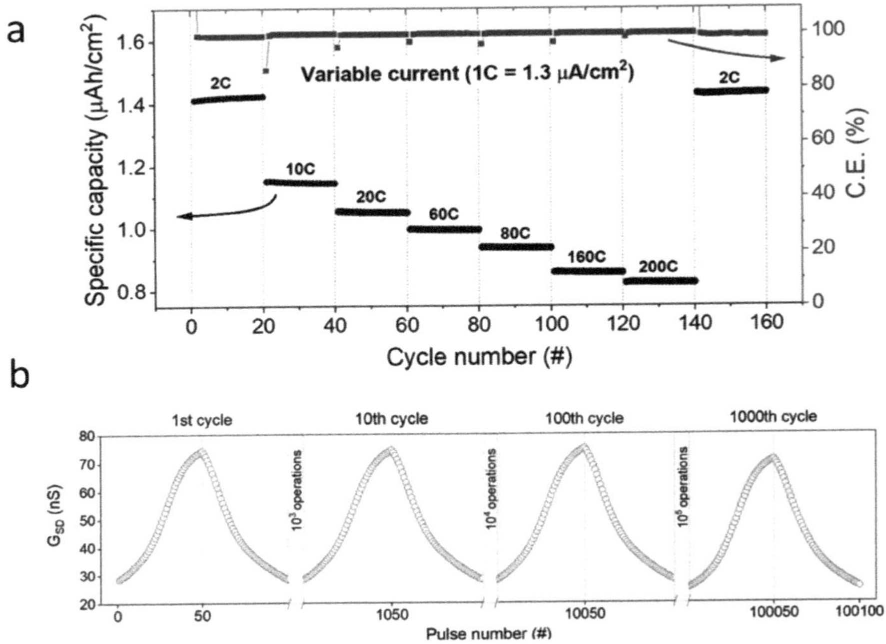

Figure 5.11 Synaptic transistor dual behavior: (a) Specific capacity and Coulombic efficiency variation with current density and cycle number. (b) Endurance test with 10^3 cycles and 10^5 potentiation/depression operations. (Property of "S. Oukassi/CEA-Leti").

5.4 Conclusions

TFBs have been developed over 30 years and several companies already commercialize products. Nevertheless, the technology is still immature and suffers from limited discharge capacity with respect to emerging applications in microelectronics and IoT. Optimization of materials such as all electrochemically active electrodes together with anode-free configuration has been recently an important upgrade toward meeting specifications. There is still room to optimize the overall electrochemical performance further by considering module (stacking) and system (power management ICs) approaches.

The pervasive aspect of TFBs, going beyond energy storage to acquire multiple additional functionalities has been discussed and numerous examples from recent reported works have been briefly described to highlight it.

Further evaluation is demanded for TFB and optimization lies mainly on the increase of energy and power densities. As a next step, new design solutions where the energy and other device functionalities are optimized could be explored, in order to bring the technology to the application level.

References

1. Xia, Q. et al. All-Solid-State Thin Film Lithium/Lithium-Ion Microbatteries for Powering the Internet of Things. *Adv. Mater.* 2200538 (2022). doi:10.1002/ADMA.202200538
2. Zhu, M. & Schmidt, O. G. Tiny Robots and Sensors Need Tiny Batteries—Here's How To Do It. *Nature* **589**, 195–197 (2021).
3. Zhu, Z. et al. Recent Advances in High-Performance Microbatteries: Construction, Application, and Perspective. *Small* **2003251**, 2003251 (2020).
4. Kyeremateng, N. A. & Hahn, R. Attainable Energy Density of Microbatteries. *ACS Energy Lett.* **3**, 11172–1175 (2018).
5. Li, Y. et al. On-Chip Batteries for Dust-Sized Computers. *Adv. Energy Mater.* **12**, 2103641 (2022).
6. Xiaohua, Yu, Bates, J. B., Jellison, G. E. & Hart, F. X. A Stable Thin-Film Lithium Electrolyte: Lithium Phosphorus Oxynitride. *J. Electrochem. Soc.* **144**, 524–532 (1997).
7. Bates, J. B., Gruzalski, G. R., Dudney, N. J., Luck, C. F. & Yu, X. Rechargeable Thin-Film Lithium Batteries. *Solid State Ionics* **70/71**, 619–628 (1994).
8. Bates, J. B. et al. Thin-Film Rechargeable Lithium Batteries. *J. Power Sources* **54**, 58–62 (1995).
9. West, W. C., Whitacre, J. F., Brandon, E. J. & Ratnakumar, B. V. Lithium Micro-Battery Development at the Jet Propulsion Laboratory. *Sixteenth Annual Battery Conference on Applications and Advances. Proceedings of the Conference (Cat. No.01TH8533)*, 237–241, doi: 10.1109/BCAA, 905131 (2001). Available at: https://ieeexplore.ieee.org/stamp/stamp.jsp?tp=&arnumber=905131. (Accessed: 8th December 2022).

10. Notten, P. H. L., Roozeboom, F., Niessen, R. A. H. & Baggetto, L. 3-D integrated all-solid-state rechargeable batteries. *Adv. Mater.* **19**, 4564–4567 (2007).

11. Trask, J. et al. Optimization of 10 μm, Sputtered, $LiCoO_2$ Cathodes to Enable Higher Energy Density Solid State Batteries. *J. Power Sources* **350**, 56–64 (2017).

12. Oukassi, S. et al. Millimeter Scale Thin Film Batteries for Integrated High Energy Density Storage. *Proc. IEEE Int. Electron Devices Meet.*, 618–621 (2019).

13. Wu, J. et al. Low-Tortuosity Thick Electrodes with Active Materials Gradient Design for Enhanced Energy Storage. *ACS Nano* **16**, 4805–4812 (2022).

14. Zhou, H., Xin, F., Pei, B. & Whittingham, M. S. What Limits the Capacity of Layered Oxide Cathodes in Lithium Batteries? *ACS Energy Lett.* **4**, 1902–1906 (2019).

15. Wu, T. et al. Thick Electrode with Thickness-Independent Capacity Enabled by Assembled Two-Dimensional Porous Nanosheets. *Energy Storage Mater.* **36**, 265–271 (2021).

16. Yang, X., Doyle-Davis, K., Gao, X. & Sun, X. Recent Progress and Perspectives on Designing High-Performance Thick Electrodes for All-Solid-State Lithium Batteries. *eTransportation* **11**, 100152 (2022).

17. Johnson, A. C. et al. Strategies for Approaching One Hundred Percent Dense Lithium-Ion Battery Cathodes. *J. Power Sources* **532** (2022).

18. Famprikis, T., Galipaud, J., Clemens, O., Pecquenard, B. & Le Cras, F. Composition Dependence of Ionic Conductivity in LiSiPO(N) Thin-Film Electrolytes for Solid-State Batteries. *ACS Appl. Energy Mater.* **2**, 4782–4791 (2019).

19. Sastre, J., Chen, X., Aribia, A., Tiwari, A. N. & Romanyuk, Y. E. Fast Charge Transfer across the $Li_7La_3Zr_2O_{12}$ Solid Electrolyte/$LiCoO_2$ Cathode Interface Enabled by an Interphase-Engineered All-Thin-Film Architecture. *ACS Appl. Mater. Interfaces* **12**, 36196–36207 (2020).

20. Lin, J. et al. Promising Electrode and Electrolyte Materials for High-Energy-Density Thin-Film Lithium Batteries. *Energy Environ. Mater.* **5**, 133–156 (2022).

21. Louli, A. J. et al. Optimizing Cycling Conditions for Anode-Free Lithium Metal Cells. *J. Electrochem* **168**, 020515 (2021).

22. Huang, C.-J. et al. Decoupling the Origins of Irreversible Coulombic Efficiency in Anode-Free Lithium Metal Batteries. *Nat. Commun.* **12**, 1452 (2021).

23. Wang, M. J., Carmona, E., Gupta, A., Albertus, P. & Sakamoto, J. Enabling "Lithium-Free" Manufacturing of Pure Lithium Metal Solid-State Batteries Through In Situ Plating. *Nat. Commun.* **11**, 5201–5209 (2020).

24. Johnson, A. C. et al. Strategies for Approaching One Hundred Percent Dense Lithium-Ion Battery Cathodes. *J. Power Sources* **532**, 231359 (2022).

25. Li, M. et al. Dense All-Electrochem-Active Electrodes for All-Solid-State Lithium Batteries. *Adv. Mater.* (2021). doi:10.1002/ADMA.202008723.

26. Xu, Y. et al. Accelerating Solid Diffusion and Suppressing Phase Transition in LiV_3O_8 Via Calcium Doping at Lithium Sites. *Nanoscale* **12**, 10205–10215 (2020).

27. Song, J. H. et al. Suppression of Irreversible Capacity Loss in Li-Rich Layered Oxide by Fluorine Doping. *J. Power Sources* **313**, 65–72 (2016).

28. Myung, S. T., Kumagai, N., Komaba, S. & Chung, H. T. Effects of Al Doping on the Microstructure of $LiCoO_2$ Cathode Materials. *Solid State Ionics* **139**, 47–56 (2001).

29. Liu, Q. et al. Approaching the Capacity Limit of Lithium Cobalt Oxide in Lithium-Ion Batteries Via Lanthanum and Aluminium Doping. *Nat. Energy* **3**, 936–943 (2018).

30. Oukassi, S., Despesse, G., & Secouard, C. US Patent 2021/0328440. Assembly of Electrical Energy Storage or Production Cells Comprising a Management Circuit for Managing the Cells. (2021).

31. Vlad, A., Singh, N., Galande, C. & Ajayan, P. M. Design Considerations for Unconventional Electrochemical Energy Storage Architectures. *Adv. Energy Mater.* **5**, 1–53 (2015).

32. Oukassi, S., Poncet, S., Frutos, J. R. & Salot R. Design, Microfabrication and Characterization of Free Form Factor, Lightweight Thin Film Battery for Powering Bioinspired Nano-Drones Based on MEMS Actuation. *Journal of Physics: Conference Series* 1407 012036 (2019). doi:10.1088/1742-6596/1407/1/012036.

33. Oukassi, S., Giroud-Garampon, C., Poncet, S. & Salot, R. Ultra-Thin Rechargeable Lithium Ion Batteries on Flexible Polymer: Design,

Low Temperature Fabrication and Characterization. *J. Electrochem. Soc.* **164**, A1785–A1791 (2017).

34. Oukassi, S. et al. Transparent Thin Film Solid-State Lithium Ion Batteries. *ACS Appl. Mater. Interfaces* **11**, 683–690 (2019).

35. Li, Y. et al. Developing a Miniature Energy-Harvesting-Powered Edge Device with Multi-Exit Neural Network. *IEEE Trans. Circuits Syst.* (2021). doi:10.1109/ISCAS51556.2021.9401799.

36. Poulin-Vittrant, G. et al. Fabrication and Characterization of ZnO Nanowire-Based Piezoelectric Nanogenerators for Low Frequency Mechanical Energy Harvesting. *Phys. Procedia* **70**, 909–913 (2015).

37. Ameziane, M., Mansell, R., Havu, V., Rinke, P. & van Dijken, S. Lithium-Ion Battery Technology for Voltage Control of Perpendicular Magnetization. *Adv. Funct. Mater.* **32** (2022).

38. Cho, J. et al. Electrochemically Tunable Thermal Conductivity of Lithium Cobalt Oxide. *Nat. Commun.* **5**, 2–7 (2014).

39. Lee, C., Akbar, S. A. & Park, C. O. Potentiometric CO_2 gas sensor with lithium phosphorous oxynitride electrolyte. *Sensors Actuators, B Chem.* **80**, 234–242 (2001).

40. Terabe, K., Tsuchiya, T. & Tsuruoka, T. A Variety of Functional Devices Realized by Ionic Nanoarchitectonics, Complementing Electronics Components. *Adv. Electron. Mater.* **8**, 1–17 (2022).

41. Zhu, Y. et al. Lithium-Film Ceramics for Solid-State Lithionic Devices. *Nat. Rev. Mater.* **6**, 313–331 (2021).

42. Trotti, P. et al. In Memory Energy Application for Resistive Random Access Memory. *Adv. Electron. Mater.* **7** (2021).

43. Nguyen, N. et al. An Ultralow Power Li_xTiO_2-Based Synaptic Transistor for Scalable Neuromorphic Computing. *Adv. Electron. Mater.*, 2200607 (2022).

Part III

Inspirations from Nature and Unifying Concepts for Future (Beyond) Technologies

Chapter 6

Memristor, Neuron, and Edge of Chaos Kernel

Leon Chua

Department of Electrical Engineering and Computer Sciences, University of California at Berkeley, Berkeley, California 94720, USA

chua@berkeley.edu

This chapter exposes Hodgkin–Huxley's "time-varying conductance" as an ill-conceived conceptual *blunder* and identifies a fundamentally new physical concept dubbed "Edge of Chaos Kernel" as nature's optimal mechanism for creating an "action potential", aka "all-or-none" response. As a bonus, it also offers a precise definition of the heretofore unresolved notion of *excitability*, which in turn finally resolves Galvani's 242-year-old enigma on the physical mechanism, which gives rise to a near-abrupt *all-or-none phenomenon*, triggered by a global saddle-node bifurcation where a stable and an unstable periodic orbit grows in size while evolving their shape, ever so gently, so as to morph onto each other, until they became a single orbit, in 4-dimensional state space. Even more intriguing is the Edge of Chaos Kernel's humble incarnation from *Chua's Riddle*, and its innate ability to exhibit both Turing instability and *Smale's paradox*.

Outlooking beyond Nanoelectronics and Nanosystems: Ultra Scaling, Pervasiveness, Sustainable Integration, and Biotic Cross-Inspiration
Edited by Simon Deleonibus

ISBN 978-981-5129-32-8 (Hardcover), 978-1-003-50990-5 (eBook)
www.jennystanford.com

6.1 Introduction

It is ironic that the state-of-the-art of the currently fashionable research area dubbed neuro-morphic, *aka* brain-like, computing did not have an electro-physiological foundation! In 1781, Luigi Galvani was dumbfounded in a frog-leg dissection experiment, when an electrical spark from a nearby Leyden Jar accidentally hit the exposed nerve of a frog leg, resulting in a sudden twitch of the leg muscle. This response was totally unexpected. Overcome with curiosity, Galvani attempted to repeat the effect and was able to convince himself that the spark alone was the cause of the muscle contraction. Further experiments showed that the size of the contraction does not increase with the strength of the discharge (spark)—a highly nonlinear response! Indeed, Galvani is probably the first scientist to recognize the fundamental difference between a linear from a nonlinear response. Galvani's acute scientific intellect led him to conclude, correctly, that the spark rather than acting directly on the muscle, had released some internal force in the muscle, as documented from the following quotation in Galvani's classic 1791 memoir:

> *"The electrical atmosphere hit, and pushed, and vibrated by the spark is that, which brought to the nerve, and similarly pushing and commuting some extremely mobile principle existing in nerves excites the action of the nerve-muscular force"* [1].

The phenomenon published by Galvani in 1791 was independently discovered 113 years later (1904) by Keith Lucas from Cambridge University, which he named *all-or-none response* [2]. The term *all-or-none* response is now used synonymously with an *action potential,* which first appeared in ref. [3]. Although Galvani correctly attributed the mechanical twitching phenomenon to be caused by a powerful electrical impulse, his insightful discovery was discredited by his peers, including Alessandro Volta. It took 83 years before Julius Bernstein invented an ingenious instrument based on a galvanometer, dubbed the *differential rheotome* that the rough outline of the *action potential* was recorded from the frog nerve in 1864, thereby proving Galvani's impulsive input is electrical in nature. However,

Bernstein's recorded *action potential* has neither "*voltage*" nor "*time*" scales. But he went on to postulate an electro-physiology of the action potential, which asserts that the physical mechanism leading to the almost abrupt emergence of the *all-or-none action potential* came from the *time-varying conductance* of the *axon*, which is selectively permeable to potassium K^+ ions at rest, but during excitation, the membrane permeability to other ions (Na^+, Ca^-, etc.) increases. The Bernstein Hypothesis was published in 1902 and became the standard theory taught at major universities until 1939, when proven wrong by Hodgkin and Huxley [4]. But Bernstein's incorrect theory continued to be taught at major universities because there was no other credible theory available, until Hodgkin and Huxley published their celebrated Hodgkin–Huxley circuit model and equations [5]. While Bernstein's erroneous hypothesis sets back research on the electro-physiology of the *action potential* for 50 *years*, Hodgkin–Huxley's circuit model is now regarded as the most successful electro-physiological model and recognized by the Nobel Prize in 1965. Notwithstanding the immense impact of the Hodgkin–Huxley circuit model, it must be pointed out that there is a fundamental error in their circuit when they incorrectly identified the potassium-ion channel conductance $G_K(t)$ and sodium-ion channel conductance $G_{Na}(t)$ as *time-varying conductances*, whose respective time-variation functions $G_K(\bullet)$ and $G_{Na}(\bullet)$ cannot be specified explicitly as required for a conductance to qualify as a *time-varying conductance* in nonlinear circuit theory [6], but must be calculated by solving one or two nonlinear differential equations, respectively. This means that the Hodgkin–Huxley circuit model and equations have no physical or biological interpretations! This is the main reason why researchers on neuro-morphic systems generally use the Hodgkin–Huxley equations only as a numerical simulation tool, and not for the immensely more powerful physical interpretations that it proffers.

We have shown in two papers [7, 8] that the Hodgkin–Huxley time-varying potassium conductance is nothing but a *time-invariant first-order memristor*; likewise, the Hodgkin–Huxley time-varying sodium conductance is nothing but a *time-invariant second-order memristor*.

The Hodgkin–Huxley circuit model and equations are merely an accurate computational model that could emulate the *all-or-none action potential* of the giant axon of a certain species of squids. It does not provide the electro-physiological mechanism of axons. In particular, Galvani's 1781 all-or-none phenomenon is still *not understood*, from the *electro-physiological* perspective.

Although the Hodgkin–Huxley model could explain *qualitatively* the *all-or-none* phenomenon, via the action of the sodium- and potassium-ion channels, it *cannot explain* the *quantitative global bifurcation dynamics* that give rise to the near-abrupt *action potential* and *nonlinear dynamical bifurcation*. In other words, it is still a *242-year-old unsolved problem*. We now have the right tool and theory to solve this problem, once and for all, by applying the latest mathematical theory of nonlinear dynamics, and the *local activity principle* [9, 10] and its gem, *the edge of chaos* calculation technique. In short, we have solved the Galvani's 242-year-old enigma. In addition, we have applied the same principle and technique to characterize neuromorphic systems, as well as volatile and nonvolatile memristors.

As for axons, it is important to understand that for an action potential to travel along the axon, it must derive the energy of its propagation not merely, from its point of excitation, but rather, from sources distributed along the entire axon. This so-called auto-wave mechanism is fundamentally different from such familiar examples as sound waves or strain waves in elastic media, where the propagation of oscillatory solutions is induced by the energy of the *initial* excitation. In the case of the axon, the energy comes from the *local activity* of the *Hodgkin–Huxley circuit cells distributed uniformly* along the entire length of the axon. In this case, the *energy is supplied locally*. A crude analog of this propagation mechanism is the firing of a train of gunpowder, where the release of the energy by the chemical change due to firing at one point raises the temperature sufficiently to cause the same changes at the next point. The propagation of the oscillations in Stan William's memristor chain follows exactly the same mechanism where the nonlinear device is a "volatile memristor" biased in a locally active domain [11].

All of the above phenomena clearly violate the second law of thermodynamics. The unknown mechanism which enables the entropy to decrease over time is precisely the "principle of local activity".

6.1.1 Galvani's 242-Year-Old Enigma

It is remarkable that the *various keen observations* reported by Galvani, since his discovery of the serendipitous spark-triggered twitching frog leg in 1781, as depicted in Figure 6.1, are still unresolved! In particular, Galvani was both astonished and perplexed to find that the frog leg could be induced to contract by using a Leyden jar almost completely discharged so as to be undetectable with the most sensitive electroscope. How it could be that such a tiny electrical force would produce muscle contraction if it were not setting in motion some internal force? Such an inexplicable phenomenon demands the discovery of some extremely *mobile principle*. According to Galvani, in the framework of the *doctrine of irritability*, external electricity was acting as an *"excitatory stimulus"* (old fashion name for a *trigger*) for the contraction, and not as the direct "efficient" cause of the observed phenomenon.

Figure 6.1 (a) Plate 1 of Galvani's Commentarius [1] shows a laboratory set up in Galvani's home with an electrostatic generator and a Leyden jar on opposite ends of the table, along with several prepared frog legs. (b) Portrait of Luigi Galvani; discoverer of the *all-or-none* phenomenon and forefather of electro-physiology. A visionary whose numerous fundamental scientific contributions were ridiculed if not rejected as reflected in the following lament from Galvani: "I am attacked by two very opposite sects—the scientists and the know nothing. Both laugh at me—calling me "the Frog's Dancing-Master", but I know that I have discovered one of the greatest forces in nature."

In this chapter, we will finally resolve Galvani's 242-year-old enigma by identifying the *"extremely mobile principle"*, which Galvani, and generations of electro-physiologists after him, had sought in vain, as the *local activity principle* [9, 10]. In particular, the *doctrine of irritability* exploited by Galvani is a direct consequence of the *Edge of Chaos* criteria, a rigorous, mathematical theory with simple tests involving only matrix algebra and ordinary differential equations. Unlike numerous theories on *complexity theory*, such as Schrödinger's *"negentropy"* [12], Prigogine's *"instability of the homogeneous"* [13], Gell-mann's *"amplification of fluctuation"* [14], Haken's *"synergetics"* [15], etc., which require excessive computations, the *local activity principle*, and its pearl, the *edge of chaos criteria* are a cinch to calculate.

6.1.2 Hodgkin–Huxley's Blunder

We have already called attention in Section 6.1 that the Hodgkin–Huxley circuit model [5] of the giant squid axon (see Figure 6.2) contains two illegitimate circuit elements, which Hodgkin and Huxley dubbed *time-varying potassium conductance* and *time-varying sodium conductance*, respectively. The two circuit elements are illegitimate in the sense that they are not recognized in the scientific discipline named *circuit theory*, which is based on a set of well-defined time-invariant circuit elements whose interconnections, called a circuit, obey a set of axioms, namely, *Kirchhoff's Voltage Law* and *Kirchhoff's Current Law* [16]. The discipline of circuit theory is based on rigorous mathematics, and physics where results are based strictly on the classical definition of time-invariant circuit elements and the two Kirchhoff Laws. Adding a *time-varying* circuit element to a *time-invariant circuit* could lead not only to nonsensical results but also to inconsistencies. In this chapter, we found a simple way to legitimize the *time-varying* potassium and sodium conductances by identifying them *not* as conductances in the classical sense, but as *time-invariant memristors*, as shown in Figure 6.3, a well-defined new *nonlinear circuit element* that has been critically vetted on the same foundation where the circuit theory discipline was based. It is a great stroke of luck that except for the change in name and element symbol, nothing else needs to be revised. What is gained is clarity, with a legitimate inclusion of the memristor-based circuit model of Figure 6.3, as an essential new building block, in electro-physiology.

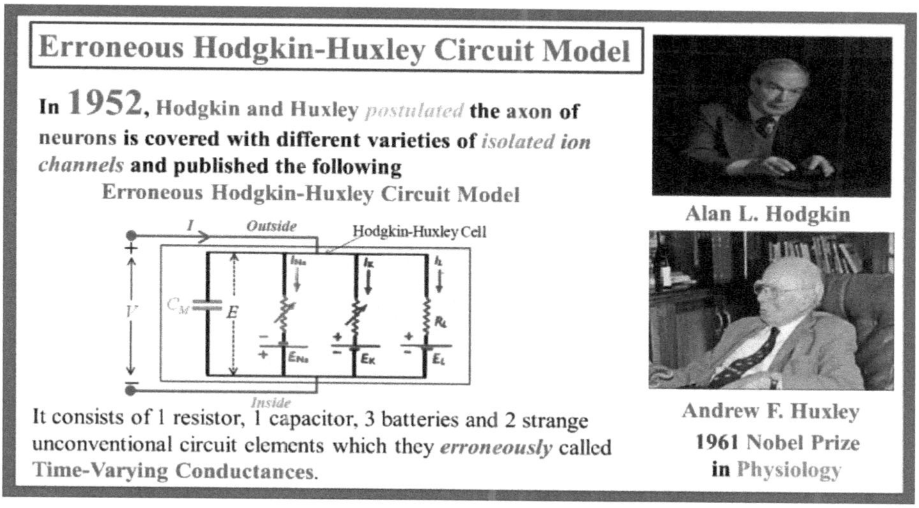

Figure 6.2 The original Hodgkin–Huxley circuit model (Figure 1 in [5]) contains two illegitimate time-varying circuit elements, which, like the infamous Bernstein hypothesis, have set back decades of progress in the electro-physiology of ion channels.

6.2 Chua's Riddle

Table 6.1 shows a *riddle* that I routinely posed in the opening lecture of a graduate course on nonlinear circuits and nonlinear dynamics that I taught at the University of California, Berkeley, at the *Chua hp Lecture* Series [18] as well as at numerous keynote lectures [9, 19]. It has attracted immense recent scientific attention and research activities in view of its deep nonlinear dynamical relevance in the electro-physiology of ion channels, in complexity theory, and above all, in demystifying the time-honored *Turing Instability* Conundrum [17, 18, 20] (unsolved since 1952) and the *Smale Paradox* [19, 21, 22] (unsolved since 1974).

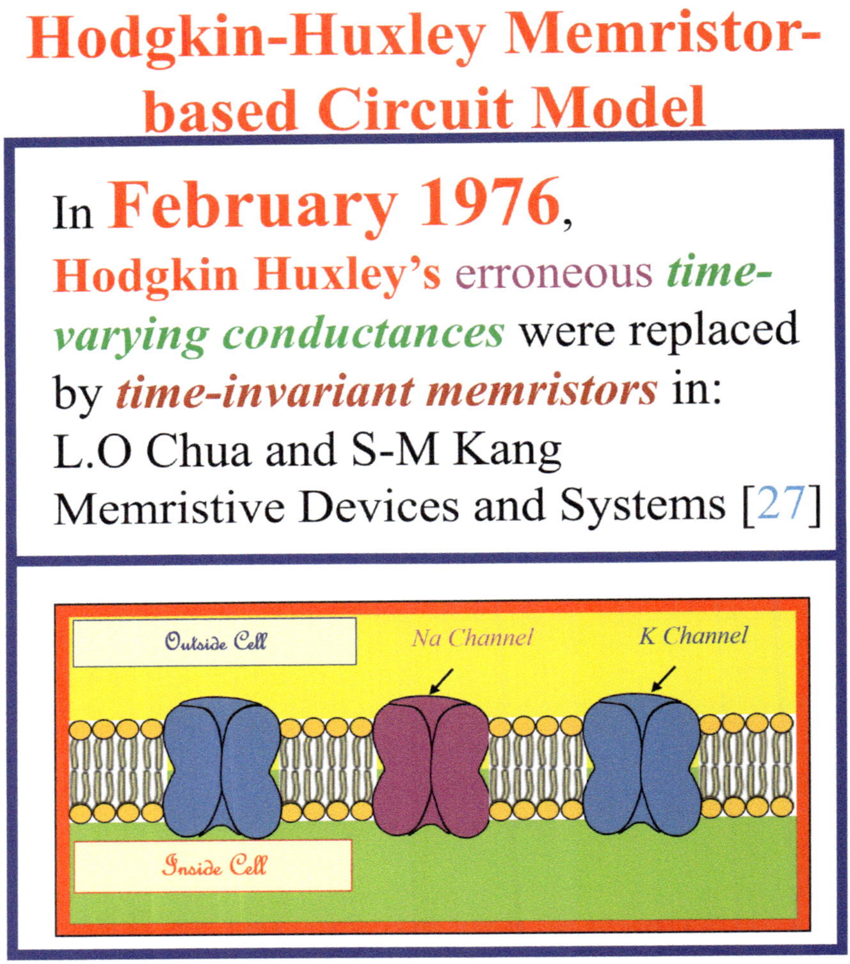

Figure 6.3 Upon replacing the original Hodgkin–Huxley potassium time-varying conductance with a potassium ion-channel memristor, and replacing the Hodgkin–Huxley sodium time-varying conductance with a sodium ion-channel memristor, we obtain a revised electro-physiologically correct Hodgkin–Huxley memristor-based circuit model. The constitutive relation defining each memristor is derived directly from the equations defining the Hodgkin–Huxley time-varying conductance in [5], and will be given in Section 6.3.

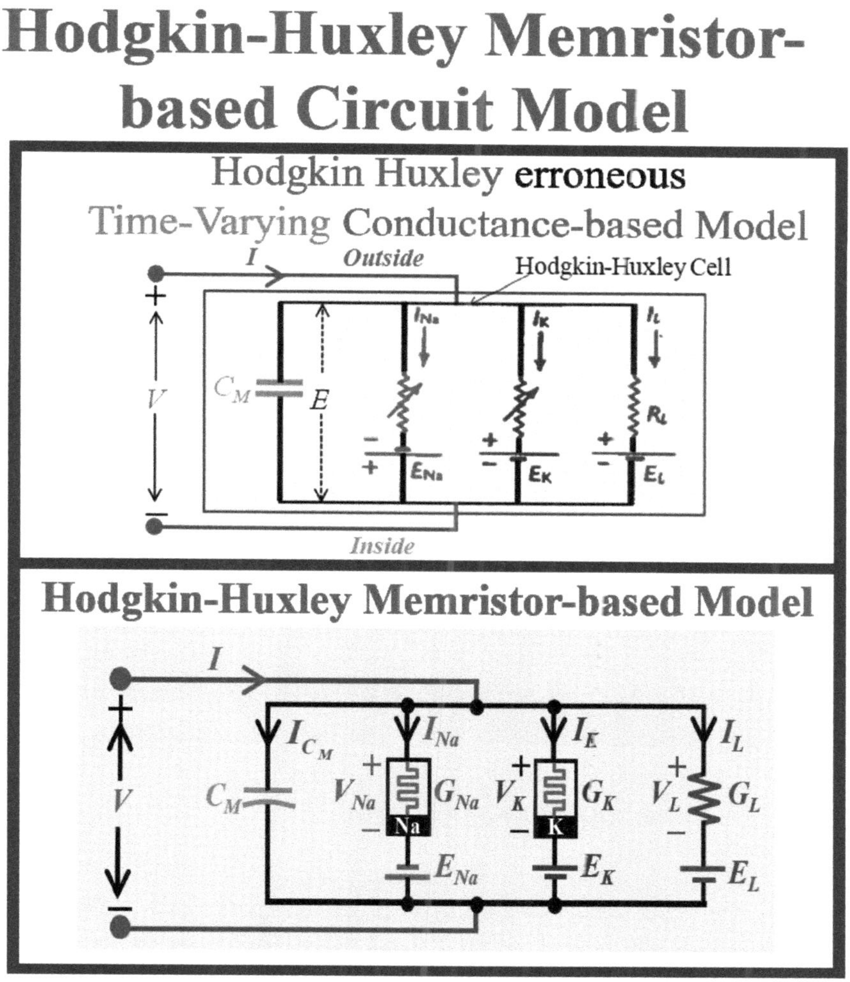

Figure 6.3 (*continued*)

Table 6.1 Gedanken experiments for illustrating Chua's Riddle. The black box N consists of a single branch, made of two circuit elements (a resistor *R* and an inductor *L*), connected in series with the two wires external to the box. In its dual version, the black box N is composed of the parallel combination between a resistor and a capacitor.

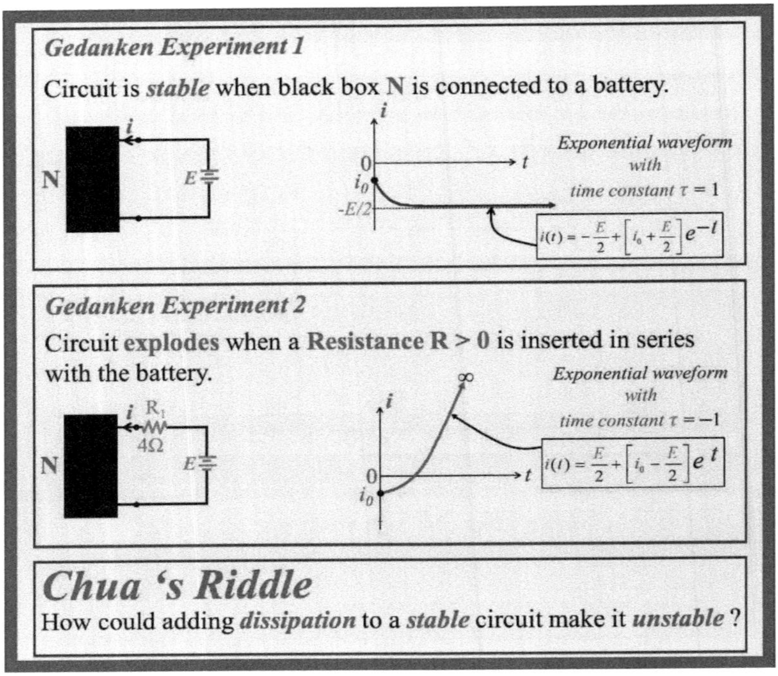

6.2.1 Resolution of Chua's Riddle

The black box **N** in Table 6.1 contains a negative resistance $R < 0$ in series with a negative inductance $L < 0$, henceforth dubbed the Chua Riddle Circuit. Without Loss of generality, let us choose $R = -2\ \Omega$ and $L = -2$ H for simplicity.

Gedanken Experiment 1 (Figure 6.4(a))

The circuit in Figure 6.4(a) is described by the following linear differential equation:

$$\frac{di}{dt} = -i - \frac{E}{2} \tag{6.1}$$

where i is the current entering **N**, and $E > 0$ is the voltage of the battery. The exact solution of Eq. (6.1) with an arbitrarily chosen initial condition $i(t) = i(0) = i_0$ at $t = 0$ is

$$i(t) = -\frac{E}{2} + \left[i_0 - \frac{E}{2}\right] e^{-\frac{t}{\tau}} \tag{6.2}$$

where $\tau = L/R > 0$ as it is easily verified by direct substitution into Eq. (6.1), and by inspection of the damped exponential waveform in Table 6.1, where $i(t) = i_0$, at $t = 0$, and $i(t) \to -E/2$ as $t \to +\infty$.

Since a positive-time constant $\tau = L/R$ is associated with the equation of a damped exponential function in Eq. (6.2), $i(t)$ gives a useful measure of how fast $i(t)$ (starting from i_0 at $t = 0$) reaches the DC steady state. In particular, after a one-time constant of $\tau > 0$ seconds, $i(t)$ would have traveled 63% of the distance between i_0 and the DC steady state (to within 1%).

Gedanken Experiment 2 (Figure 6.4(b))

The circuit in Figure 6.4(b) is described by the following linear differential equation:

$$\frac{di}{dt} = i - \frac{E}{2} \tag{6.3}$$

The exact solution of Eq. (6.3) with the same initial condition $i(0) = i_0$ is

$$i(t) = \frac{E}{2} + \left[i_0 - \frac{E}{2}\right] e^{-\frac{t}{\tau}} \tag{6.4}$$

where $\tau = \dfrac{L}{R + R_1} = -1 < 0.$

Observe that $i(0) = i_0$ at $t = 0$, and $i(t) \to \infty$, since $\left(i_0 - \frac{E}{2}\right) > 0$, by assumption. If $i_0 < E/2$, the exponential waveform would tend to $-\infty$, and the circuit would explode as well.

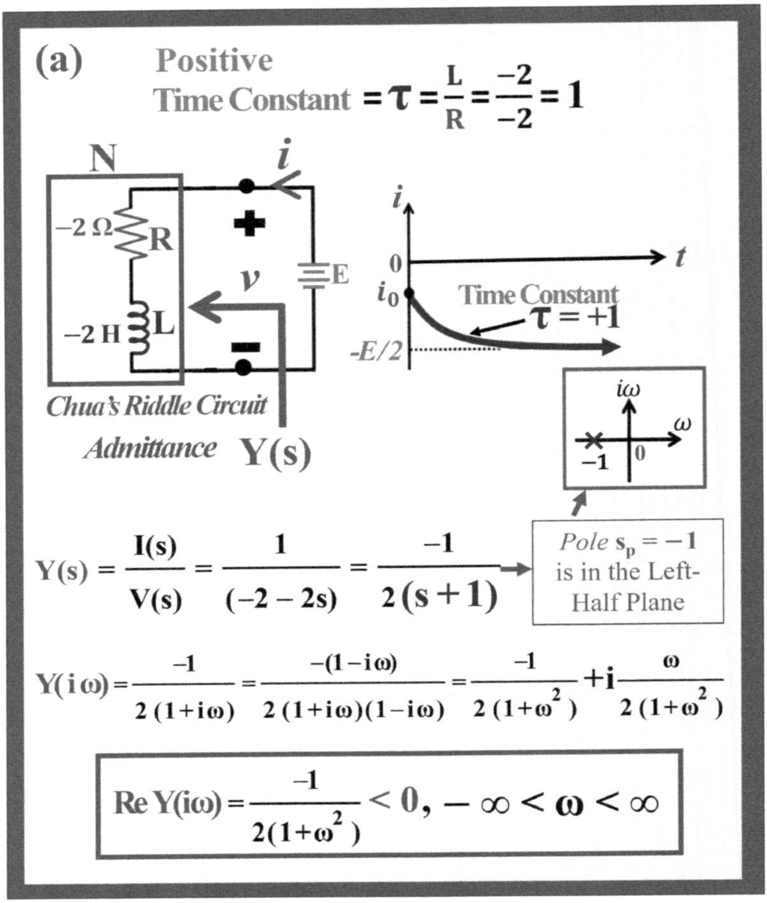

Figure 6.4 Resolution of Chua's Riddle. The black box N in Table 6.1 contains a −2 Ω resistance in series with a −2 H inductance. (a) The admittance of the 2-terminal black box N on the left is $Y(s) = -\frac{1}{2}(s+1)^{-1}$, which has a pole $s_p = -1 < 0$ in the open left half plane, thereby guaranteeing the circuit is asymptotically *stable*, when connected across a battery. The pole s_p and the *time constant* τ are related by $s_p = -\frac{1}{\tau}$. [16] (b) When a 4 Ω resistance is inserted in series with the battery in the black box-cum battery circuit on the left, the resulting composite box N′ has an equivalent circuit consisting of a +2 Ω resistance in series with a −2 H inductance whose composite admittance is $Y'(s) = -\frac{1}{2}(s-1)^{-1}$. Observe that $Y'(s)$ has a pole $s_p = 1 > 0$ in the open right-half-plane, thereby destabilizing the Chua's Riddle Circuit by adding dissipation via the positive resistance $R_1 > 0$, a shocking counter-intuitive phenomenon!

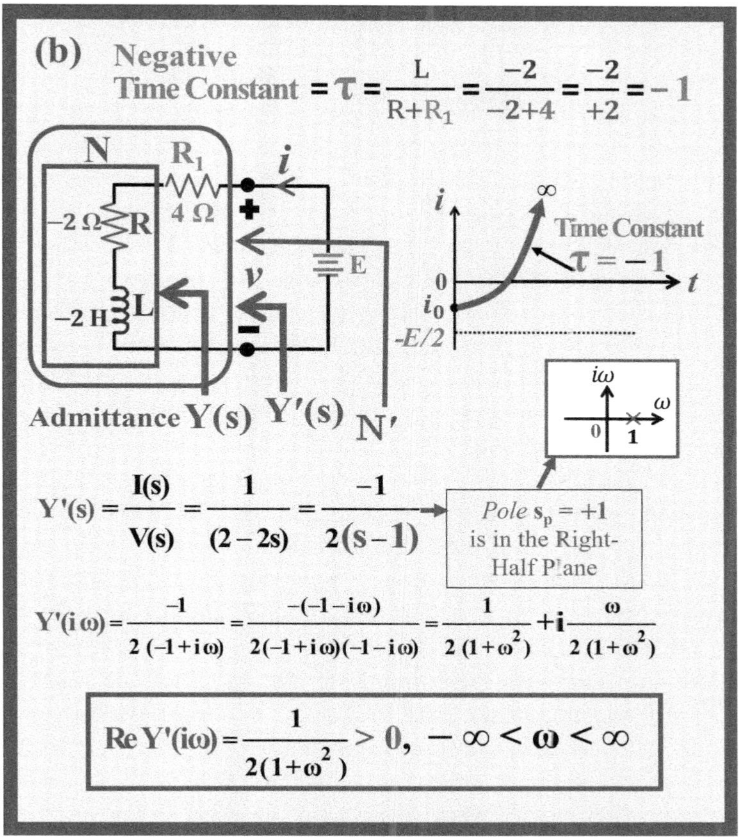

Figure 6.4 (*continued*).

6.2.2 Circuit-Theoretic Properties of Chua's Riddle Circuit

Applying basic circuit theory [6] to the Chua's Riddle circuit inside the black box N in the Gedanken Experiments in Table 6.1, i.e., the magenta box in Figure 6.4(a) and the composite box N′ in Figure 6.4(b), we obtain the admittance $Y(s) = I(s)/V(s)$ given in Figure 6.4(a) and $Y'(s) = I(s)/V(s)$ in Figure 6.4(b), respectively. Observe the admittance $Y(s)$ of N has a pole $s_p = -1$ in the open left half plane, which decodes into the stable damped exponential waveform in Figure 6.4(a), with a positive time constant $\tau = L/R = 1$. In sharp contrast, the admittance $Y'(s)$ of N′ has a pole $s_p = 1$ in the open right-half plane, which decodes into the unstable exponential waveform in Figure 6.4(b) with a negative time constant $\tau = L/R = -1$. By varying the value of the positive resistance R_1 inserted in series with the battery, in Figure 6.4(b), we can move the pole from the left-half plane in Figure 6.4(a) to the right-half plane. In particular, the pole in Figure 6.4(a) moves smoothly to the right as we increase R_1 from $R_1 = 0$ to $R_1 = 2\ \Omega$ where the pole arrives at $s_p = 0$. Increasing R_1 further would move the pole to the right-half plane where the circuit becomes unstable.

In electrical circuit theory parlance, any enclosure or box **N** (real or conceptual) containing an arbitrary interconnection of electrical circuit elements, including voltage and current sources, which has two accessible external electrical wires is called a *1-port*. The port voltage v is labeled with an arbitrary voltage reference polarity, and the associated port current reference direction entering the "plus" terminal is labeled i with an arrowhead entering **N**. The explicit formula for the admittance $Y(s)$ for the 1-port **N** and $Y'(s)$ for the 1-port **N′** is given in Figure 6.4(a) and Figure 6.4(b) respectively, as a function of the complex frequency $s = \sigma + i\omega$ in the complex s-plane for $R = -2\ \Omega$, $L = -2$H, and $R_1 = 4\ \Omega$, respectively. While $Y(s)$ and $Y'(s)$ are needed to find the pole s_p of $Y(s)$ and $Y'(s)$, we will find the real part $\text{Re}[Y(i\omega)]$ and imaginary part $\text{Im}[Y(i\omega)]$ of the 1-port **N**, as well as the real part $\text{Re}[Y'(i\omega)]$ and imaginary part $\text{Im}[Y'(i\omega)]$ of the 1-port **N′** to be more revealing. They are shown in Figure 6.5 and Figure 6.6, respectively, along with their loci in the $\text{Im}[Y(i\omega)]$ vs. $\text{Re}[Y(i\omega)]$ plane (resp., $\text{Im}[Y'(i\omega)]$ vs. $\text{Re}[Y'(i\omega)]$ plane), dubbed the Nyquist plot [16].

Following is a poetic interpretation of *Chua's Riddle*.

Ode to *Chua's Riddle*

Tame when open circuited
Tame when short circuited
Explode when dissipated

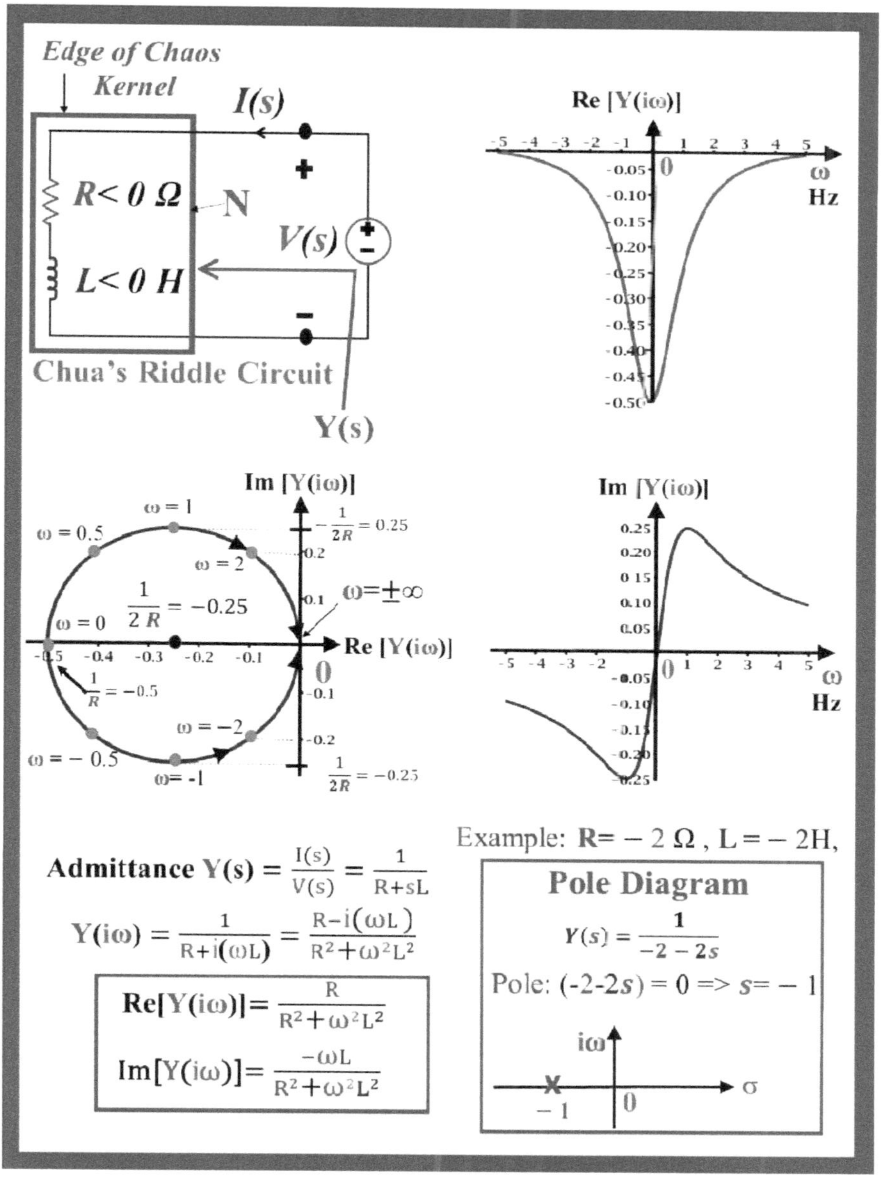

Figure 6.5 The *Nyquist plot* of the admittance $Y(s)$ of the *Edge of Chaos Kernel* lies in the *left-half plane* $\text{Re}[Y(i\omega)] \leq 0$, for $-\infty < \omega < +\infty$. It is a circle of radius ρ = $\mathbf{|2R|^{-1} = 0.25}$, whose center is located at ($\mathbf{Re}[Y(i\omega)]$, $\mathbf{Im}[Y(i\omega)]$) = $\mathbf{((2R)^{-1}, 0)}$.

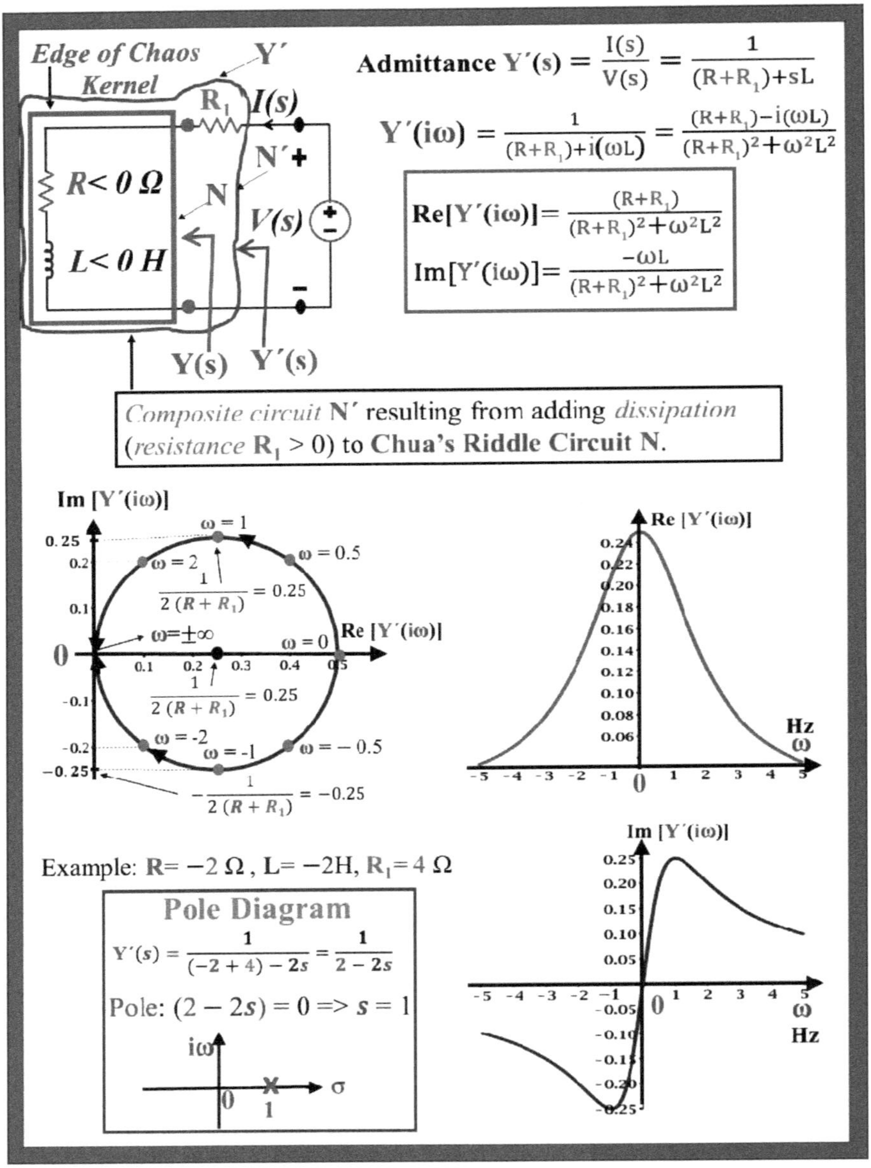

Figure 6.6 When a positive resistance $R_1 > |R|$ is inserted in series with the *Chua's Riddle Circuit*, the *Nyquist plot* of the admittance $Y'(s)$ of the composite 1-port N′ flips over the *right-half* plane $\text{Re}[Y'(i\omega)] \geq 0$, for $-\infty < \omega < +\infty$.

Figure 6.7 The holy name pertaining to a 1-port made of a *negative resistance* in series with a *negative inductance*, *aka* the *Chua's Riddle Circuit*.

6.2.3 Edge of Chaos Kernel

The Chua's Riddle Circuit in Figure 6.4(a) consists of a 1-port **N** connected across a DC voltage source E, namely, a battery. When the voltage source is specified by the Laplace transform $V(s) = \mathcal{L}\{v(t)\}$ of an input voltage time function $v(t)$, then $I(s) = \mathcal{L}\{i(t)\}$ is the Laplace transform of the port current response $i(t)$. The corresponding Chua's Riddle Circuit in the s-domain is shown in Figures 6.5 and 6.6. In the most common situations, where we will be concerned with the neuro-morphic systems, we will switch to the standard notation $V(i\omega)$ and $I(i\omega)$, respectively.

To highlight the main result of this chapter, we have extracted the content of Chua's Riddle 1-port **N** and christened it the *Edge of Chaos Kernel*, as exhibited in Figure 6.7. The *Edge of Chaos Kernel* is the simplest electrical 1-port that exhibits the edge of Chaos phenomenon of being stable yet potentially unstable. But where does the name Edge of Chaos come from? Is it a philosophical hypothesis that is impossible to quantify via mathematical calculations, or is it just a jargon bereft of any scientific basis, as is quite typical in such an over-hyped field dubbed complexity theory? The answer is none of the above. "Edge of Chaos," is one of four mathematically rigorous and easily testable conditions on complexity theory which I developed in 1998[1] [9, 10, 19]. The tests are *quantitative* involving only elementary ordinary differential equations and sophomore-level matrix analysis. The theory that I developed is called the Principle of Local Activity [9, 10, 17–22]. To determine whether a 1-port **N** is locally active, or not, requires executing four independent constructive mathematical tests. Satisfaction of *any one* of the four tests guarantees **N** is locally active. Often several tests would be satisfied even though satisfaction of only one of the four tests would imply **N** is locally active. The 1-port **N** is on the Edge of Chaos domain if, and only if, the condition

Edge of Chaos Criterion

$$Re\,[Y(i\omega)] < 0 \text{ for some } \omega \in [0, \infty] \tag{6.5}$$

is satisfied.

We cannot over-emphasize that the single condition (Eq. (6.5)) for testing whether a 1-port Admittance $Y(s)$ is on the Edge of Chaos domain is both a necessary and a sufficient condition. This implies that the other three sufficient conditions for $Y(s)$ to be locally active should **not** be satisfied. For example, one such sufficient condition for $Y(s)$ to be locally active is for $Y(s)$ to have a pole in the right-half plane (RHP), which would make $Y(s)$ unstable. This, in turn, would disqualify the Admittance $Y(s)$ of the 1-port candidate **N** from earning the Edge of Chaos accolade. Yes, the Edge of Chaos club is very exclusive indeed. In general, the Edge of Chaos parameter domain of a 1-port device D is a very tiny subset of the parameter space of D. This is the reason Edge of Chaos is universally acclaimed to be the *Pearl of Local Activity*, as dramatized in Figure 6.8.

---

[1]For a cursory glimpse at the *Local Activity Principle*, and its pearl, the *Edge of Chaos*, see lecture 10 (*Local Activity and Edge of Chaos*) from the Chua Lectures [18], and part III (Local Activity: The Genesis of Complexity) from CNN: A Paradigm for Complexity [19].

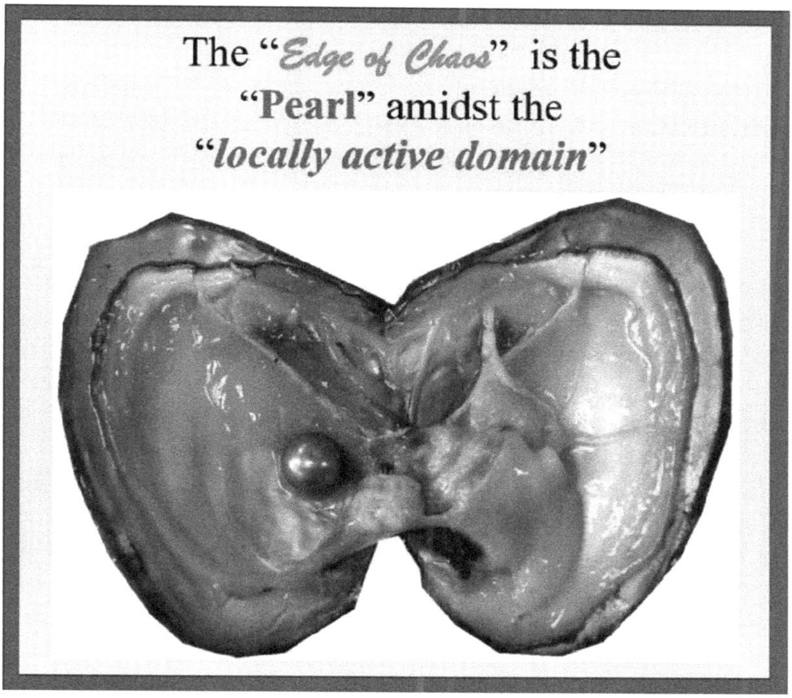

Figure 6.8 If *Edge of Chaos* is the *pearl of Local Activity*, then *Edge of Chaos Kernel* is the *elan vital* of *life*.

6.3 Hodgkin–Huxley's Mistaken Identity: Their Erroneous *Time-Varying Conductance* Are *Time-Invariant Memristors*

The *Hodgkin–Huxley equations* are widely regarded as the most important model across all the physiological literature. It is studied and applied in multiple disciplines, including neuro-physiology, biology, physics, mathematics, and electrical engineering, where the Hodgkin–Huxley equations were originally conceived and derived empirically from an electrical circuit, shown in Figure 6.2, where two of the components are unconventional in the sense that they resemble *electrical conductance* whose values *cannot* be prescribed *a priori*, but had to be calculated by solving several complicated *nonlinear differential equations*, whose solution not only varies with time, but also with *the input voltage* $V(t)$, or *input current* $I(t)$, as exhibited in Table 6.2, and in Figure 6.2, where two arrows are overlaid on top of the conventional symbol of a conductance. There are three other variables n, m, and h, which make the Hodgkin–Huxley equations a system of four highly nonlinear differential equations, whose solutions can only be calculated numerically.

In view of the *strong nonlinearity*, such as the *exponential* function of four dynamical variables V, n, m, and h, respectively, and its high-dimensional state space, very few in-depth theoretical analyses of the original Hodgkin–Huxley equations *enframed* in Table 6.2(a) has been published. Among the few that were published, the *Hodgkin–Huxley equations* were reduced by the publications' authors, via various assumptions, to 3- or 2-dimensional systems of nonlinear ordinary differential equations [23, 24]. Part of the reason for the stagnation in understanding the true scientific, rather than empirical, principle still hidden within the abyss of the *Hodgkin–Huxley equations* is that no one has succeeded in finding the route to this *"all-or-nothing"* wall. By naming, without anchoring upon a well-established nonlinear circuit theoretical foundation, the potassium and sodium ion channels as *time-varying* conductance, Hodgkin and Huxley have led generations of neuro-physiologists into a dark alley to nowhere. The issue here is not a matter of *naming semantics*. We will show in the following in-depth analysis, with rigorous physical and mathematical circuit theory, that the Hodgkin–Huxley equations are blessed with an *Edge of Chaos Kernel* (EOCK), which we have introduced, albeit lightly, in the preceding Section 6.2.3. Instead of reducing the *Hodgkin–Huxley equations* in Table 6.2, usually by wantonly mutilating and disfiguring various components of the equations, we will embark on a precise reverse engineering journey, for the sole purpose of identifying, *the nonlinear circuit-theoretic* nature of the *time-varying* potassium conductance $G_K(t)$, and the *time-varying* sodium conductance $G_{Na}(t)$, in the original 1952 Hodgkin–Huxley circuit model reproduced from their Nobel Prize winning research [5]. For ease of comparison, unless otherwise stated, we will follow the same *voltage polarity* and *current direction reference* adopted by Hodgkin and Huxley in their monumental seminal 1952 paper [5].

Table 6.2 The Hodgkin–Huxley equations. The symbols V and I denote the port voltage and the port current of the *one-port* (yellow box in Figure 6.2) reproduced in plot (b) below. All other symbols are circuit element parameters: C_M denotes the membrane capacitance; $\bar{g}_K$, $\bar{g}_{Na}$, $\bar{g}_l$ are associated with the potassium, sodium, and leakage channels; E_K, E_{Na}, and E_l are battery voltages associated with the potassium, sodium, and leakage channels; $V = V_m$ denotes the voltage across the membrane capacitance.

(a)

$$\frac{dV}{dt} = -\frac{1}{C_M}\left[\bar{g}_K n^4 (V - E_K) + \bar{g}_{Na} m^3 h (V - E_{Na}) + \bar{g}_l (V - E_l) - I\right]$$

$$\frac{dn}{dt} = \left[\frac{0.01\,(V+10)}{\left(\exp\frac{V+10}{10} - 1\right)}(1-n)\right] - \left[0.125\exp\left(\frac{V}{80}\right)n\right]$$

$$\frac{dm}{dt} = \left[\frac{0.1\,(V+25)}{\left(\exp\frac{V+25}{10} - 1\right)}(1-m)\right] - \left[4\exp\left(\frac{V}{18}\right)m\right]$$

$$\frac{dh}{dt} = \left[0.07\exp\left(\frac{V}{20}\right)(1-h)\right] - \left[\frac{1}{\left(\exp\frac{V+30}{10} + 1\right)}h\right]$$

Hodgkin-Huxley Model

(b)

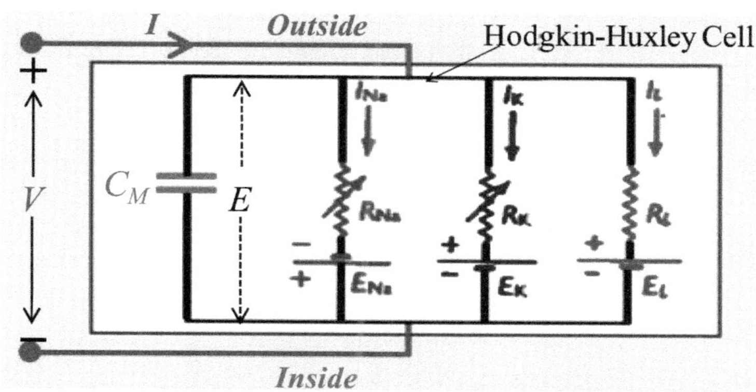

6.3.1 Potassium Time-Varying Conductance Is Not a Resistor but a Time-Invariant Memristor

By inspection of the Hodgkin–Huxley circuit model in Table 6.2(b), it is a simple exercise to extract from the Hodgkin–Huxley equations in Table 6.2(a), the subset of equations shown in Figure 6.9(b), along with Hodgkin–Huxley's symbol of a time-varying conductance $G_K(t)$, in Figure 6.9(a). This subset of equations has never seen the light of day in the circuit theory literature and Hodgkin and Huxley would need to invent a name for future publication of their impressive research. It would have been safer to name it a time-varying element and pass the naming rights to other circuit theory authorities. Unfortunately, by naming it a "time-varying conductance", Hodgkin and Huxley had inadvertently misled four generations of researchers into many long winters of unproductive research on ion-channel dynamics. Just as Bernstein's infamous hypothesis on the all-or-none mechanism was continuously taught at major universities for nearly 40 years after his publication in 1902, the erroneous Hodgkin–Huxley "time-varying conductance" hypothesis is still being taught 60 years after their 1952 paper. It is time to call a spade a spade and neuro-physiology departments should banish time-varying conductance from their courses on electro-physiology and commence teaching memristors forthwith as the scientifically confirmed circuit element responsible for the onset of the all-or-none phenomenon. We will explain and justify the above recommendation in the ensuing sections.

Indeed, if we simply insert "State-Dependent Ohm's Law" above the first equation in Figure 6.9(b), and insert "State-Equation" above the second equation in Figure 6.9(b), we would recognize the Hodgkin–Huxley time-varying conductance defined in Figure 6.9, and which reappears in its dress-up version in Figure 6.10 is simply a memristor as defined formally in Figure 6.11 [25–32]. It makes sense, therefore, to replace the erroneous time-varying conductance symbol in Figure 6.9 with the standard memristor symbol, which we did in Figure 6.10.

For those "doubting Thomases" who found it difficult to accept the notion that the key component of the Hodgkin–Huxley circuit model is not a time-varying conductance, but a relatively recently discovered circuit element dubbed the memristor [25–29], the electrical engineer's Litmus test is shown in Figure 6.12. In sharp contrast, there does not exist a corresponding test for certifying Hodgkin–Huxley's time-varying conductance.

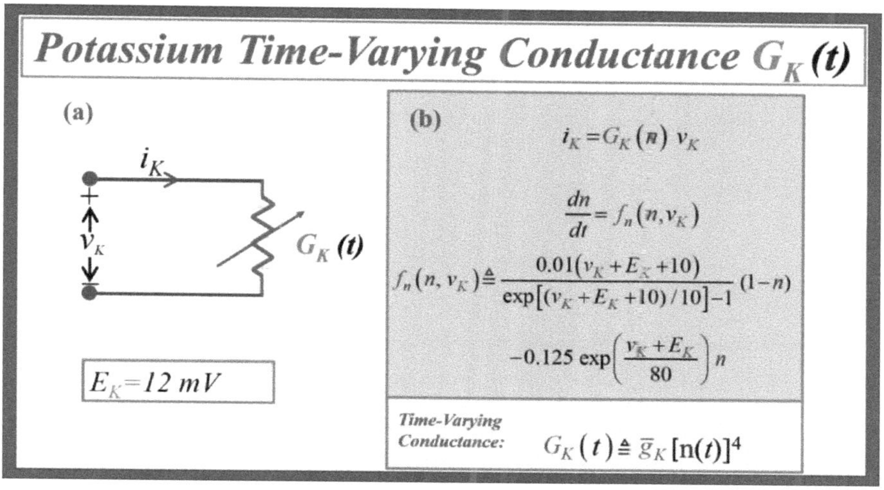

Figure 6.9 (a) Hodgkin–Huxley's symbol of a potassium time-varying conductance $G_K(t)$. (b) Equations defining $G_K(t)$. These equations are a subset of the Hodgkin–Huxley equations in Table 6.2(a).

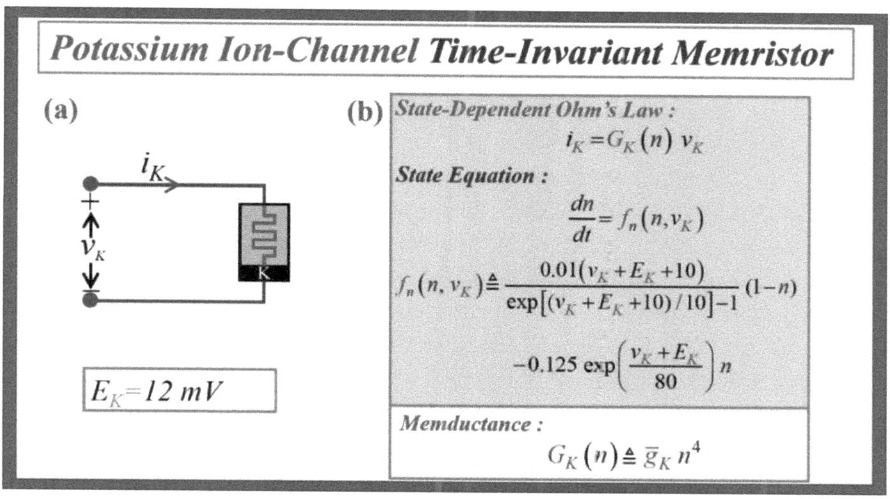

Figure 6.10 The *same equation* defining a *time-varying conductance* by Hodgkin and Huxley in Figure 6.9(b) fits exactly the standard definition of a *time-invariant voltage-controlled memristor*, which is reproduced in Figure 6.11.

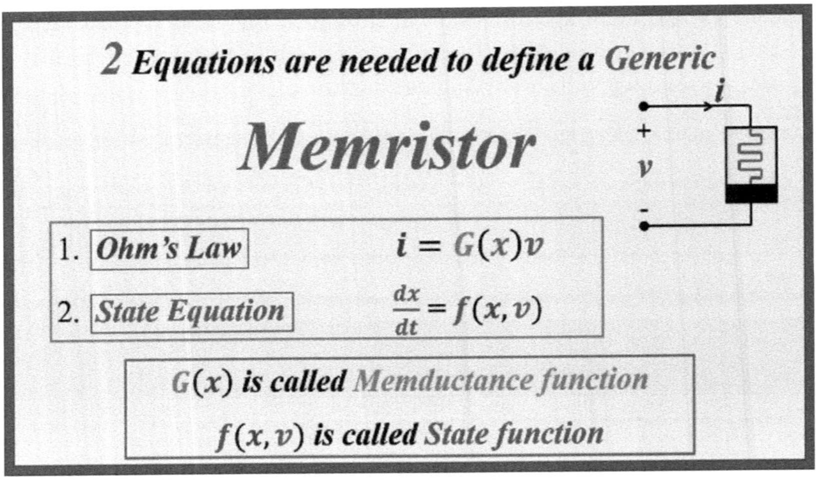

Figure 6.11 Standard definition of a generic memristor [28, 30–32].

Another distinct property of a memristor is that it has equilibrium points under DC excitation, which in electrical engineer's parlance, means that it has a DC voltage-current characteristic, aka DC *V-I* plot, or simply a DC *V-I* curve, if it can be drawn by a contiguous stroke of a pen. For the potassium ion-channel memristor, its DC V_K–I_K curve is calculated and reproduced in Figure 6.13. Moreover, the existence of a DC *V-I* plot, in a one-port electrical device $\mathfrak{D}$ implies that $\mathfrak{D}$ is time-invariant. It follows that Hodgkin and Huxley's choice of the name "time-varying conductance" is unscientific and ambiguous.

One of many fundamental applications of the existence of a DC *V-I* plot is that it allows the application of the Taylor series expansion about certain DC operating point *Q* with special significance, to which the *Edge of Chaos Kernel* (EOCK) stands out as the crowning jewel. In addition to each DC-operating point *Q* on the DC *V-I* curve, one can derive, via sophomore-level mathematics and basic circuit theory [6], a small-signal equivalent circuit of the one-port device $\mathfrak{D}$, about a prescribed DC operating point *Q*. We have carried out this exercise for both the potassium ion-channel memristor and the sodium ion-channel memristor, in reference [7, 8]. The relevant results, namely, a dossier of $\mathfrak{D}$, are summarized for the potassium ion-channel memristor in part (e) of Table 6.3, for a sample set of DC operating points *Q*.

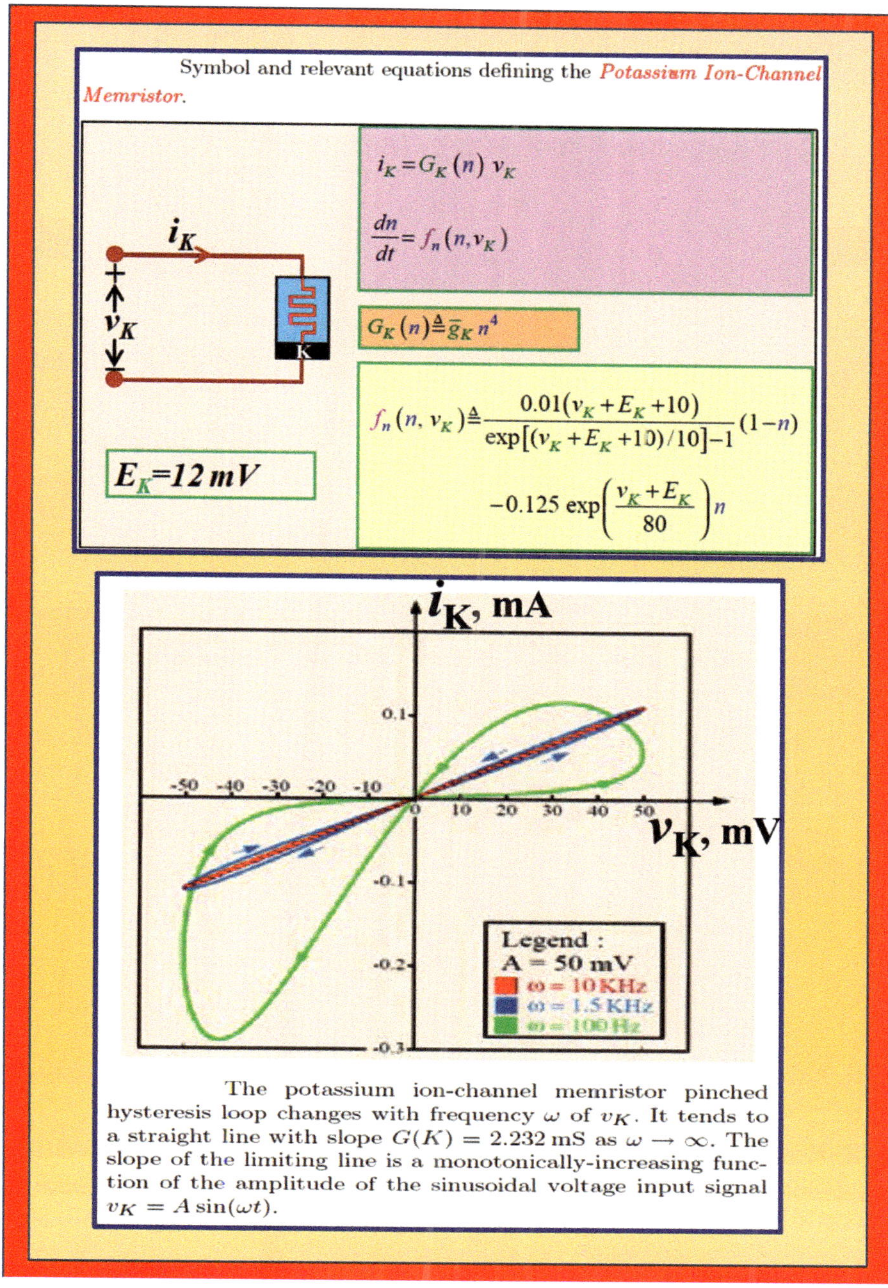

Figure 6.12 Three fingerprints of potassium *ion-channel memristor* [31, 33].

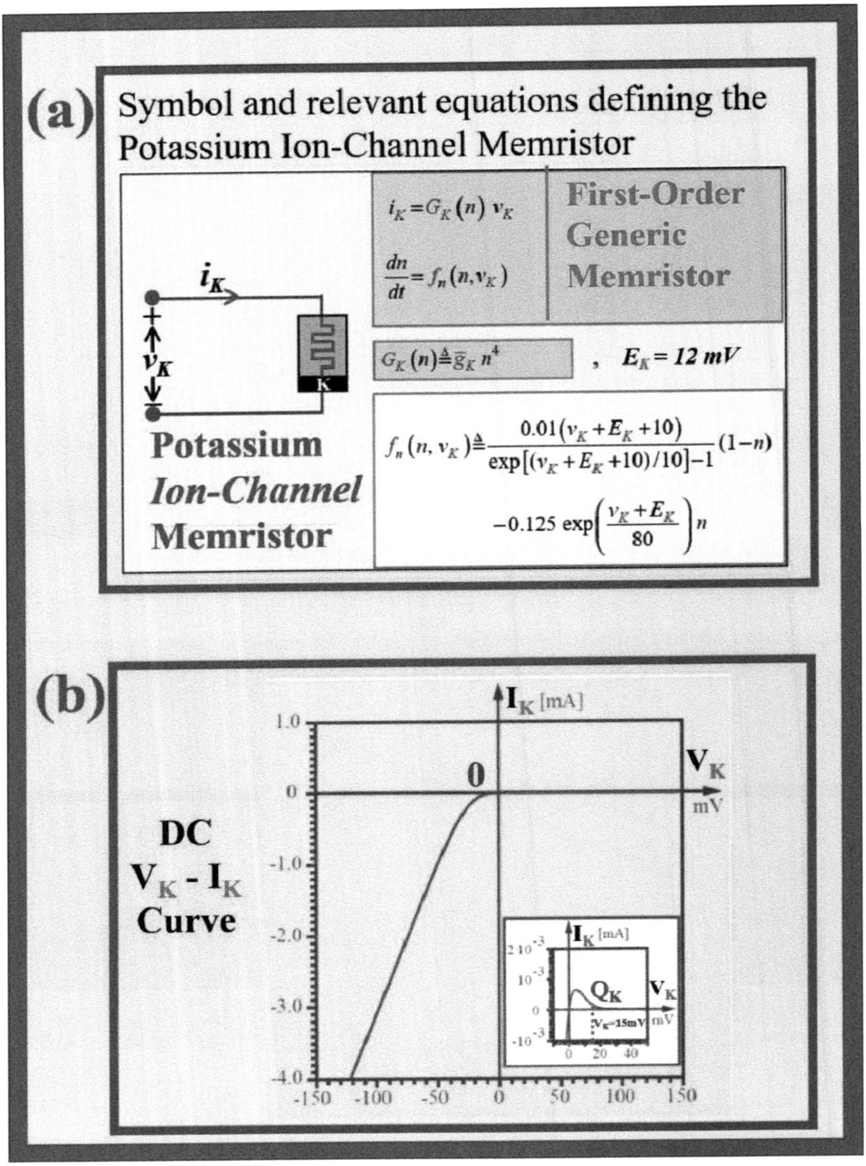

Figure 6.13 DC V_K-I_K curve of the potassium ion-channel memristor.

Table 6.3 Small-signal equivalent circuit of potassium ion-channel memristor about a DC operating point Q ($v_K = V_K$) via external DC bias ($V = V_m$).

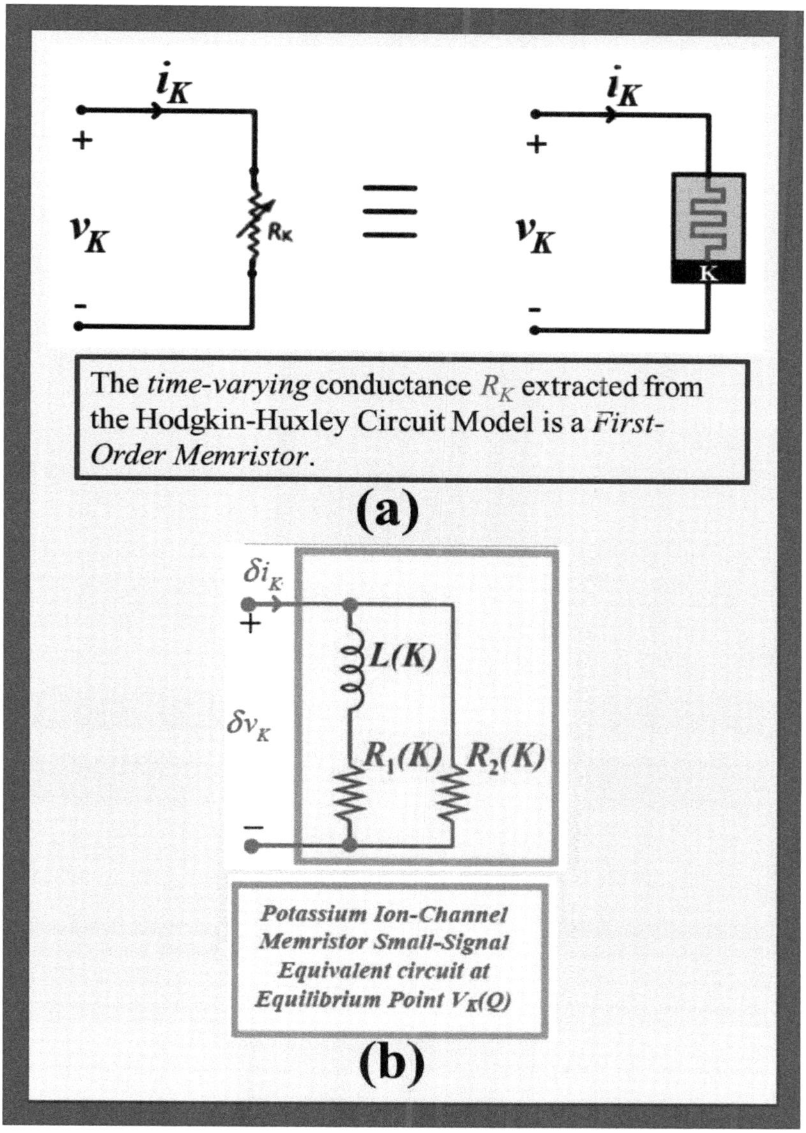

Table 6.3 (*continued*).

(c)

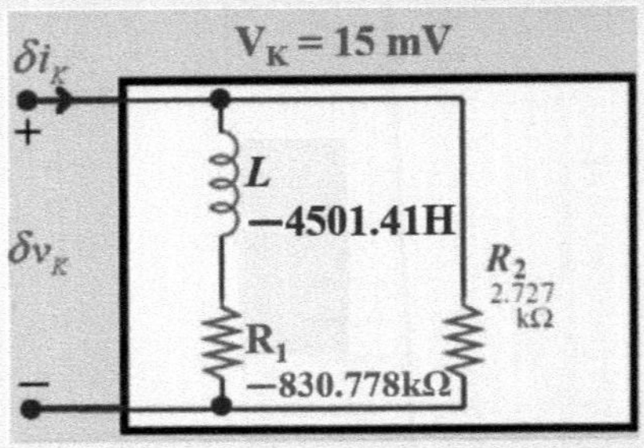

Small-signal equivalent circuit of the potassium ion channel memristor about the equilibrium point $Q_K(V_K, I_K)$. The values of L, R_1 and R_2 are calculated at $V_K = +15$ mV. In this case, the (L-R_1) series 1-port is distinguished as an *Edge of Chaos Kernel*

(d)

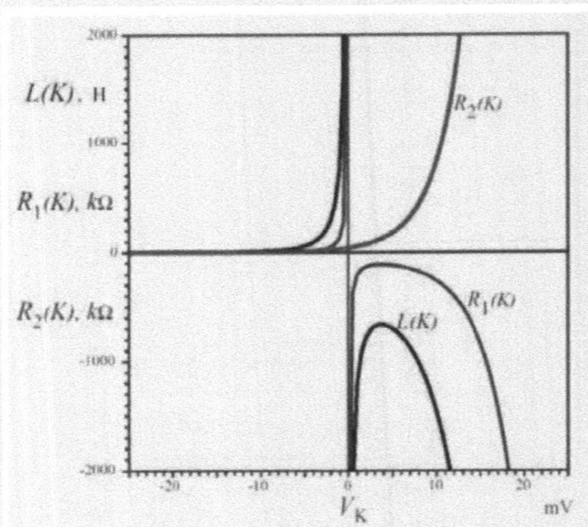

Inductance $L(K)$, resistance $R_1(K)$, and resistance $R_2(K)$ in the small-signal equivalent circuit of the potassium ion-channel memristor at equilibrium voltage V_K.

Table 6.3 (*continued*).

(e)

V_K, mV	L, H	R_1, kΩ	R_2, kΩ	V_m, mV
-25	0.589568	0.130885	0.37588	-13
-20	1.23073	0.250377	0.714556	-8
-15	3.19081	0.604497	1.57402	-3
-10	10.9705	1.97344	4.06444	2
-5	59.4587	10.3713	12.3348	7
0	-9.50947e+015	-1.64175e+015	43.6909	12
5	-696.443	-121.215	178.012	17
10	-1454.43	-259.128	818.572	22
15	-4501.41	-830.778	4163.96	27
20	-17088.0	-3295.8	22997.4	32
25	-74195.6	-15048.3	135707.0	37

Circuit element parameters of potassium memristor admittance $Y_K(s\,;V_K)$.

Readers interested in the three key parameter values (inductance $L(\mathrm{K})$, series resistance $R_1(\mathrm{K})$, and load resistance $R_2(\mathrm{K})$) at DC operating points Q that are not listed in Table 6.3, can either extrapolate them from the three colored curves in part (d) or calculate the exact values from the explicit formulas given in reference [7, 8].

6.3.2 Sodium Time-Varying Conductance Is Not a Resistor but a Time-Invariant Memristor

Following the same reasoning, *mutatis mutandis*, as Section 6.3.1, to the Hodgkin–Huxley's time-varying conductance $G_{\mathrm{Na}}(t)$, we obtain the corresponding Figures 6.14–6.17, the corresponding Table 6.4, the DC V_{Na}-I_{Na} curve in Figure 6.17, as well as the small-signal equivalent circuit at $Q_{\mathrm{Na}}(V_{\mathrm{Na}} = 115\text{ mV})$ in Figure 6.18, where the values of L_1 and R_1 are extracted from columns 1 and 3 of Table 6.4 (row 2 from bottom) at $V_{\mathrm{Na}} = 115$ mV. Observe that just like the potassium ion-channel memristor in Table 6.3, this sodium ion-channel memristor is blessed with an *Edge of Chaos Kernel* (refer to Table 6.4).

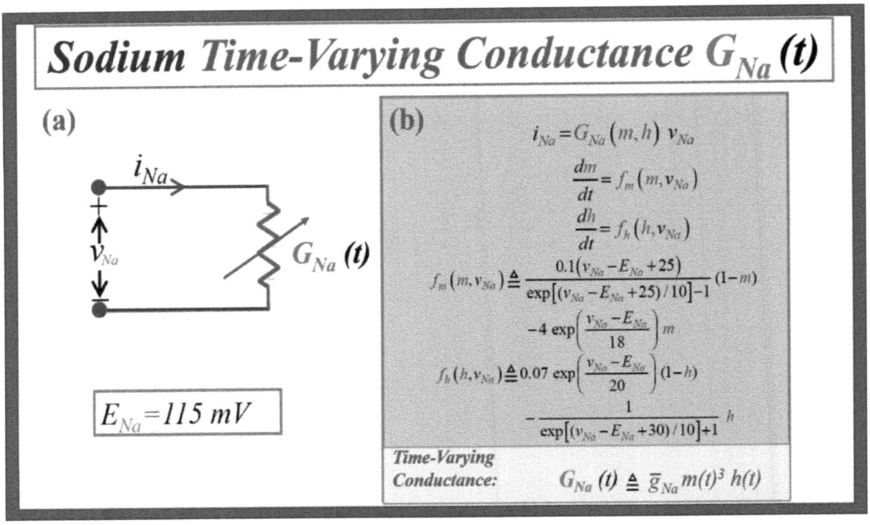

Figure 6.14 (a) Hodgkin–Huxley's symbol of a *sodium time-varying conductance* $G_{Na}(t)$. (b) Equations defining $G_{Na}(t)$. These equations are a subset of the Hodgkin–Huxley equations in Table 6.2(a).

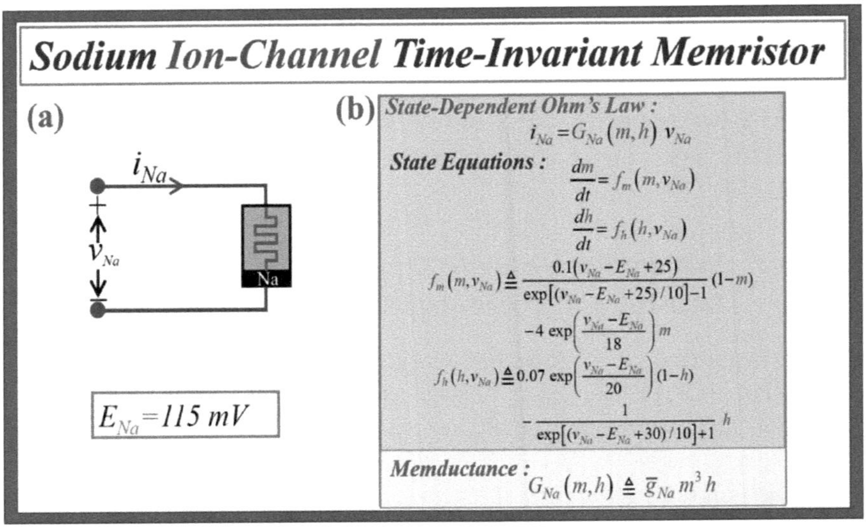

Figure 6.15 The same equation defining a *time-varying conductance* by Hodgkin and Huxley in Figure 6.14(b) fits exactly the standard definition of a *time-invariant voltage-controlled memristor*, as shown in Figure 6.11, where $\boldsymbol{x} = (m, h)$ is a 2-dimensional vector.

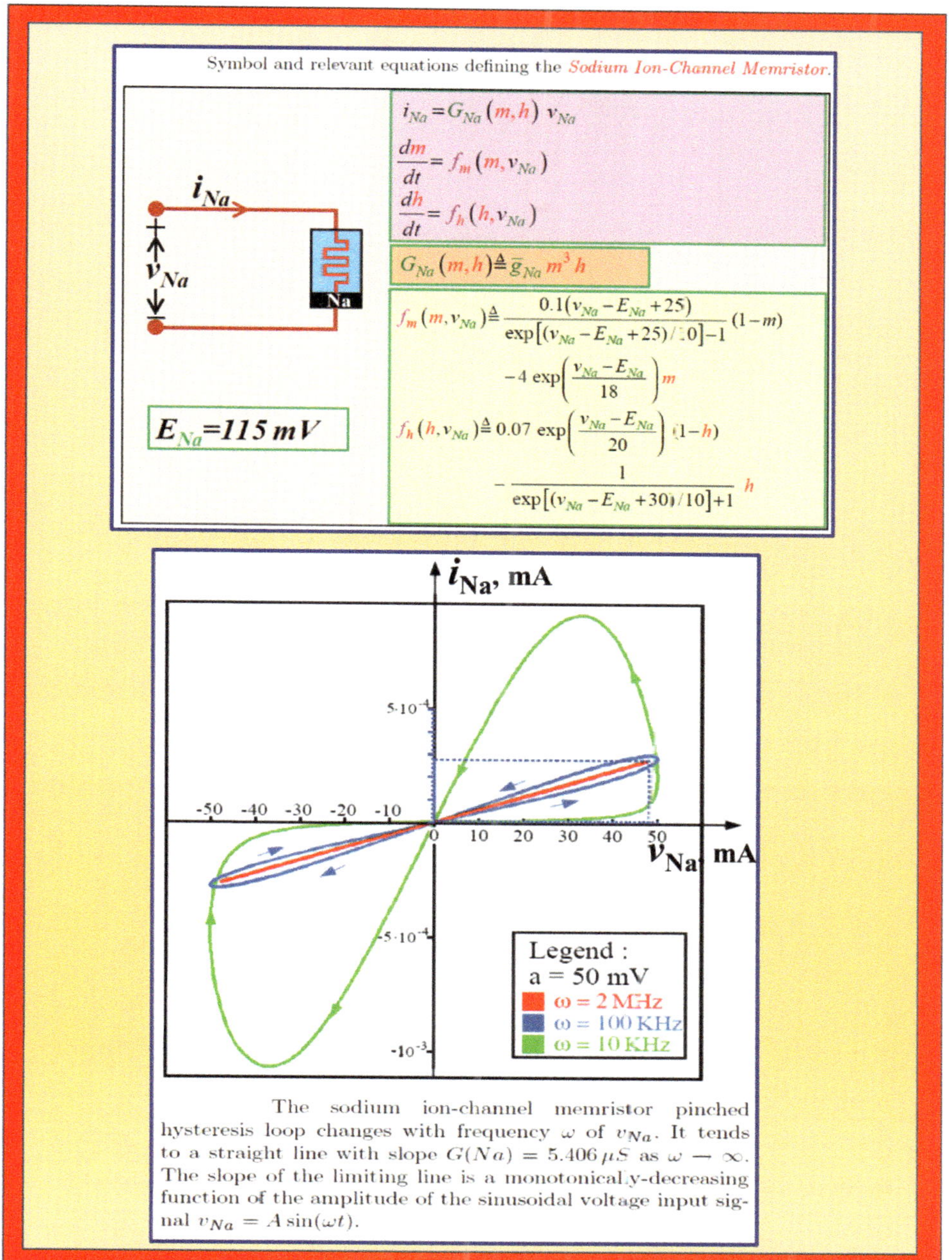

Figure 6.16 Three fingerprints of the sodium ion-channel memristor [31, 33].

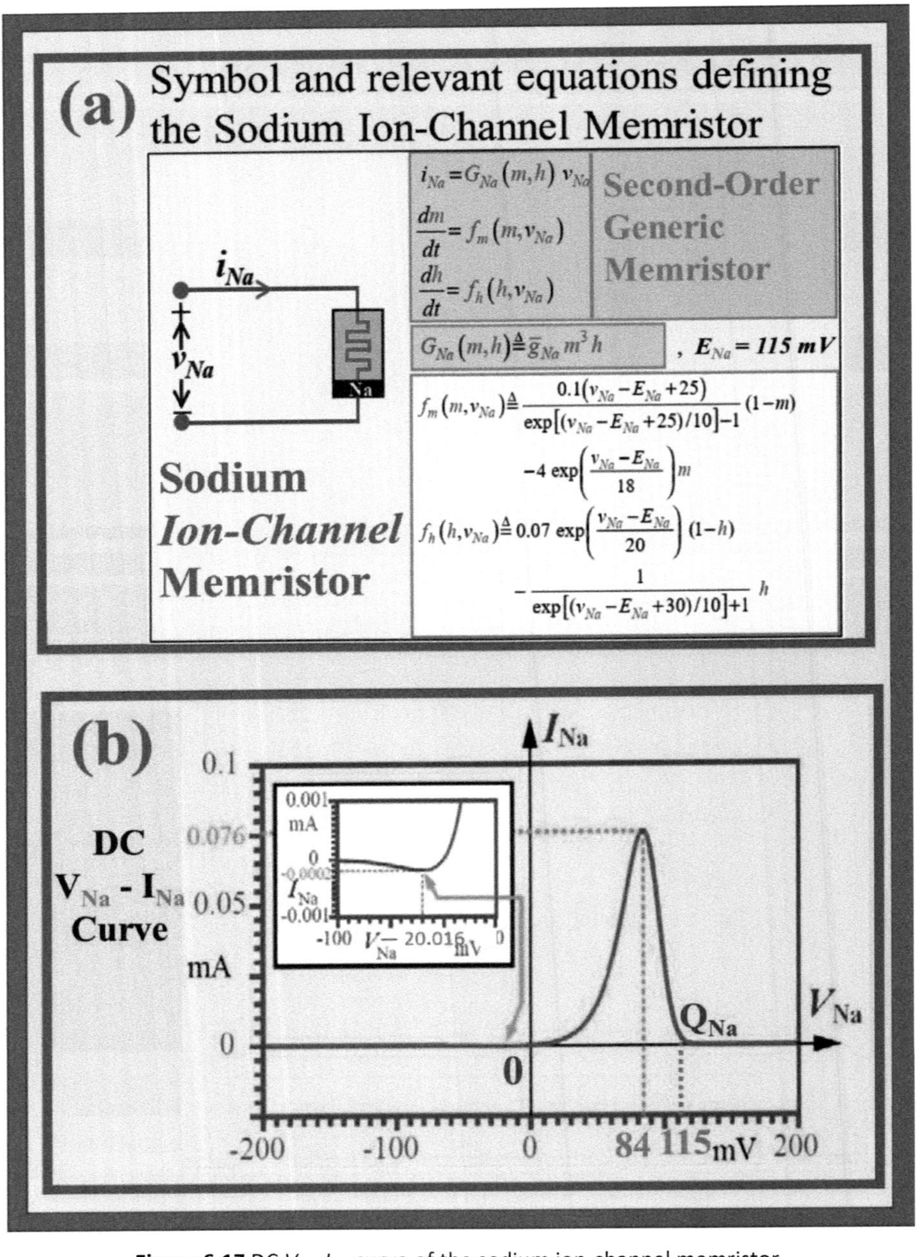

Figure 6.17 DC V_{Na}-I_{Na} curve of the sodium ion-channel memristor.

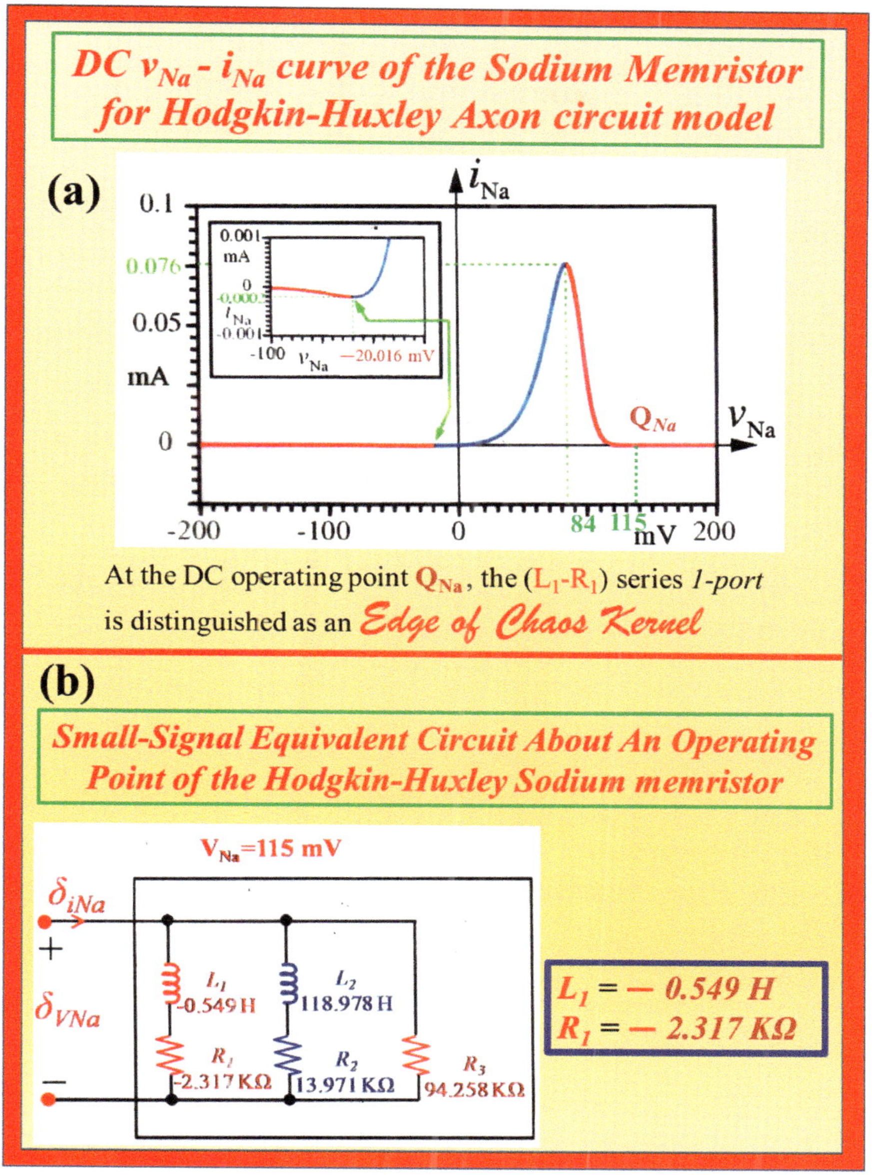

Figure 6.18 The small-signal equivalent circuit of the sodium ion-channel memristor about a DC (V_{Na}-I_{Na}) operating point Q_{Na} contains two inductors L_1 and L_2, and three resistors R_1, R_2, and R_3. Only the left-most L_1-R_1 series one-port qualifies to be an Edge of Chaos Kernel for the operating point Q_{Na}.

6.3.3 Hodgkin–Huxley Circuit Model Is Made of Memristors Blessed with an *Edge of Chaos Kernel*

It follows from Sections 6.3.1 and 6.3.2 that the original Hodgkin–Huxley circuit model based on time-varying potassium conductance $G_{\mathrm{K}}(t)$ and sodium conductance $G_{\mathrm{Na}}(t)$ is flawed from a scientific perspective. In the name of science, we have no choice but to bite the bullet and henceforth teach physiology with the physically sound circuit model shown in Figure 6.19, where the time-varying potassium conductance $G_{\mathrm{K}}(t)$ is replaced by the time-invariant potassium ion-channel memristor, defined in Figure 6.10(b), and where the time-varying sodium conductance $G_{\mathrm{Na}}(t)$ is replaced by the time-invariant sodium ion-channel memristor defined in Figure 6.15(b). The DC *V-I* curve of the memristor-based circuit model can be obtained by a straightforward graphical method presented in [7, 8], and reproduced in Figure 6.20. The small-signal equivalent circuit at any DC operating point Q on the DC *V-I* curve is easily obtained, as shown in Figure 6.21. Observe from the parameter value table for the sodium ion-channel memristor in part (e) of Table 6.4 that there exist DC operating points where L_1 (Na) < 0 and R_1 (Na) < 0. This proves that the Hodgkin–Huxley equations are endowed with a memristor blessed with an *Edge of Chaos Kernel*.

Table 6.4 Small-signal equivalent circuit of sodium ion-channel memristor about a DC operating point Q ($v_{\mathrm{Na}} = V_{\mathrm{Na}}$) via external DC bias ($V = V_m$).

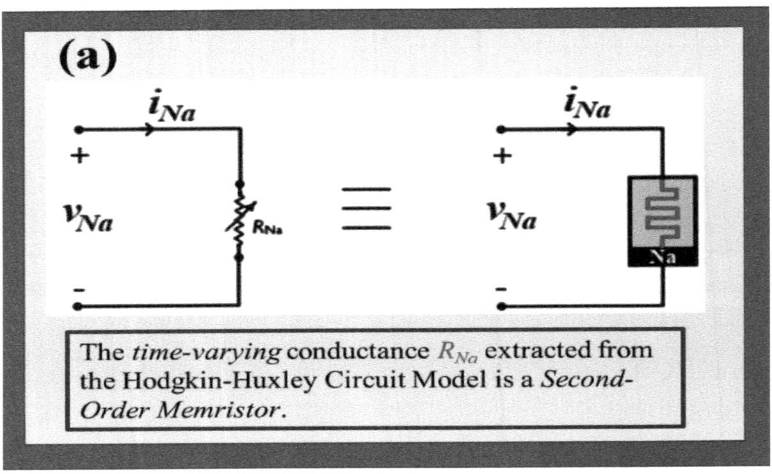

Table 6.4 (*continued*).

(b)

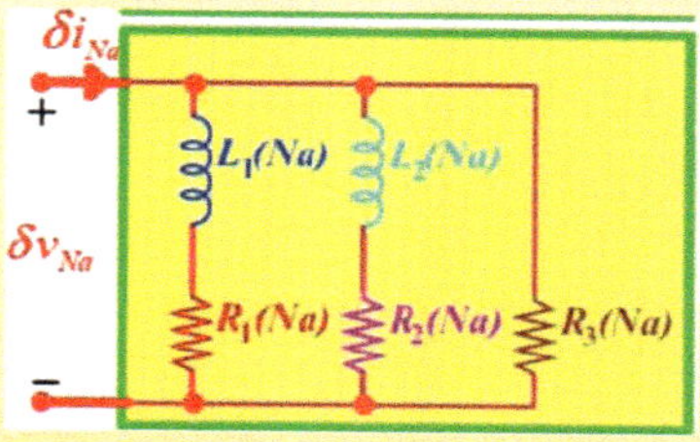

Sodium Ion-Channel Memristor Small-Signal Equivalent circuit at Equilibrium Point $V_{Na}(Q)$

(c)

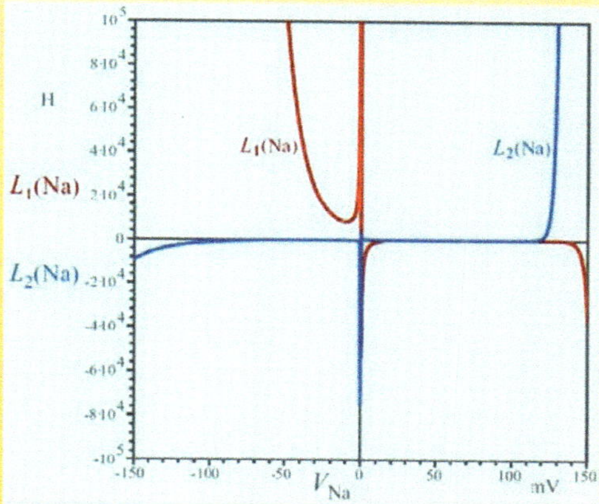

Inductance $L_1(Na)$ and $L_2(Na)$ in the small-signal equivalent circuit of the sodium ion-channel memristor at equilibrium voltage V_{Na}.

(d)

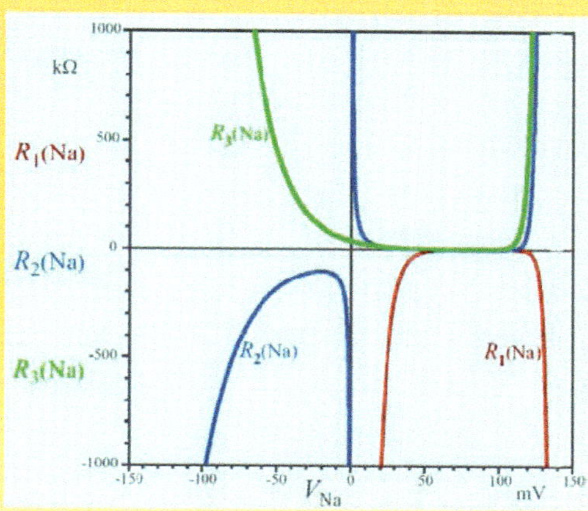

Resistance $R_1(Na)$, resistance $R_2(Na)$, and resistance $R_3(Na)$ in the small-signal equivalent circuit of the sodium ion-channel memristor at equilibrium voltage V_{Na}.

Table 6.4 (*continued*).

(e) Circuit element parameters of sodium memristor admittance $Y_{Na}(s\,;V_{Na})$.

V_{Na}, mV	L_1 H	L_2 H	R_1 kΩ	R_2 kΩ	R_3 kΩ	V_m, mV
0	-1.96302e+018	5.27493e+016	-1.76825e+019	5.27503e+016	37.4882	-115
5	-3262.38	116.84	-27764.8	116.834	29.2198	-110
10	-953.116	45.5388	-7638.65	45.5304	22.782	-105
15	-370.987	23.6716	-2789.68	23.6612	17.7699	-100
20	-162.321	13.8475	-1140.6	13.8351	13.8682	-95
25	-75.6962	8.644	-494.806	8.62935	10.8316	-90
30	-36.7434	5.62315	-222.316	5.60588	8.46878	-85
35	-18.3342	3.76428	-102.113	3.74391	6.63098	-80
40	-9.33623	2.57354	-47.5768	2.5495	5.20238	-75
45	-4.8309	1.78808	-22.3788	1.7597	4.09294	-70
50	-2.53394	1.2584	-10.5988	1.22493	3.23282	-65
55	-1.34636	0.895416	-5.05109	0.85607	2.5681	-60
60	-0.725427	0.644425	-2.42697	0.598424	2.05768	-55
65	-0.397837	0.471397	-1.18248	0.417913	1.67112	-50
70	-0.223838	0.355284	-0.591241	0.293092	1.38757	-45
75	-0.131151	0.283677	-0.310081	0.210072	1.196	-40
80	-0.0821023	0.251228	-0.176868	0.159435	1.09783	-35
85	-0.057089	0.262424	-0.115677	0.135311	1.1153	-30
90	-4.62226e-015	0.35333	-9.16143e-015	0.140483	1.34759	-25
95	-0.0460213	0.602394	-0.0960704	0.177521	1.8947	-20
100	-0.058022	1.43207	-0.134633	0.308599	3.44212	-15
105	-0.0945203	4.65009	-0.257647	0.751734	8.03651	-10
110	-0.200267	20.4233	-0.669472	2.66268	24.2703	-5
115	-0.548625	118.978	-2.31715	13.9711	94.2579	0
120	-1.88833	884.521	-10.2687	105.43	457.571	5

Table 6.4 (*continued*).

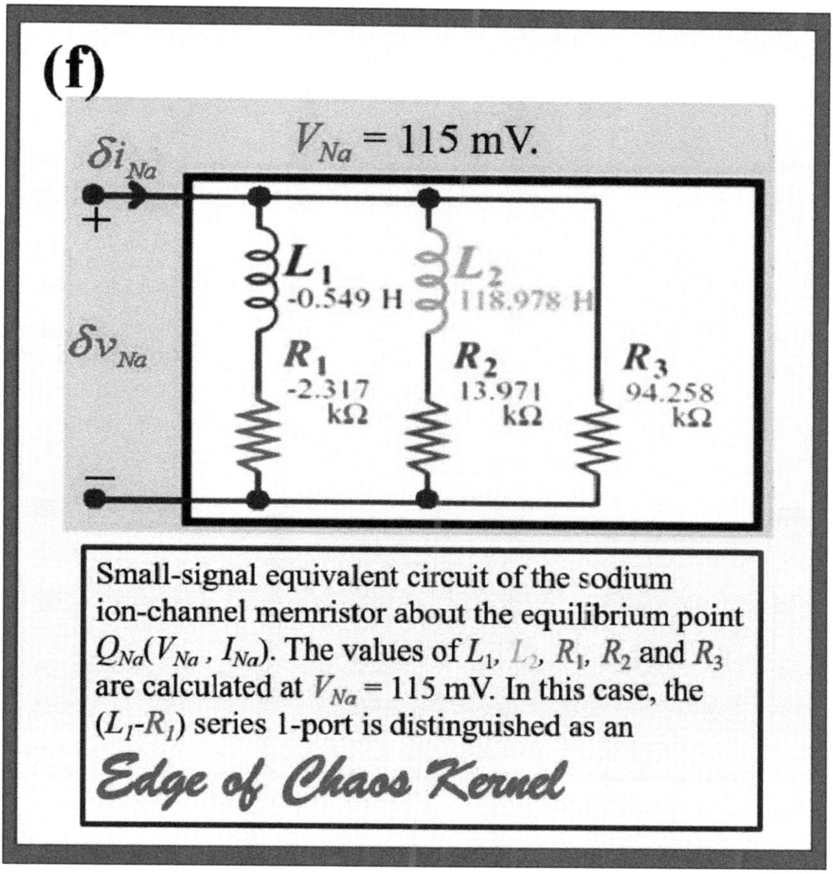

6.3.4 Life Is Impossible Without *Edge of Chaos Kernel*

In this subsection, we present a bird's eye view of four memristors extracted from physiological models of living systems, all of which are blessed with an *Edge of Chaos Kernel*.

6.3.4.1 Memristor model of the heart's Purkinje fibers harbors an *Edge of Chaos Kernel*

Figure 6.22 shows the location of the Purkinje fibers in the human heart where the heartbeat action potentials are emulated by

D. Noble [36, 37], using a modified erroneous time-varying circuit model shown in Figure 6.23(a), along with its scientifically-based time-invariant memristive-based circuit model, shown in Figure 6.23(b). Under a current source input, the cardiac Purkinje fiber one-port (with a yellow background) is blessed with an *Edge of Chaos Kernel*, as shown in Figure 6.24.

6.3.4.2 Two-memristor model of the Balanus Nubilus Harbors two *Edge of Chaos Kernels*

Figure 6.25 shows a colony of barnacles, "a clam-like" shellfish attached to the rocks. It has been erroneously modeled with a time-varying calcium conductance $g_{\mathrm{Ca}}(t)$ and a time-varying potassium conductance $g_{\mathrm{K}}(t)$, respectively [23]. Its scientifically correct circuit elements are a time-invariant calcium ion-channel memristor and a potassium ion-channel memristor, respectively. Both their small-signal equivalent circuits about a DC operating point host *Edge of Chaos Kernels*, as shown in Figure 6.27.

6.3.4.3 One-memristor model of the Balanus Nubilus Harbors only one *Edge of Chaos Kernel*

The reduced single potassium ion-channel memristor model of Balanus Nubilus 1-port in Figure 6.28 has a single *Edge of Chaos Kernel* at all DC operating points where $V_{\mathrm{K}} < 0$.

6.3.4.4 The Chua corsage memristor: Simplest electronic device blessed with an *Edge of Chaos Kernel*

Figure 6.29 shows the schematic circuit diagram of a one-port containing inexpensive electronics components, such as resistors, inductors, capacitors, op-amps, and analog multipliers along with two perspectives of the actual circuit board implementation of this diagram. The resulting one-port is christened the Chua Corsage Memristor (*CCM*) and was first defined and published in [18]. Its DC *V-I* curve consists of a contiguous string that looks like a corsage, as shown in Figure 6.29(b) [38, 43, 44]. The small-signal equivalent circuit calculated at a DC operating point Q is shown in Figure 6.30. This poor man's memristor is blessed with an *Edge of Chaos Kernel* and is capable of emulating most of the neuromorphic phenomena reported in the literature [38, 43, 44].

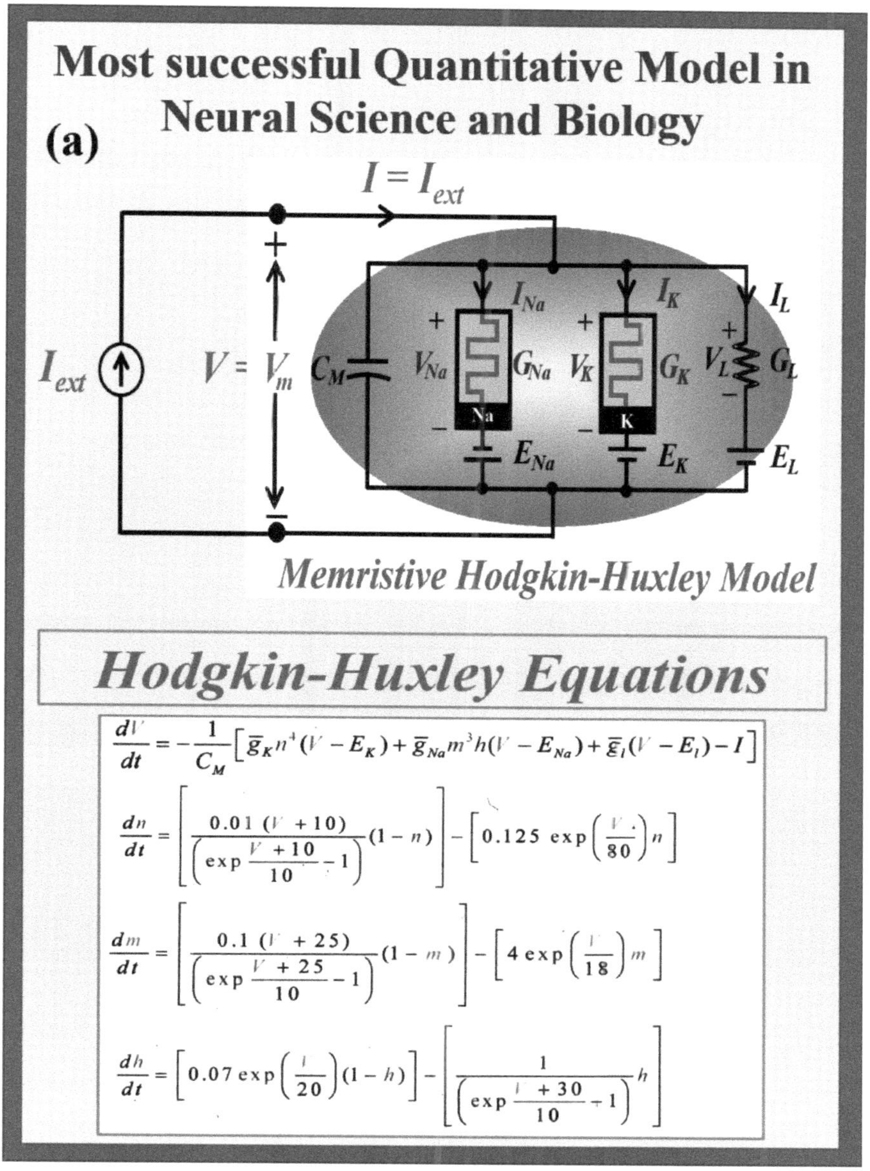

Figure 6.19 (a) Exact memristor-based circuit model for Hodgkin–Huxley equations. (b) Sketch of the *giant axon* of a squid specie *Loligo forbesi* whose extra large diameter makes it fairly easy to stick a capillary (shown on the right) down the axon and record the potential differences across the surface membrane.

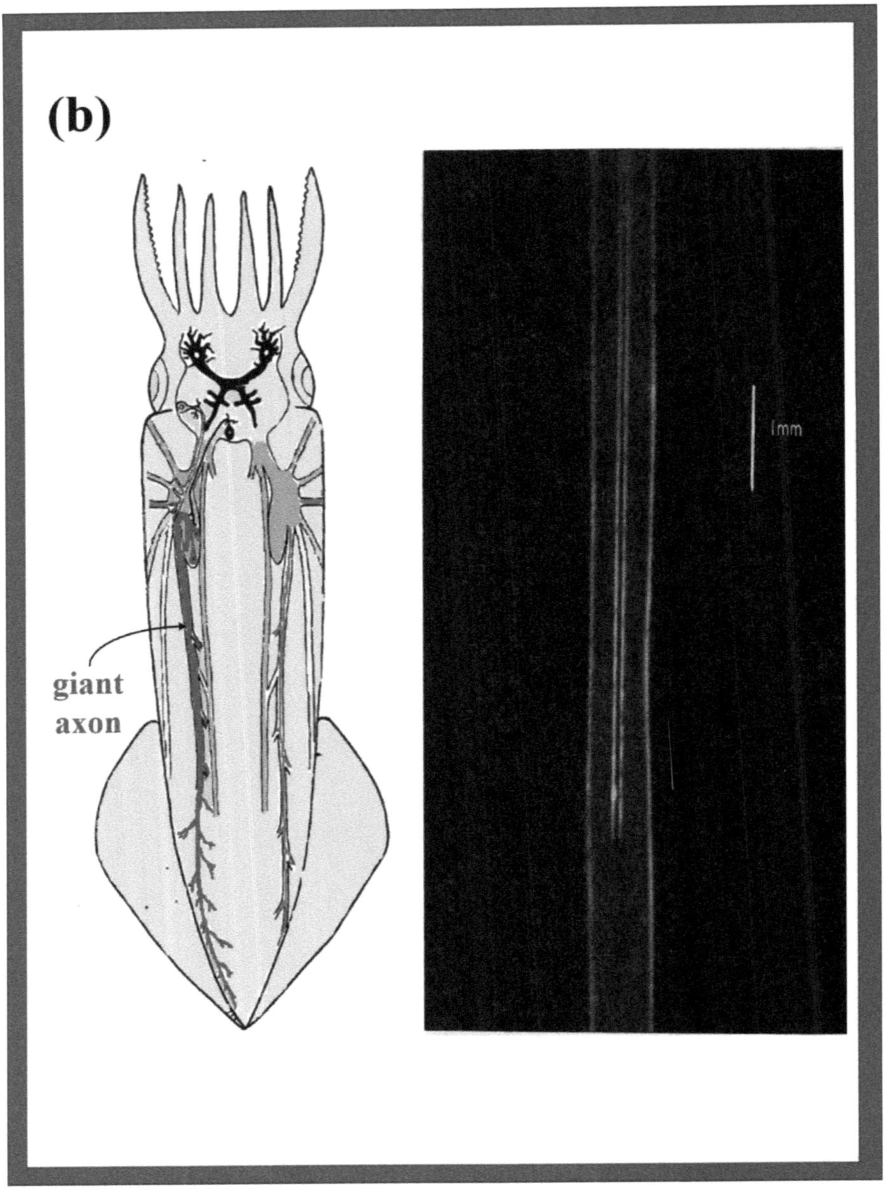

Figure 6.19 (*continued*).

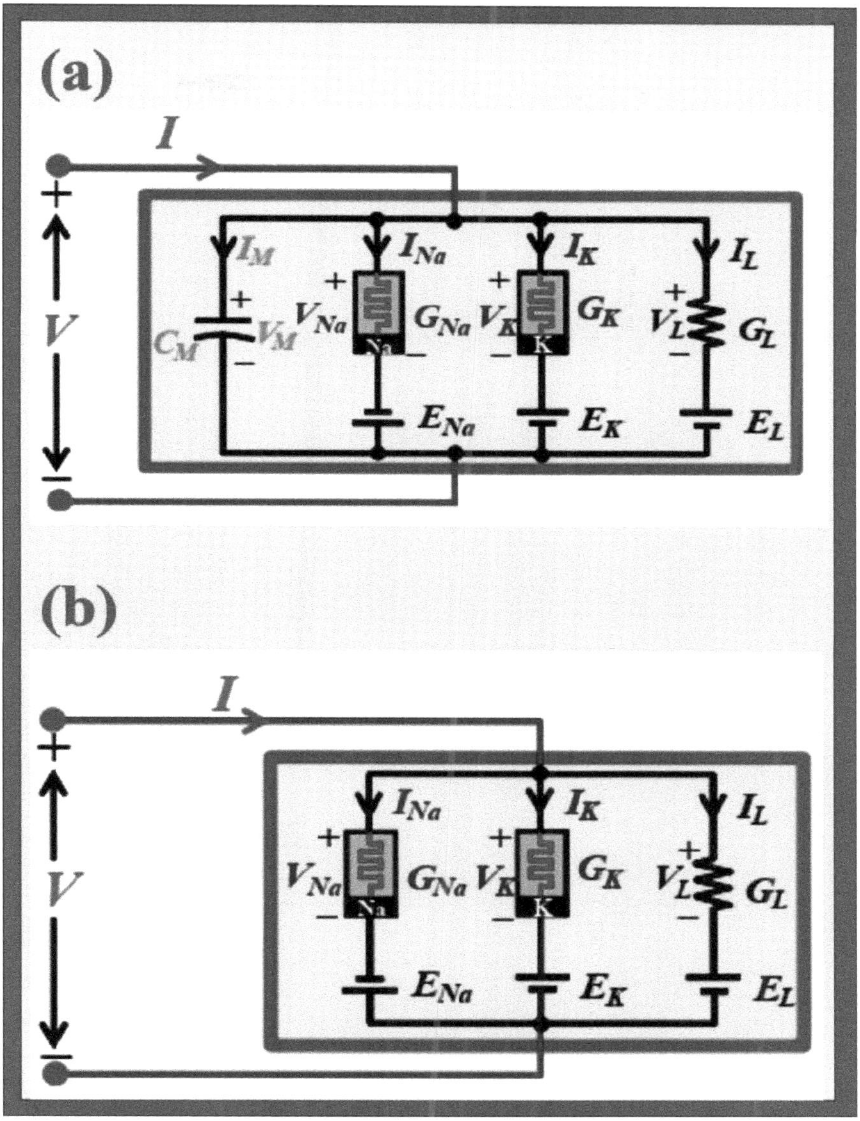

Figure 6.20 Graphical method [6] for deriving the DC *V-I* curve of the Hodgkin–Huxley circuit model. (a) Memristor-based Hodgkin–Huxley model. (b) Delete the capacitor C_M and obtain the DC Hodgkin–Huxley circuit where the *sodium ion-channel memristor* is modeled by its DC V_{Na}-I_{Na} curve in Figure 6.17(b), and the *potassium ion-channel memristor* is modeled by its DC V_K-I_K curve in Figure 6.13(b).

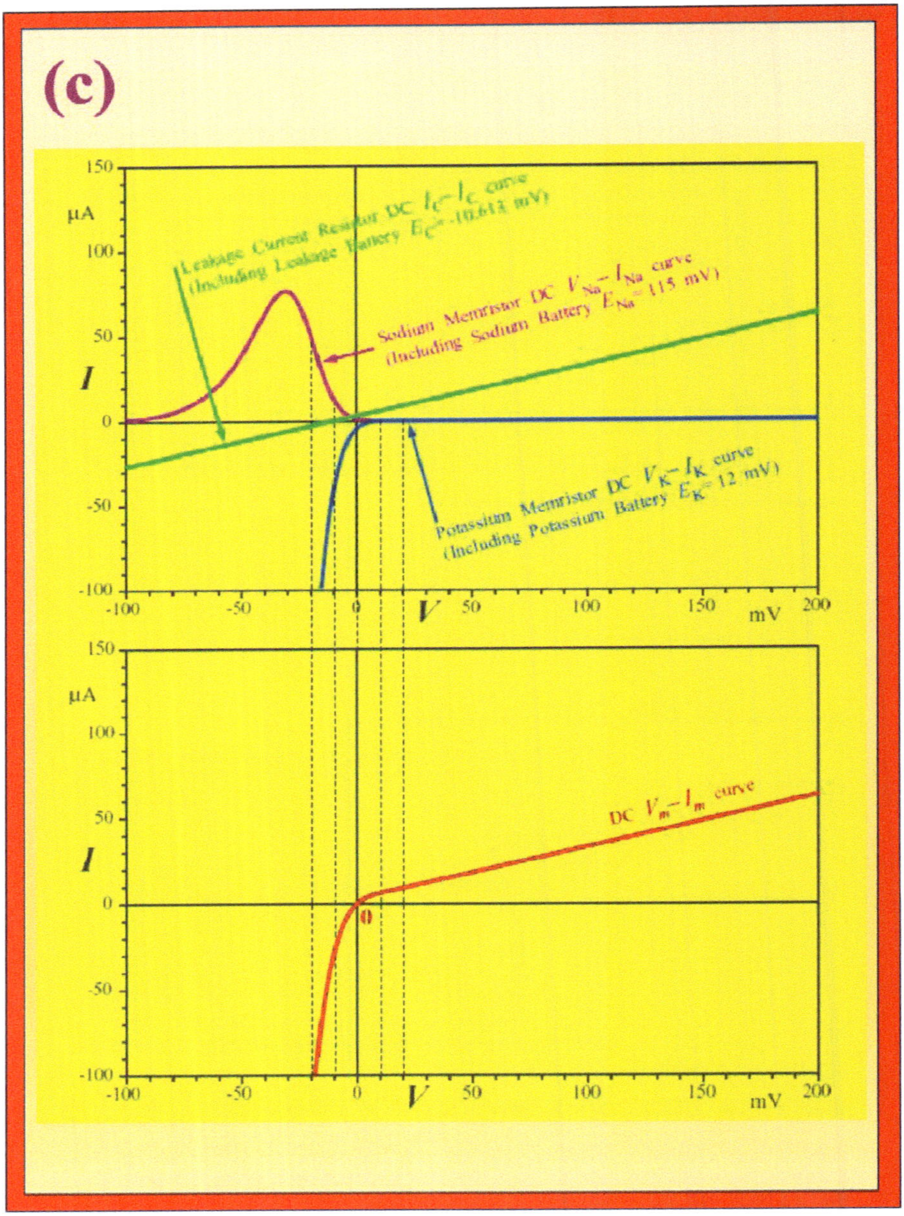

Figure 6.20 (*continued*).

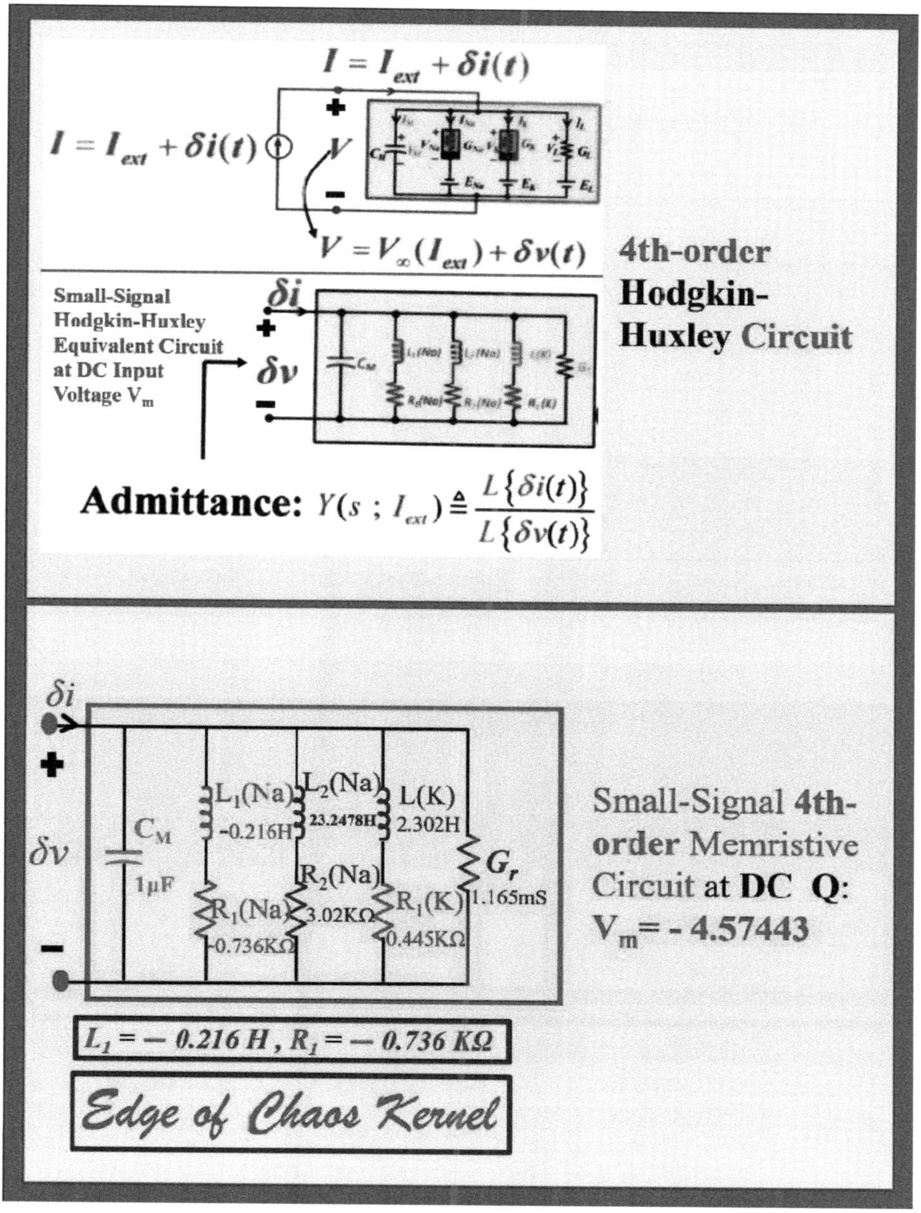

Figure 6.21 Small-signal equivalent circuit about an equilibrium point Q of the Hodgkin–Huxley equations.

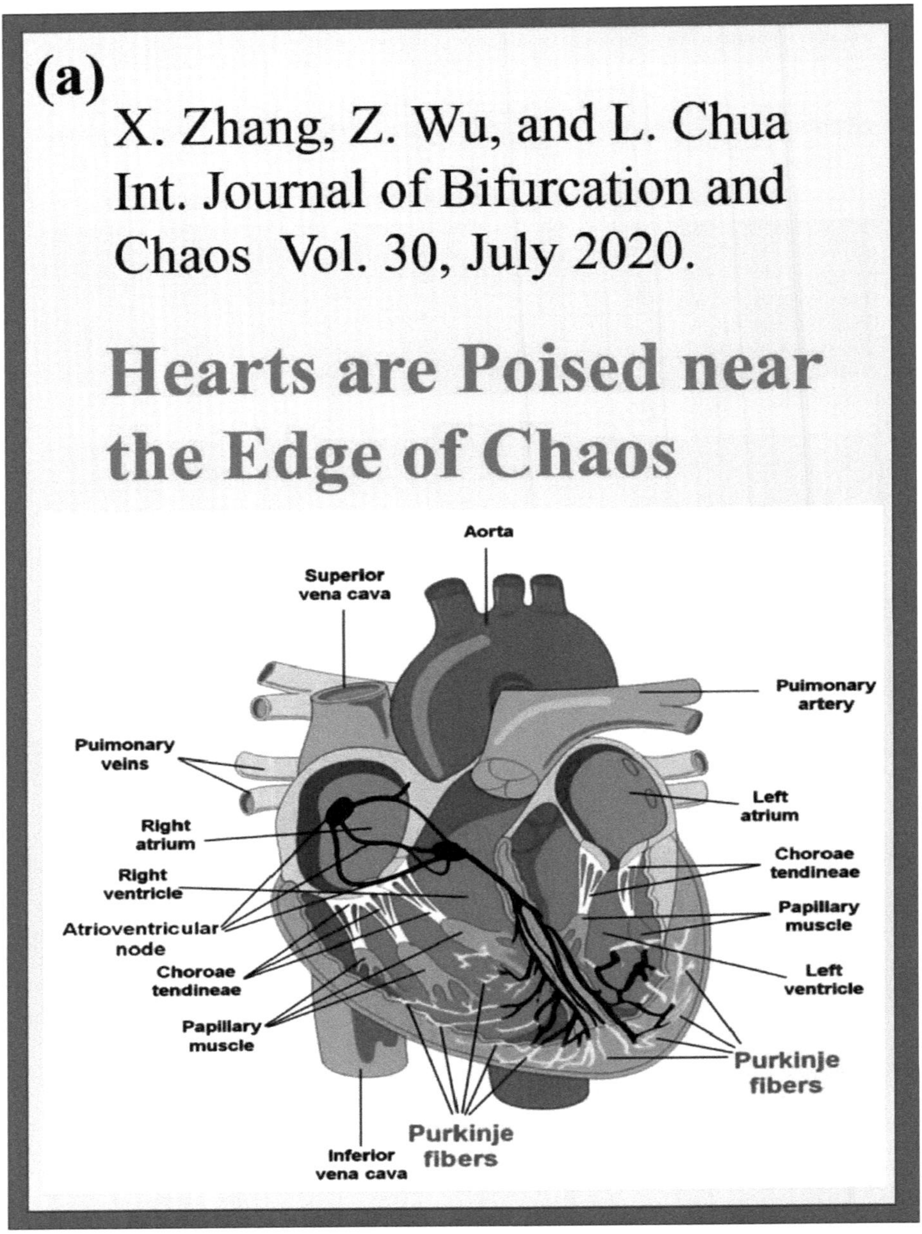

Figure 6.22 The Purkinje fibers of the human heart can be modeled by a set of four nonlinear differential equations, very similar to the Hodgkin–Huxley equations.

(b)

Action Potentials from the Heart Are Generated by Memristors

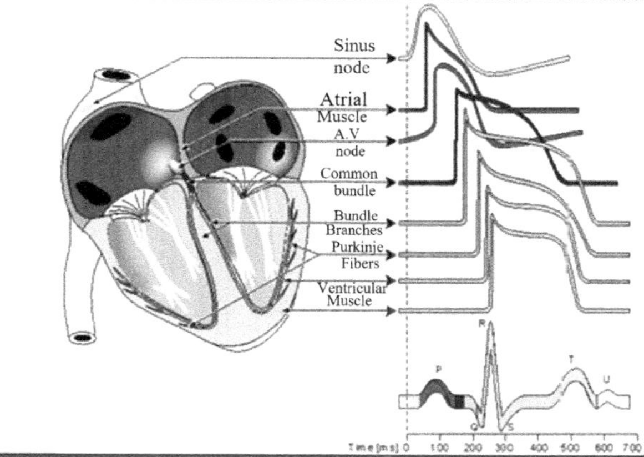

Hodgkin-Huxley "Cardiac" Equations

$$\frac{dV}{dt} = -\frac{1}{C_m}\Big\{[400m^3h + 0.14](V - a) + \big[1.2\exp((-V-90)/50) + 0.015\exp((V+90)/60) + 1.2n^4\big](V + b)\Big\} + D\nabla V$$

$$\frac{dm}{dt} = \alpha_m(V)(1-m) - \beta_m(V)m$$

$$\frac{dh}{dt} = \alpha_h(V)(1-h) - \beta_h(V)h$$

$$\frac{dn}{dt} = \alpha_n(V)(1-n) - \beta_n(V)n$$

D: Noble, D. [1962] A modification of the Hodgkin-Huxley equations applicable to Purkinje fibre action and peacemaker potentials, *J. Physiol.* 160, 317-352.

Figure 6.22 (*continued*).

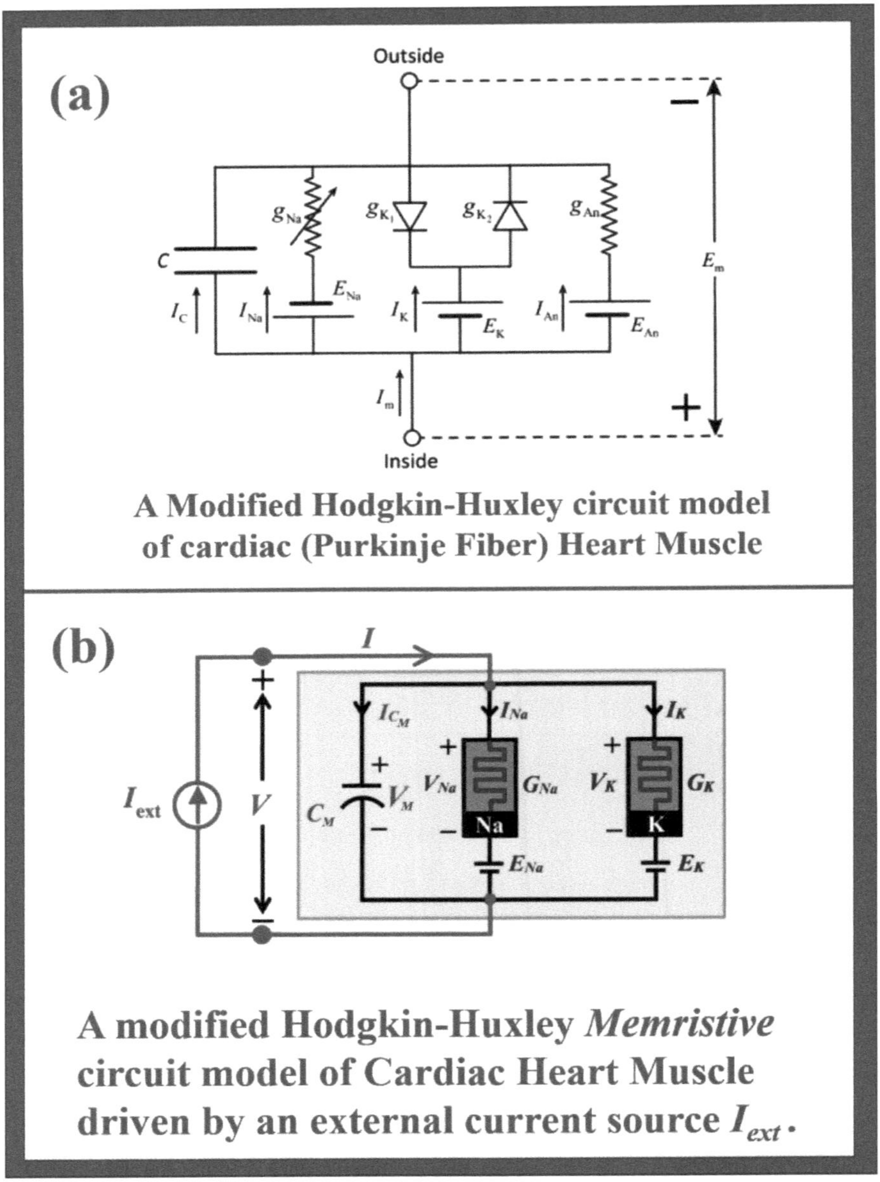

Figure 6.23 A modified Hodgkin–Huxley circuit model of the cardiac Purkinje fiber heart muscle which uses a *time-varying* sodium ion-channel conductance g_{Na}, and its equivalent *time-invariant* model using only *time-invariant* sodium and potassium ion-channel memristors.

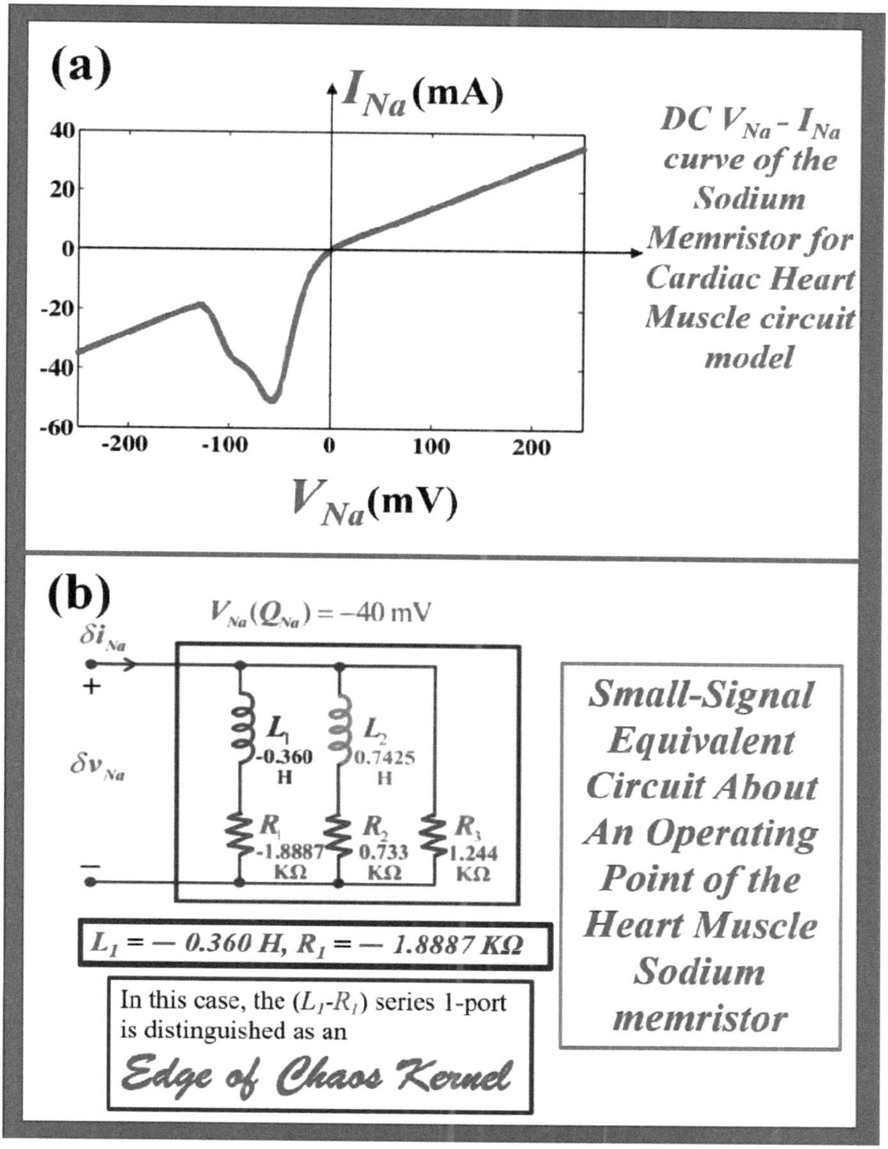

Figure 6.24 DC V_{Na}-I_{Na} curve (this curve differs from Figure 6.18 because a different $G_{Na}(m,h)$ is used [37]) of the sodium ion-channel *memristor* and its *small-signal equivalent circuit* calculated at the DC operating point $V_{Na}(Q_{Na}) = -40$ mV. At this equilibrium point, the sodium ion-channel memristor harbors an *Edge of Chaos Kernel*.

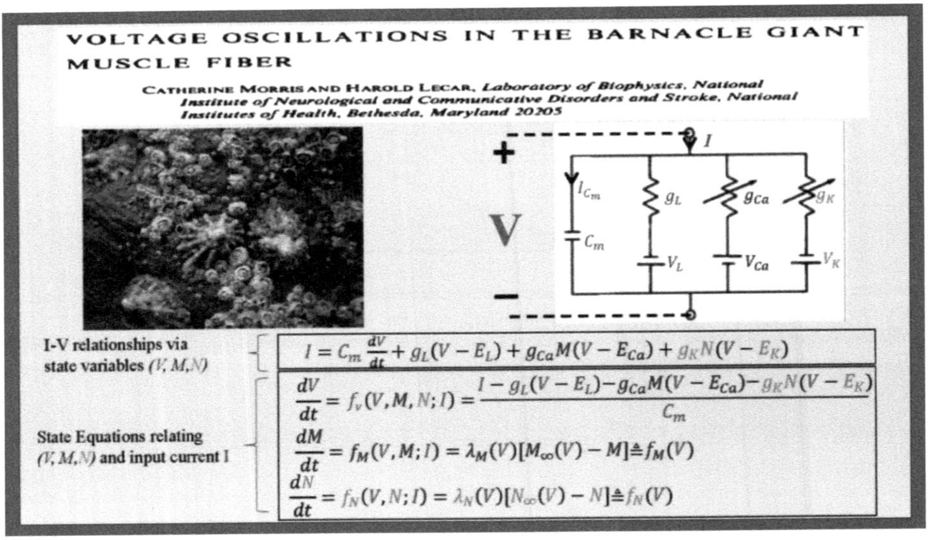

Figure 6.25 A group of barnacles (upper-left) in which each giant muscle fiber was modeled by Morris and Lecar [23] via a Hodgkin–Huxley-like circuit composed mainly of a time-varying calcium conductance $g_{Ca}(t)$ and a time-varying potassium conductance $g_K(t)$.

Figure 6.26 Close-up view of a *Balanus Nubilus* barnacle.

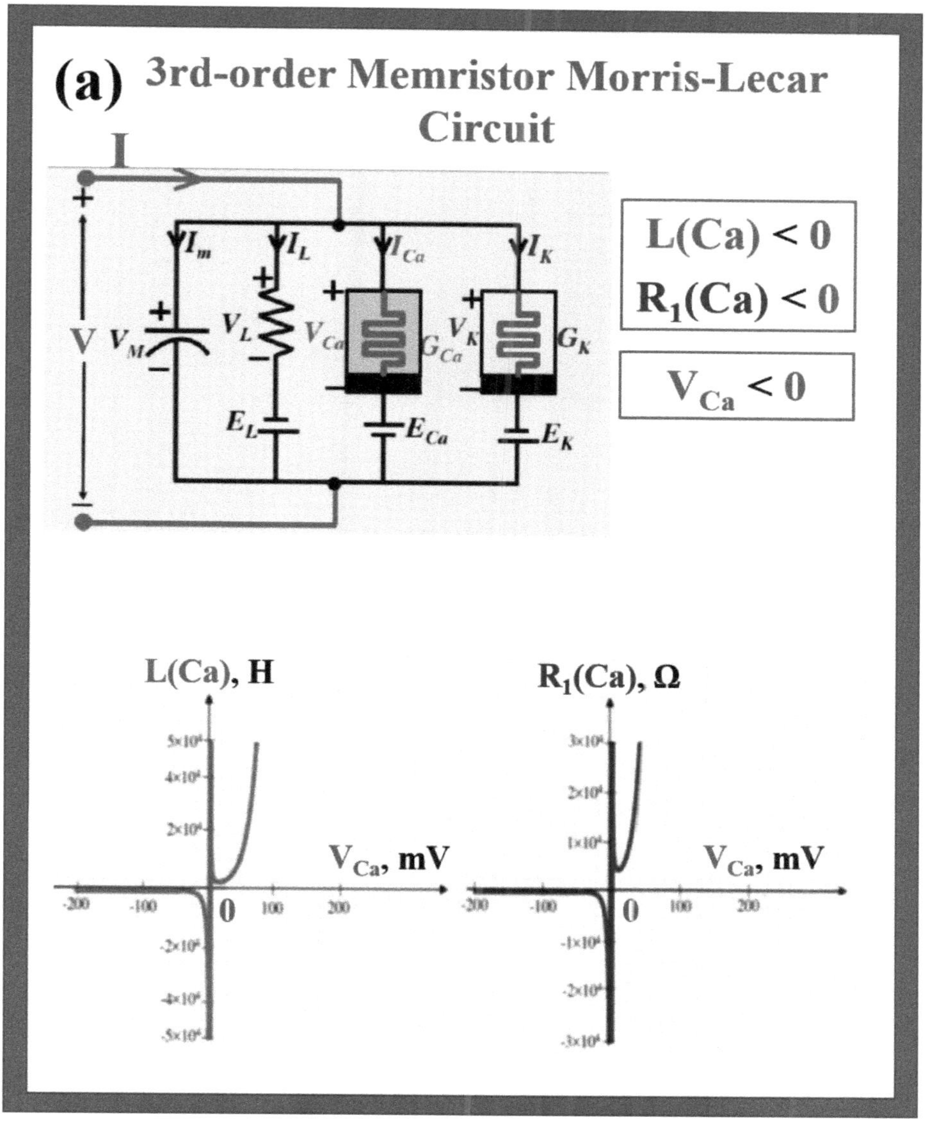

Figure 6.27 Modelling the giant muscle fiber of the Balanus Nubilus in Figure 6.26 with a calcium ion-channel memristor and a potassium ion-channel memristor.

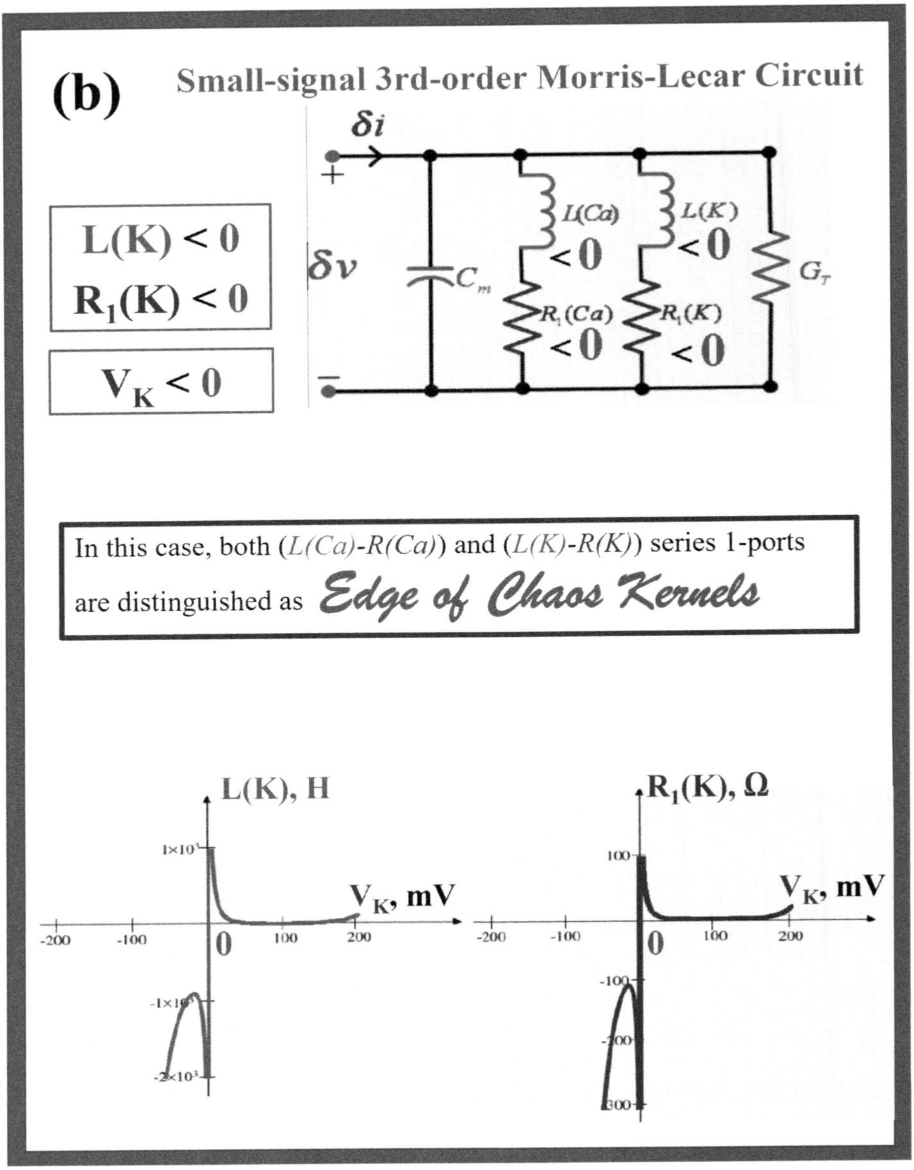

Figure 6.27 (*continued*).

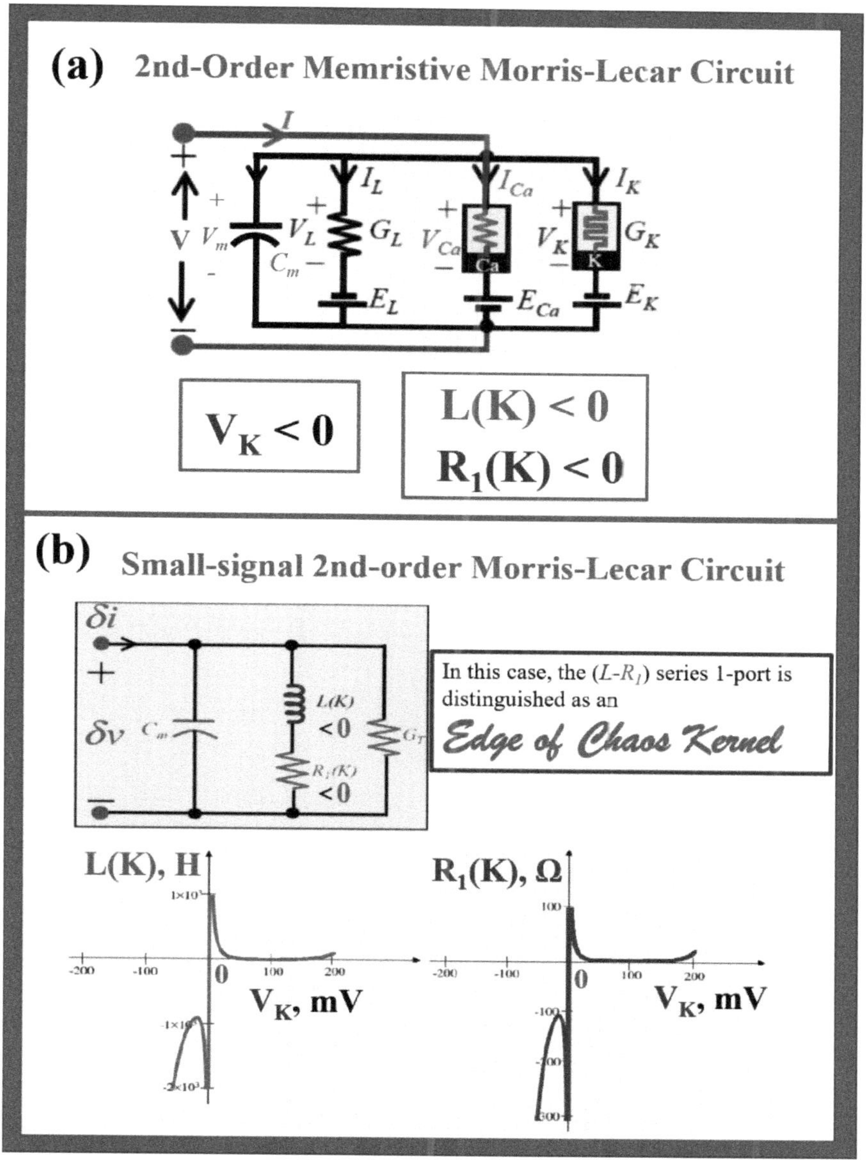

Figure 6.28 Due to the significant difference in the time scales of the two memristors in the 3rd-order Morris–Lecar circuit in Figure 6.27, the latter is occasionally simplified by replacing the calcium memristor in Figure 6.27 with a *nonlinear resistor*. The simplified circuit is called a 2nd-order Morris–Lecar circuit.

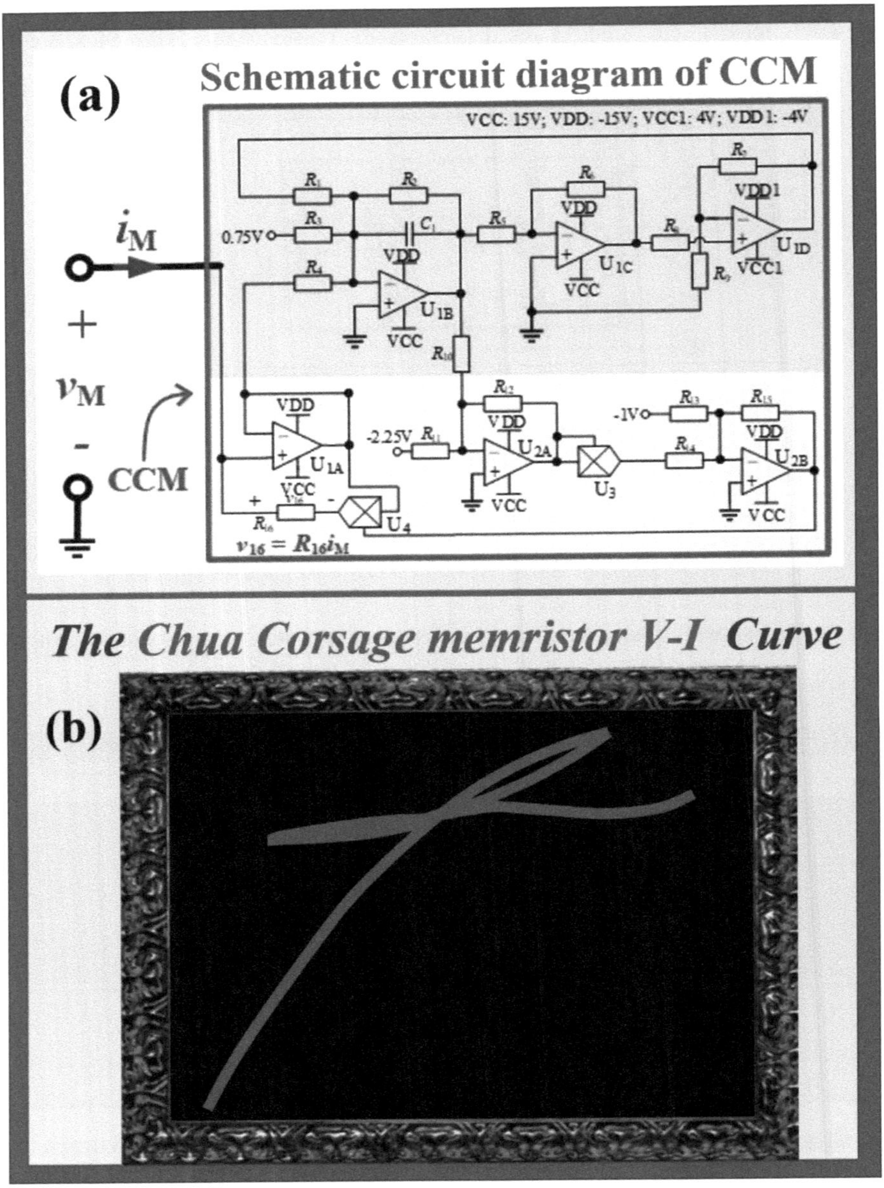

Figure 6.29 Schematic circuit diagram for building the *Chua Corsage Memristor*. All components are standard off-the-shelf electronic elements, such as resistors, capacitors, op-amps, analog multipliers, and batteries.

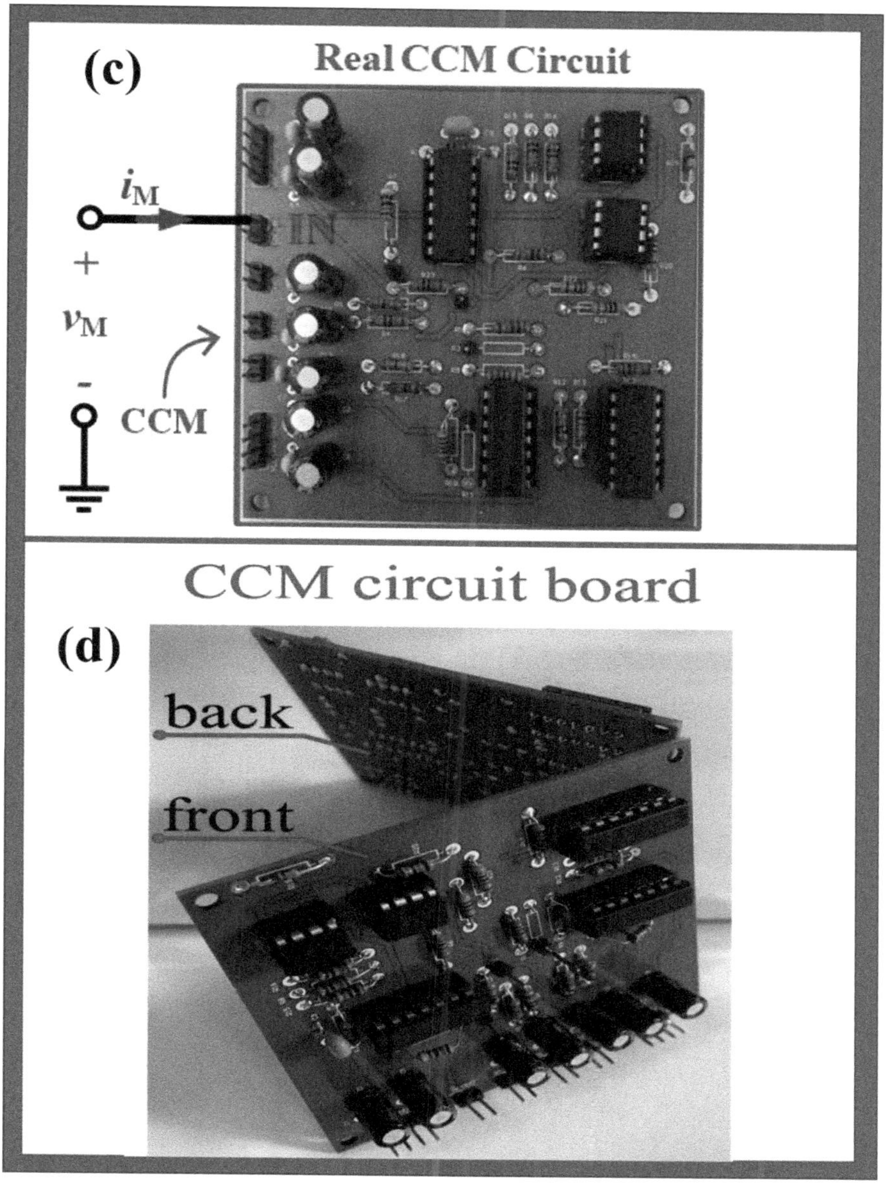

Figure 6.29 (*continued*).

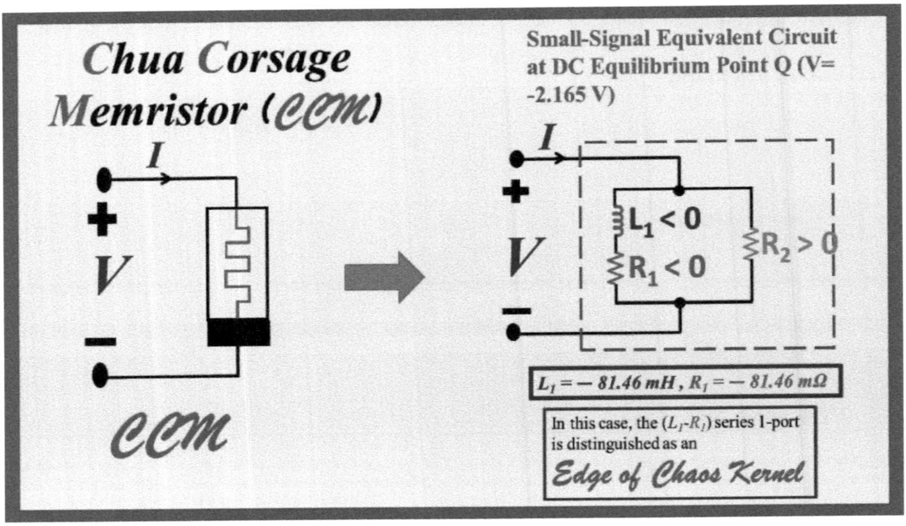

Figure 6.30 The *Chua Corsage Memristor* can be synthesized using *off-the-shelf* circuit components [38, 43, 44].

6.4 Galvani's 242-Year-Old "All-Or-None" Enigma Finally Resolved

How does the neuron generate action potentials (spikes)? An easy answer is that it is due primarily to the interplay between sodium ions and potassium ions via some positive feedback interactions. But this is mainly a handwaving *qualitative* answer. A mathematician could offer to predict the waveform by solving the Hodgkin–Huxley equations numerically on his PC and claim to have provided a *quantitative* answer. But neither would satisfy Luigi Galvani because neither could explain the actual global nonlinear dynamical bifurcations that made this miracle possible. By exploiting the global nonlinear dynamical complexity of the *Edge of Chaos Kernel*, we have succeeded in resolving Galvani's 242-year-old unresolved enigma. Since a *rigorous resolution* via high-dimensional *Poincare* maps [34, 35] of this enigma would require a substantial increase in the length of this chapter, we have opted to publish it as a separate paper. In this final section, we will offer the first-ever close-up view of the global nonlinear dynamics responsible for this fascinating *all-or-none* phenomenon, via a sequence of phase-portraits of the 4-dimensional Hodgkin–Huxley equations in Table 6.2 plotted in the 3-dimensional (V_m, n, m)-subspace. They are shown in Figures 6.31–6.39. The fascinating dynamics revealed by these nine-*phase portrait perspectives*[2] are the solution to Galvani's unsolved enigma, which will be explained in detail in Section 6.4.1.

6.4.1 Apocalypse Via Vector Field Metamorphosis

The global bifurcation phenomenon dubbed *Saddle-Node* that has been identified as the dynamical mechanism for the demise of the action potential [39, 40] is somewhat misleading because it implies that at some critical excitation current $I_{ext} = I^*$, the *stable limit cycle collides* with the *unstable limit cycle* in the 4-dimensional (V_m, n, m, h)-state-space, as in an anti-missile defense system, as illustrated in Figure 6.39 (a) and (b) of [8], obtained by numerical solution of the original Hodgkin–Huxley equations given in Table 6.2. A more careful simulation over the range $0 < I_{ext} < 8.5$ μA[3] reveals a distinctly different mechanism involving no collision at all! To illustrate how a *stable periodic limit cycle* in the 4-dimensional (V_m, n, m, h)-state-space can suddenly disappear, Figures 6.31–6.39 exhibit a *gallery* of nine-*phase portrait perspectives* in the 3-dimensional (V_m, n, m)-subspace for nine values of I_{ext}.

Each phase portrait perspective in this exhibition is presented in two parts. The **upper part** contains, in addition to the *stable* and *unstable limit-cycle* (printed in red and green color, respectively), a typical trajectory *converging* to the *red* stable periodic orbit, and a typical trajectory *diverging* from the *green* unstable periodic orbit, while converging to the *stable-limit cycle*, or to the stable equilibrium point Q, depending on whether the *initial state* lies in the *basin of attraction* of the *red stable limit cycle*, or of the *blue stable equilibrium point Q*.

[2]By definition, the phase portrait is the vector field in the 4-dimensional (V_m, n, m, h) state space R^4, where a trajectory from any initial point P_0 is tangent to the vector of the vector field located at P_0. In this paper, we plot only the first three coordinates (V_m, n, m) of the trajectories and refer to the picture that emerges as the *phase portrait perspective* [41].

[3]I_{ext} in these *phase portrait perspectives* is defined with the *standard current reference convention*, namely, the opposite of that adopted by Hodgkin and Huxley in their 1952 paper, for ease of comparison.

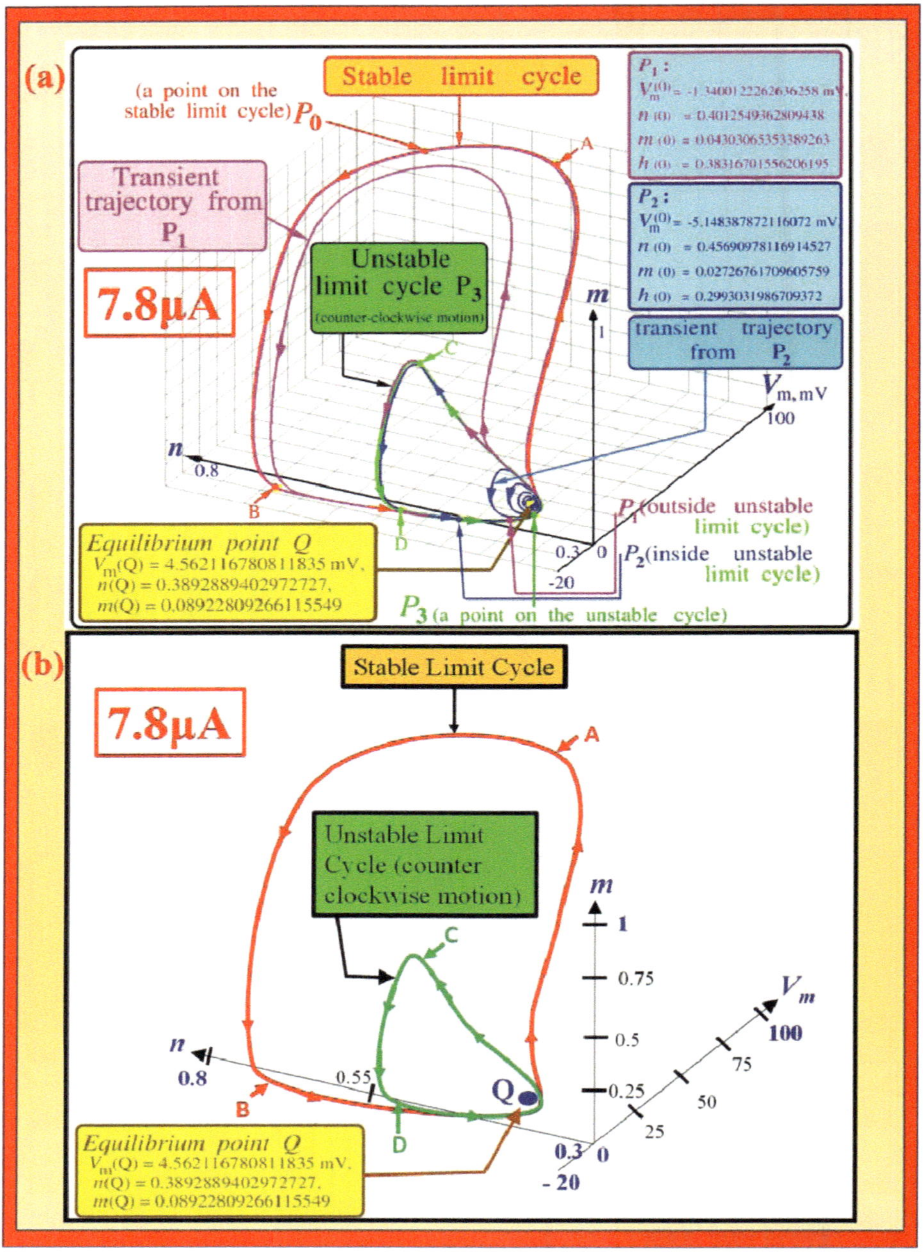

Figure 6.31 (V_m, n, m) *phase portrait perspective* at I_{ext} = 7.8 μA.

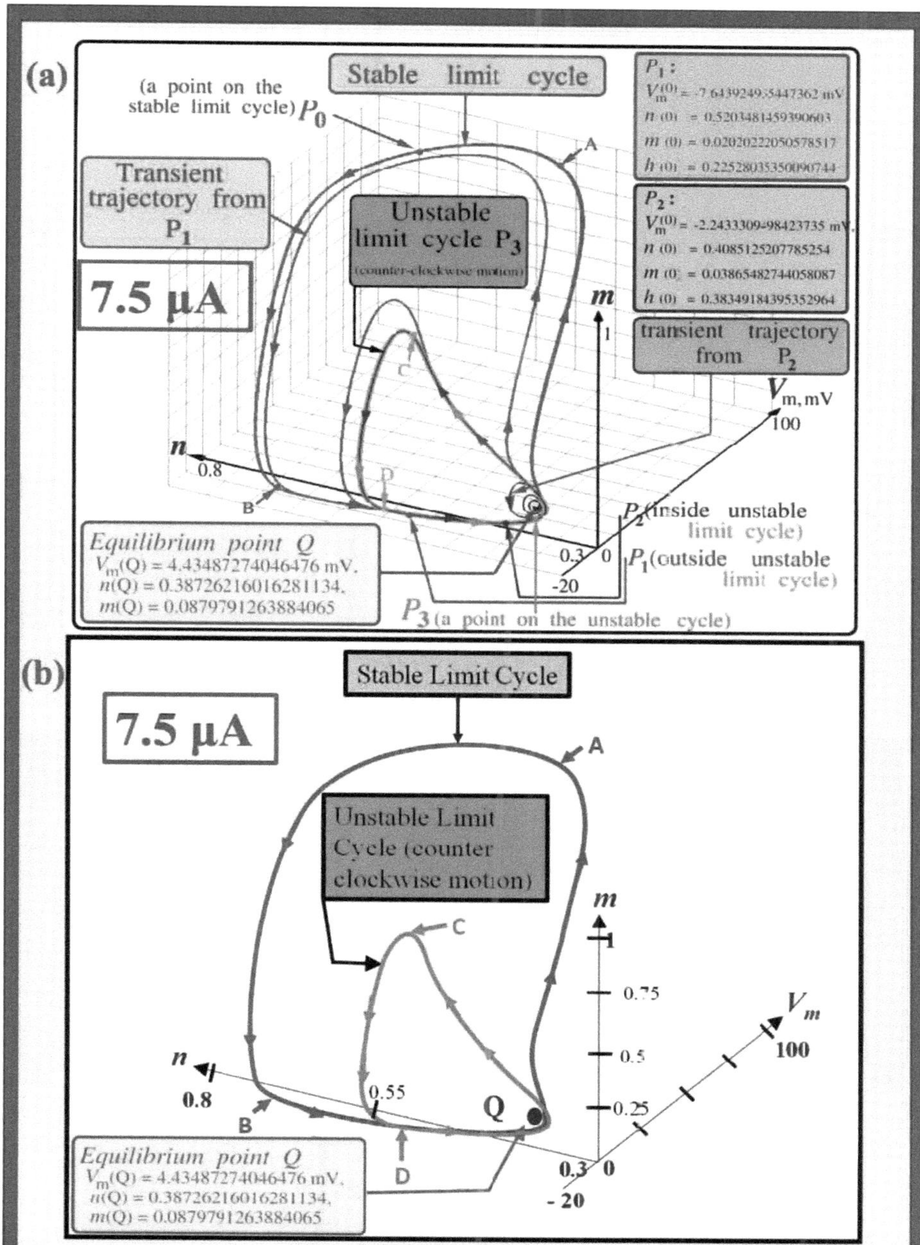

Figure 6.32 (V_m, n, m) *phase portrait perspective* at I_{ext} = 7.5 µA.

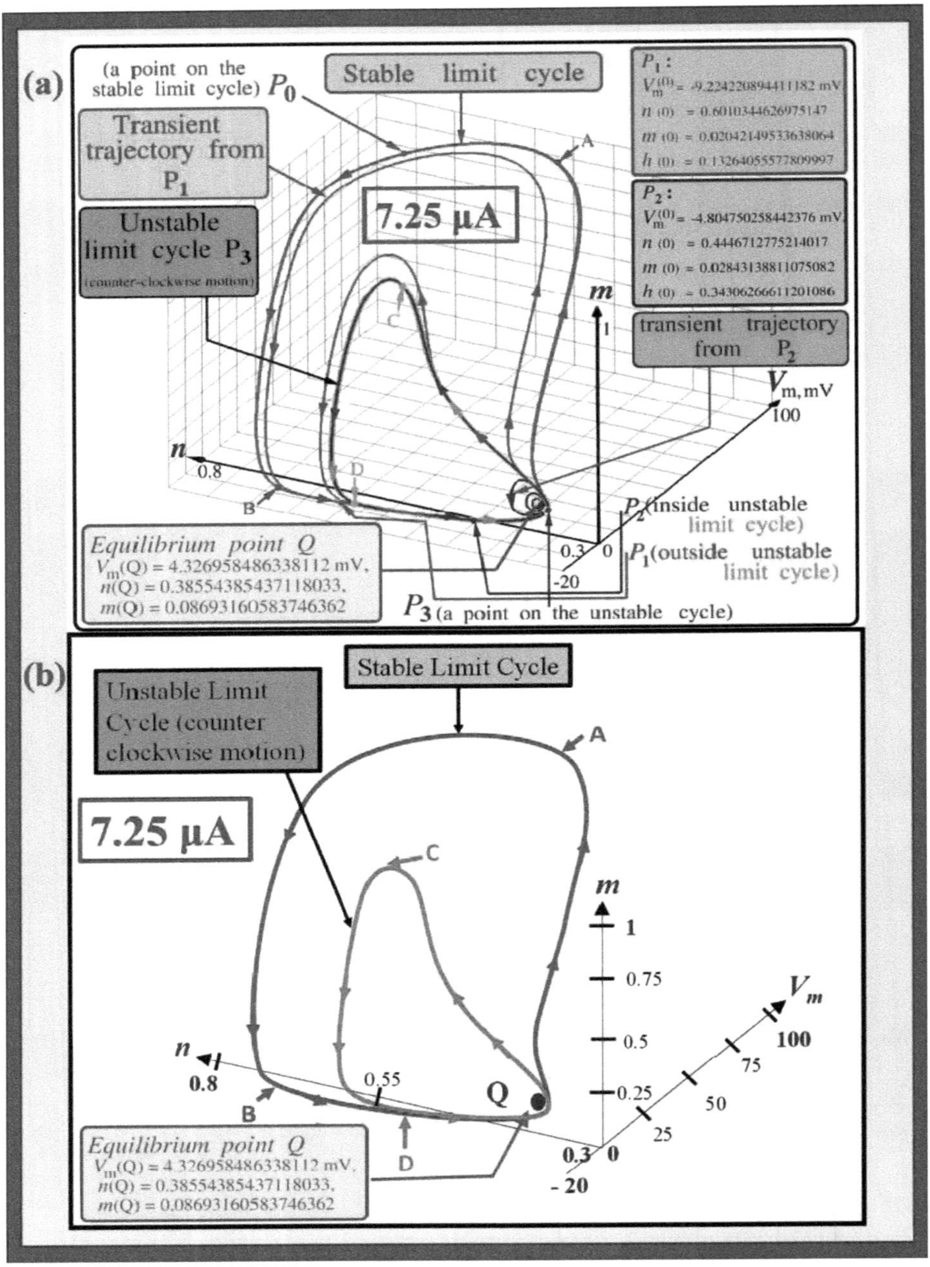

Figure 6.33 (V_m, n, m) *phase portrait perspective* at I_{ext} = 7.25 μA.

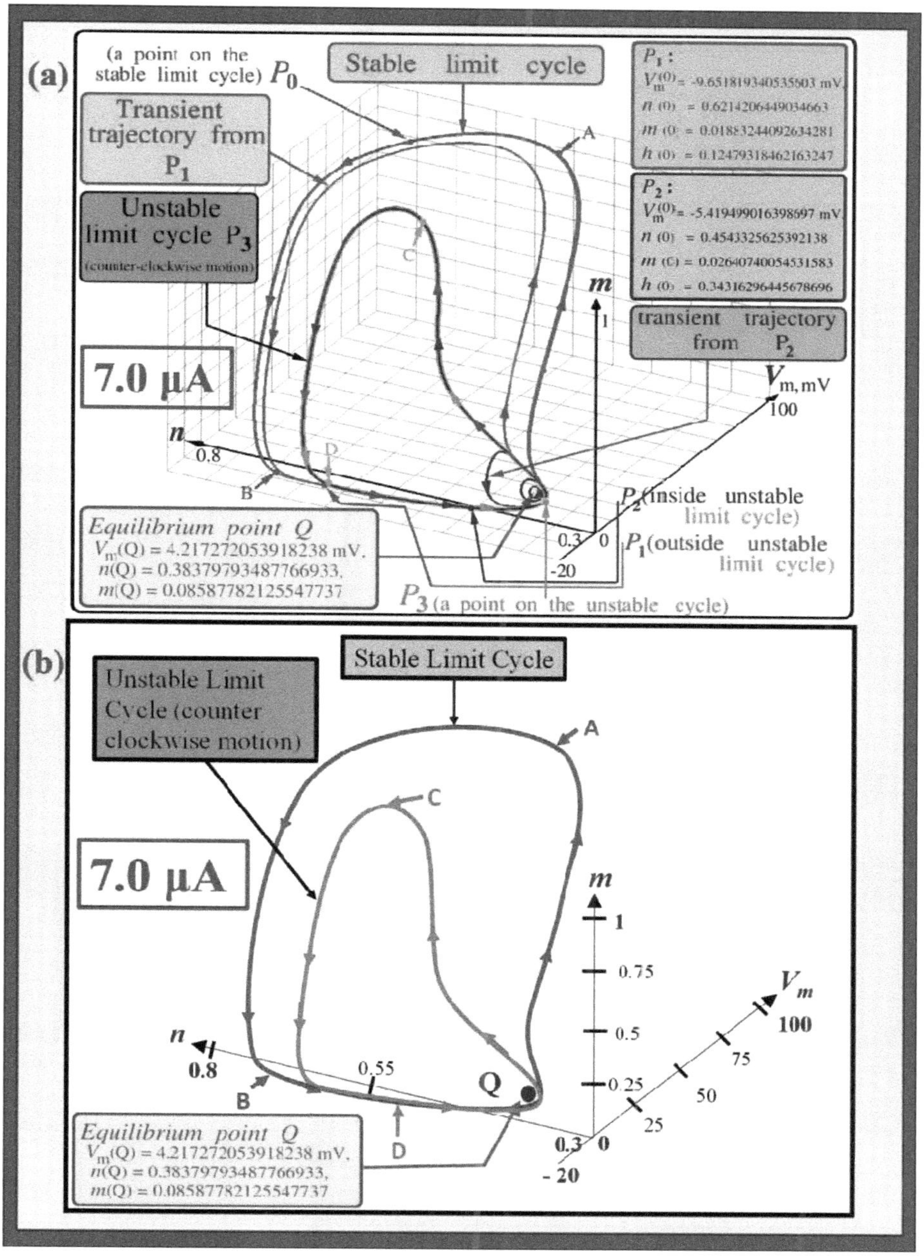

Figure 6.34 (V_m, *n*, *m*) *phase portrait perspective* at I_{ext} = 7.0 µA.

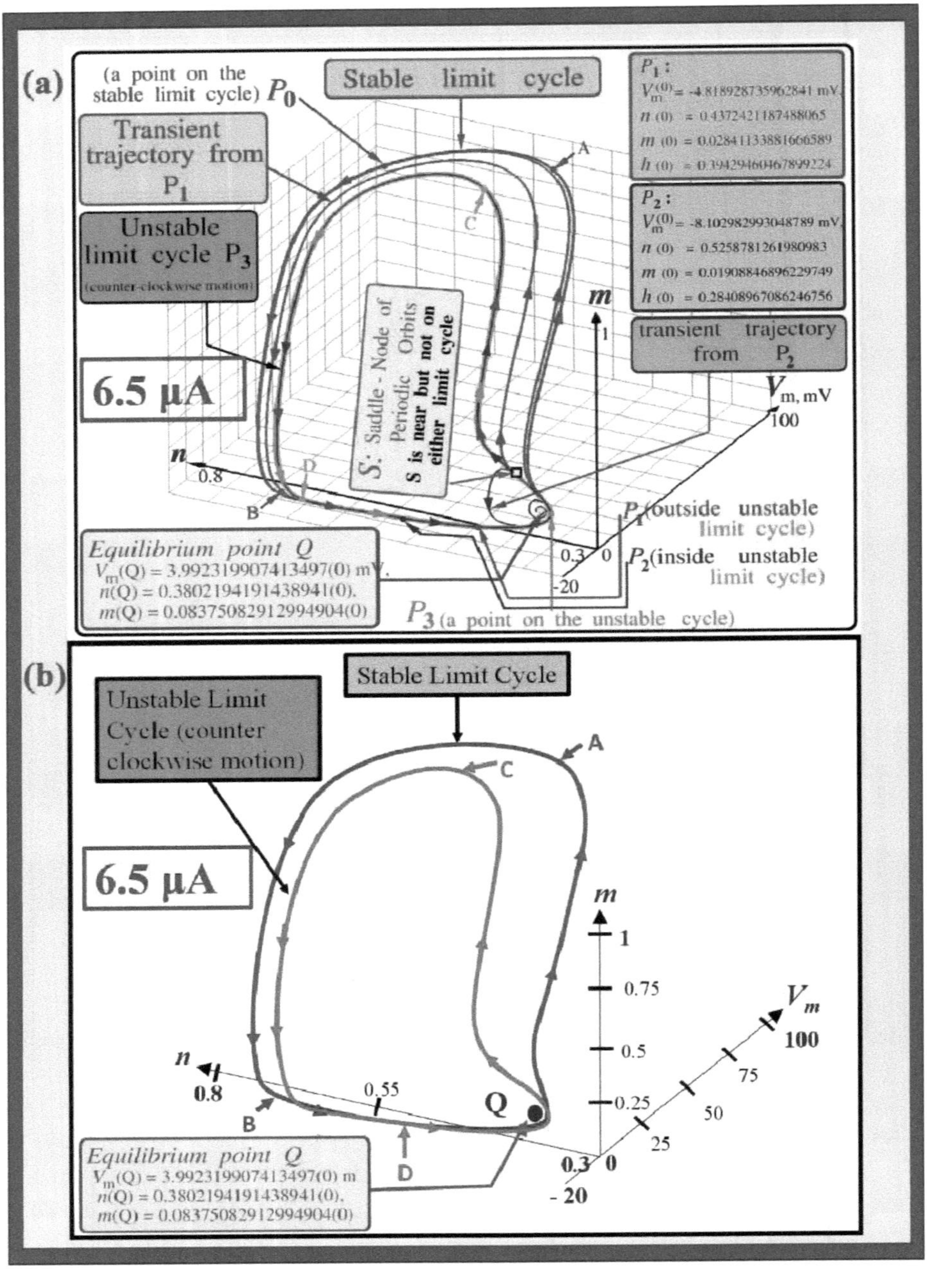

Figure 6.35 *(V_m, n, m) phase portrait perspective* at I_{ext} = 6.5 µA.

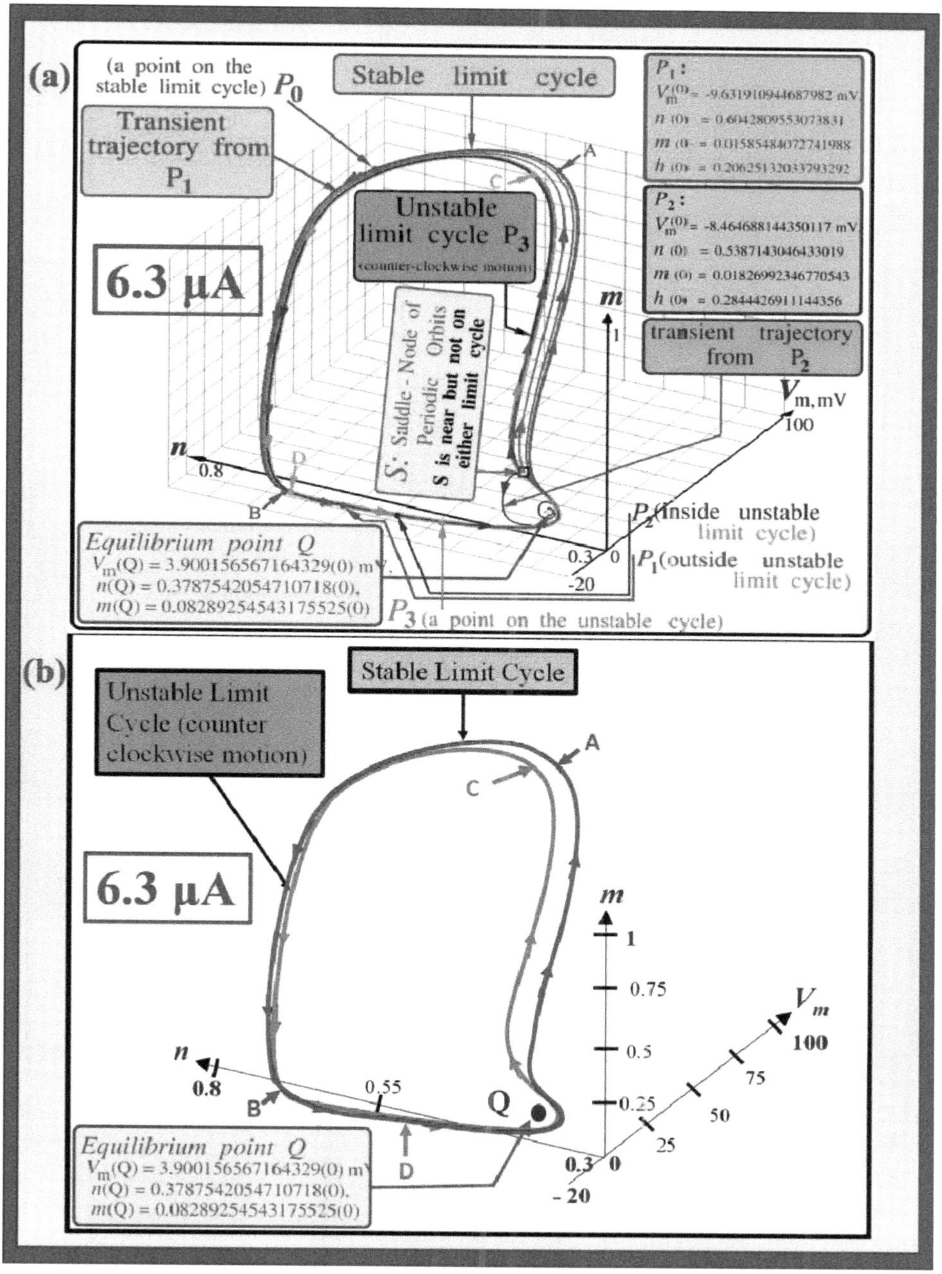

Figure 6.36 (V_m, n, m) *phase portrait perspective* at I_{ext} = 6.3 μA.

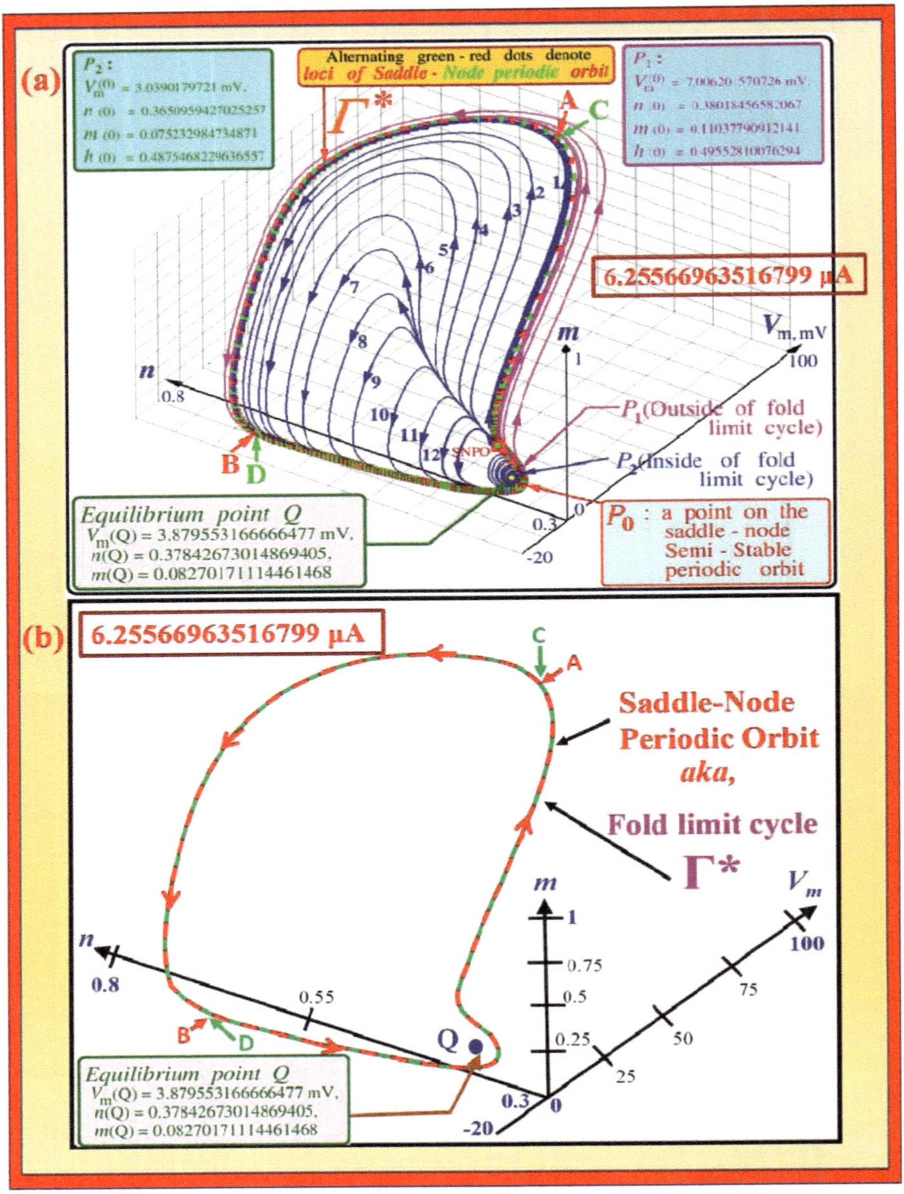

Figure 6.37 *(V_m, n, m) phase portrait perspective* at I_{ext} = 6.25566963516799 μA = I^*. All trajectories with an initial point inside Γ^* converge to Q. All trajectories with an initial point outside Γ^* converge to Γ^*. Here inside and outside are loosely defined with respect to the plane spanned by all pairwise combinations of the four state variables V_m, n, m, and h.

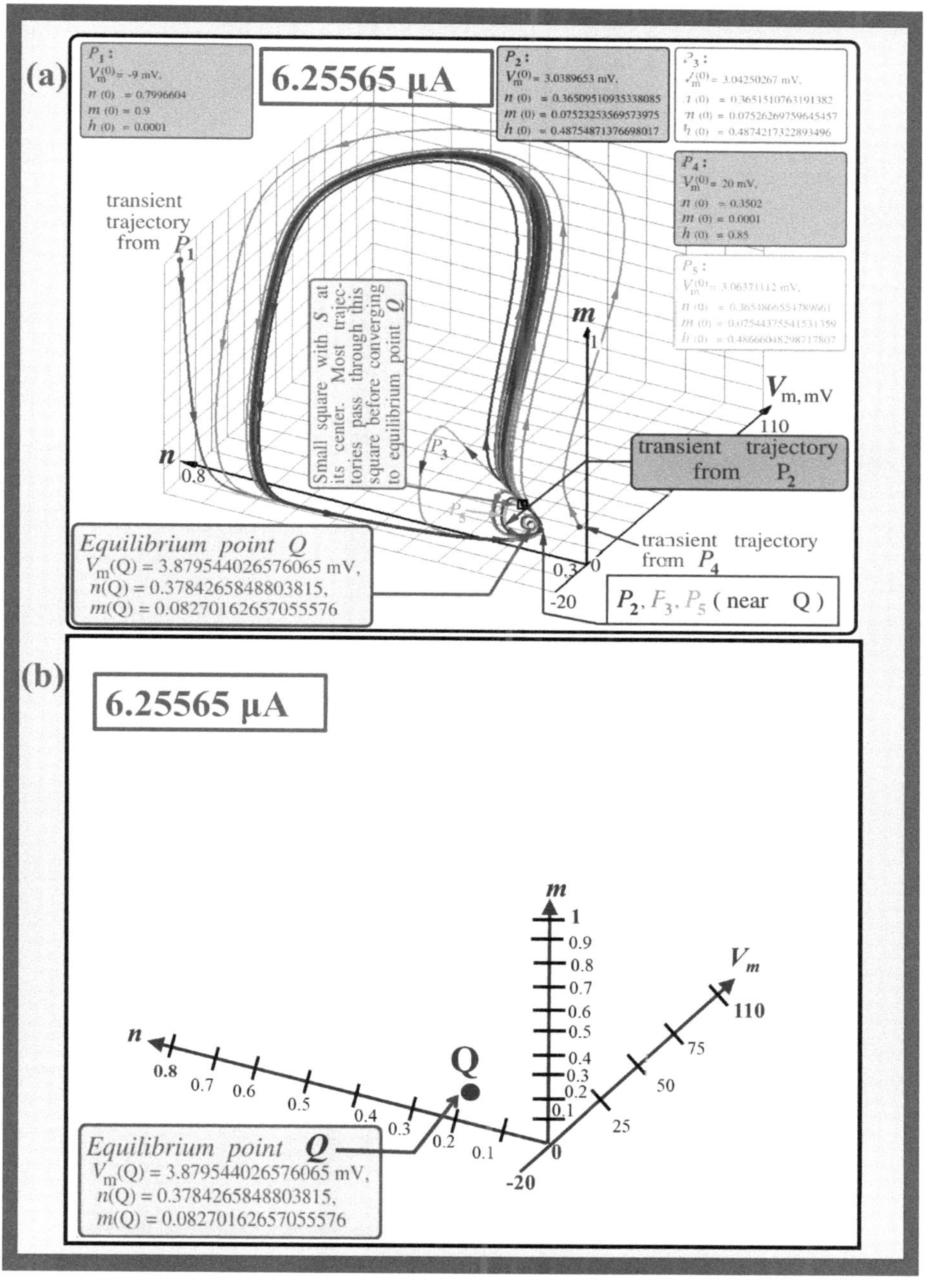

Figure 6.38 (V_m, n, m) *phase portrait perspective* at I_{ext} = 6.25565 μA.

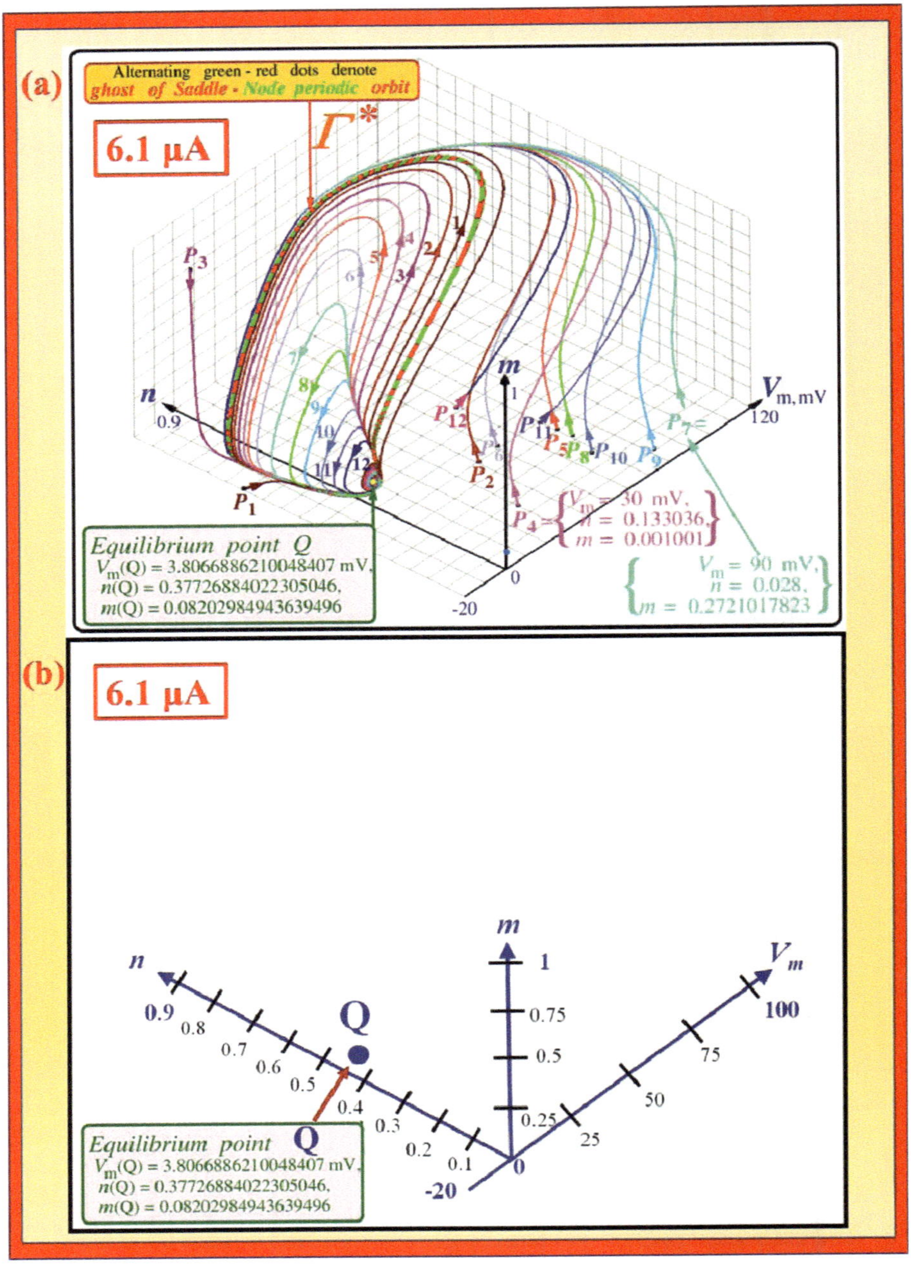

Figure 6.39 *(V_m, n, m) phase portrait perspective* at I_{ext} = 6.1 μA.

In each of the nine-phase portrait perspectives in Figures 6.31–6.39, the *maximum point* and the *minimum point* on the *red stable limit cycle* are identified and labeled by a red letter, A and B, respectively. The distance between A and B is therefore *the peak-to-peak* voltage of the associated ***stable action potential*** $V_m(t)$. Similarly, the *maximum* point and the *minimum* point on the *green unstable* limit cycle are identified and labeled by a green letter C and D, respectively. The distance between C and D is therefore the *peak-to-peak voltage* of the *unstable* limit cycle.

The **lower part** shows three *subsets* from the *upper part*, namely, the *stable limit cycle* in red color (without extraneous transient components), the *unstable limit cycle* in green color (without extraneous transient components), and the *stable DC equilibrium point* $\boldsymbol{Q}$ in blue color. The reader is encouraged to observe that while the outer large *stable limit cycle* retains, approximately, both its shape and size, in each of the nine exhibits (Figures 6.31–6.39), the inner *unstable limit cycle* grows in size along with a change in *shape* (i.e., morphology) that tends to emulate the shape of the corresponding large stable outer orbit as I_{ext} decreases until I_{ext} tends to $\boldsymbol{I^* = 6.25566963516799\ \mu A}$ in Figure 6.37, where the *unstable limit-cycle morphs* exactly into the *stable limit cycle*, as depicted in alternating red–green dashes.

The reader is encouraged to witness the next extraordinary event, where the red–green alternating loop disappears upon an infinitesimally small further reduction in I_{ext}, say $I_{\text{ext}} = I^* - \varepsilon$, for an arbitrarily small positive value of ε. We have a full large-amplitude action potential when $I_{\text{ext}} = I^*$, but the latter melts into nothing after an arbitrarily small perturbation ε, as depicted in the corresponding phase portrait projection in Figure 6.38(b), when we reduce the value of $I_{\text{ext}} = I^* = 6.25566963516799\ \mu\text{A}$ to $I_{\text{ext}} = 6.255650\ \mu\text{A}$, where $\varepsilon = 0.00001963516799$, and witness the **Galvani's all-or-none enigma, an example *par excellence*!**

6.4.2 From All to None Gently

To reassure the reader that all trajectories from any initial point in Figure 6.38(a) converge to the equilibrium point Q, we pick, arbitrarily, three points P_2, P_3, and P_5 from Figure 6.38(a) and show, in Figure 6.40, their corresponding waveform $V_m(t)$ indeed converge to $V_m(Q)$ = 3.879544026576065 mV at the equilibrium point $\boldsymbol{Q}$. The same scenario repeats itself at $I_{\text{ext}} = 6.1$ μA, as shown in Figure 6.39, and for all $I_{\text{ext}} < 6.1$ μA.

A cursory examination of the evolution of the stable (red) and unstable (green) limit cycles in the nine-phase portrait perspectives exhibited in the gallery displayed in Figures 6.31–6.39 reveals that while the peak-to-peak voltage of the red stable limit cycle varies only slightly, the green unstable limit cycle grows rapidly in size between $I_{\text{ext}} = 7.8$ μA in Figure 6.31 and $I_{\text{ext}} = 6.5$ μA in Figure 6.35, with a continuous change in shape to try to emulate its larger stable brother, to avoid an eventual head-on collision.

It can be shown that the Principle of Local Activity, [9, 10, 19] and its pearl, the Edge of Chaos, may explain Galvani's all-or-none **enigma**, as well as the dogma of irritability, which we now know is the old-fashioned name of what modern physiologist dub excitability, an intuitive but still undefined concept!

The above transition from all to none is depicted in a caricature shown in Figure 6.41 (modified from Figure 6.22 of [42]). Yes, the transition suggests some violent reactions. Yet the perceived Apocalypse was benign. The miracle is a direct consequence of the uniqueness theorem from the theory of differential equations which decrees that no two trajectories can ever cross each other. This law requires therefore that the vector field in the neighborhood of the stable limit cycle, and of the unstable limit cycle, must reorient itself gently, such that the two cycles become eventually identical, i.e., coalescing one into the other, at some critical value $I_{\text{ext}} = I^*$. By virtue of the continuity property of the vector field of differential equations, an infinitesimal perturbation of I_{ext} from I^* would allow a smooth splitting into two distinct trajectories, each one coasting along the same mother vector field, and attracted to the same equilibrium point Q.[4]

We end this chapter by reminding the reader that none of the above magical events could occur in the Hodgkin–Huxley equations without the presence of the *Edge of Chaos Kernel* hidden inside the small-signal equivalent circuit of the sodium ion-channel memristor of the 4th-order Hodgkin–Huxley circuit in Figure 6.21. Yes indeed, *Edge of Chaos Kernel* is the **elan vital** of life.

[4]The above scenario can be rigorously proved in R^4 using the rather deep mathematical tools developed in Chapter 12 (Global bifurcations at the disappearance of saddle-node equilibrium states and periodic orbits) of Shilnikov, et al. [35].

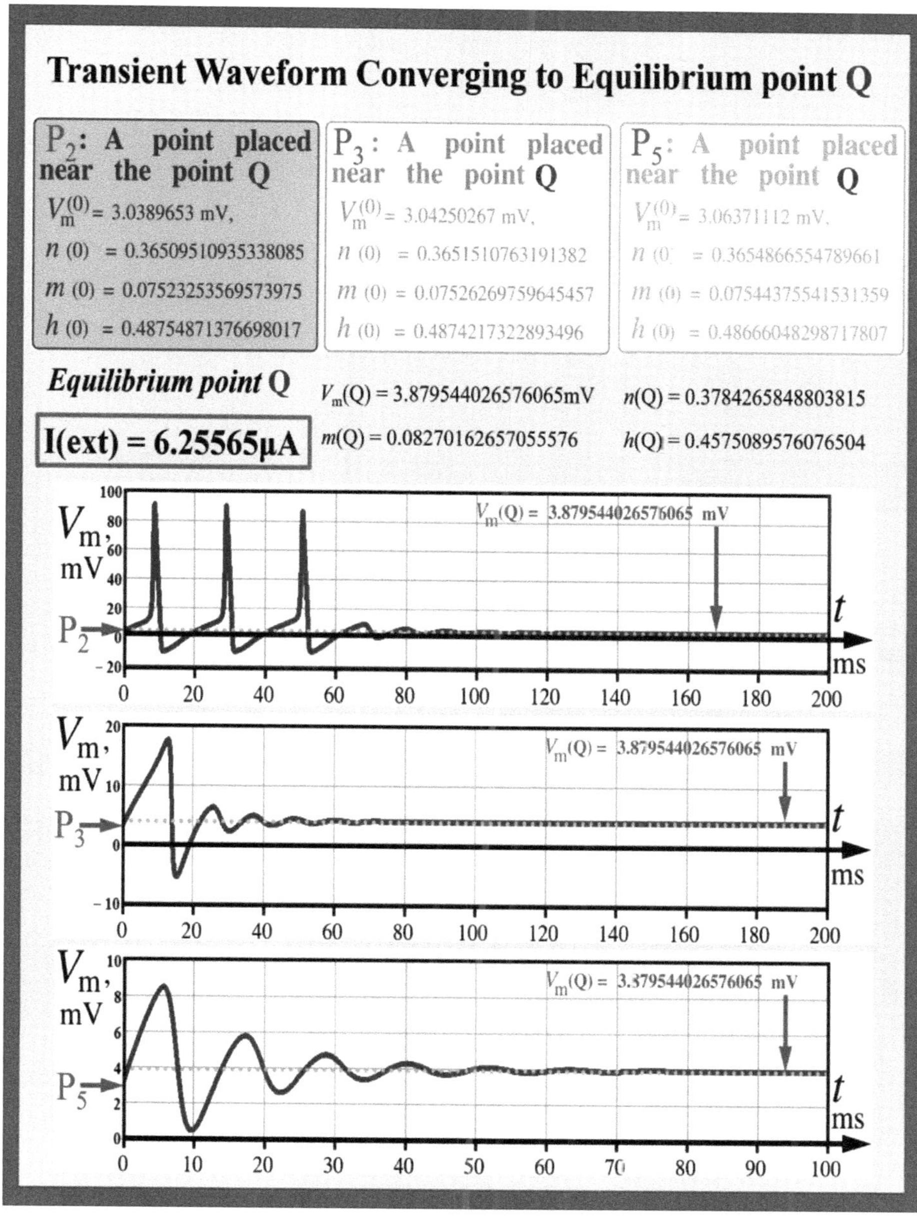

Figure 6.40 Time function $V_m(t)$ of trajectory from three initial points P_2, P_3, and P_5 near the *Equilibrium point Q* identified in the lower-right corner of Figure 6.38(a). Observe that they all converge to Q in less than 80 ms. For all I_{ext} < 6.25566963516799 μA = I^* trajectories originating from *any* initial point in R^4 converge to the asymptotically stable equilibrium point Q.

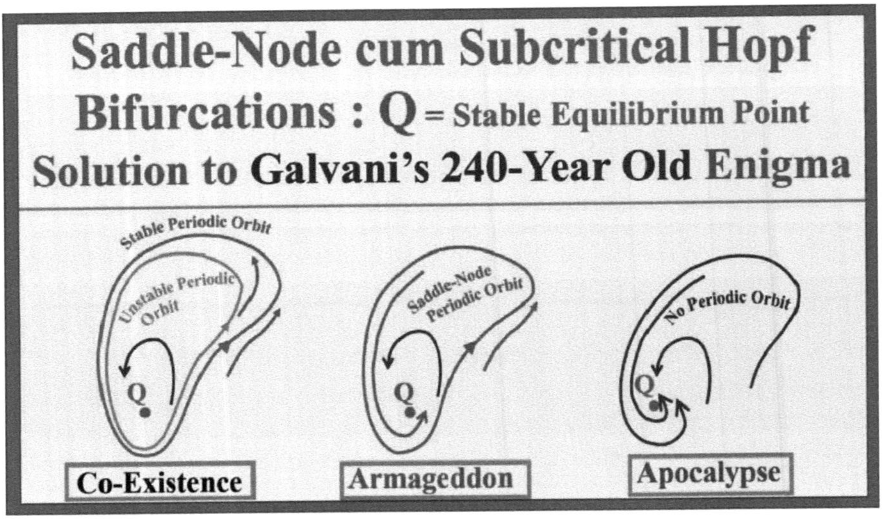

Figure 6.41 As I_{ext} decreases, starting from I_{ext} = 7.8 μA in Figure 6.31, where a stable and unstable periodic orbit, co-existed and dubbed the *co-existence* I_{ext}-domain (left). At I_{ext} = 6.25566963516799 μA = I^*, the unstable periodic orbit morphs smoothly onto the stable periodic orbit, in the so-dubbed *Armageddon* I^*-domain (middle). For $I_{ext} < I^*$ μA, we reach the *Apocalypse* I_{ext}-domain, where no periodic orbits exist any longer, and the "*all*" state has turned into a "*none*" state (right).

Acknowledgments

The research is supported by the USA Office of Scientific Research grant FA9550-18-1-0016. The author would like to thank Professor Hyongsuk Kim, Dr. Valery Sbitnev, and Muhammad Umraiz for generating most of the figures and for the production of the chapter, and Dr. Alon Ascoli for his incisive review of this chapter.

References

1. Galvani, L. (1791). De viribus electricitatis in motu musculari, Commentarius, *De Bononiensi Scientiarum et Artium Instituto atque Academia Commentarii*, 7, pp. 363–418.
2. Lucas, K. (1909). The 'all or none' contraction of the amphibian skeletal muscle fibre, *The Journal of physiology*, 38(2–3), pp. 113–133.

3. Forbes, A. and Thacher, C. (1920). Amplification of action currents with the electron tube in recording with the string galvanometer, *American Journal of Physiology-Legacy Content*, 52(3), pp. 409–471.
4. Hodgkin, A.L. and Huxley, A.F. (1939). Action potentials recorded from inside a nerve fibre, *Nature*, 144(3651), pp. 710–711.
5. Hodgkin, A.L. and Huxley, A.F. (1952). A quantitative description of membrane current and its application to conduction and excitation in nerve, *The Journal of physiology*, 117(4), p. 500.
6. Chua, L. (1969). *Introduction to Nonlinear Network Theory*, McGraw-Hill.
7. Chua, L., Sbitnev, V., and Kim, H. (2012). Hodgkin–Huxley axon is made of memristors, *International Journal of Bifurcation and Chaos*, 22(03), pp. 12300111–123001148.
8. Chua, L., Sbitnev, V., and Kim, H. (2012). Neurons are poised near the edge of chaos, *International Journal of Bifurcation and Chaos*, 22(04), pp. 12500981–125009849.
9. Chua, L. O. (2005). Local activity is the origin of complexity, *International journal of bifurcation and chaos*, 15(11), pp. 3435–3456.
10. Mainzer, K. and Chua, L. (2013). *Local Activity Principle*, Imperial College Press.
11. Messaris, I., Brown, T. D., Demirkol, A. S., Ascoli, A., Al Chawa, M. M., Williams, R. S., Tetzlaff, R., and Chua, L. O. (2021). NbO 2-Mott memristor: A circuit-theoretic investigation, *IEEE Transactions on Circuits and Systems I: Regular Papers*, 68(12), pp. 4979–4992.
12. Schrödinger, E. (1948). *What is Life?* Cambridge University Press, Cambridge.
13. Prigogine, I. (1980). *Exploring Complexity*, Freeman, San Fransisco.
14. Gell-Mann, M. (1994). *The Quark and the Jaguar*, Freeman, New York.
15. Haken, H. (1993). *Advanced Synergetics*, Springer, New York.
16. Chua, L. O., Desoer, C. A., and Kuh, E. S. (1987). *Linear and Nonlinear Circuits*, McGraw-Hill Book Company.
17. Ascoli, A., Demirkol, A.S., Tetzlaff, R., and Chua, L. (2022). Edge of Chaos is sine qua non for Turing Instability, *IEEE Transactions on Circuits and Systems I: Regular Papers*, 69(11), pp. 4596–4609.
18. Chua, L. (2021). *The Chua Lectures* (in 4 volumes), World Scientific, Singapore.
19. Chua, L. (1998). *CNN: A Paradigm for Complexity*, World Scientific.

20. Turing, A. (1952). The chemical basis of morphogenesis, *Philos. Trans. Roy. Soc. London, Series* B, 237, pp. 37–72.
21. Smale, S. (1974). A Mathematical model of two cells via Turing's equations, *Lectures in Applied Mathematics 6, American Mathematical Society*, pp. 15–26.
22. Ascoli, A., Demirkol, A. S., Tetzlaff, R., and Chua, L. (2022). Edge of chaos theory resolves Smale paradox, *IEEE Transactions on Circuits and Systems I: Regular Papers*, 69(3), pp. 1252–1265.
23. Morris, C. and Lecar, H. (1981). Voltage oscillations in the barnacle giant muscle fiber, *Journal of Biophysics*, 35, pp. 193–213.
24. FitzHugh, R. (1961). Impulses and physiological states in theoretical models of nerve membrane, *Journal of Biophysics*, 1, pp. 445–446.
25. Chua, L. (1971). Memristor: The missing circuit element, *IEEE Trans. On Circuit Theory*, 18, pp. 507–519.
26. Strukov, D. B., Snider, G. S., Stewart, D. R., and Williams, R. S. (2008). The missing memristor found, *Nature*, 453(7191), pp. 80–83.
27. Chua, L. and Kang, S. (1976). Memristive devices and systems, *Proc. of IEEE*, 64, pp. 209–223.
28. Chua, L. (2011). Resistance switching memories are memristors, *Applied Physics A*, 102, pp. 765–783.
29. Chua, L. (2015). Everything you wish to know about memristors but are afraid to ask, *Radioengineering*, 24, pp. 319–368.
30. Chua, L. (2013). Memristor, Hodgkin–Huxley, and Edge of Chaos, *Nanotechnology*, 24, pp. 383001–3830014.
31. Chua, L. (2014). If it's pinched it's a memristor, *Semiconductor Science and Technology*, 29, pp. 104001–1040042.
32. Chua, L. (2018). How we predicted the memristor, *Nature Electronics*, 1.5, p. 322.
33. Adhikari, S.P., Sah, M.P., Kim, H., and Chua, L. (2013). Three fingerprints of memristor, *IEEE Transactions on Circuit and Systems*, 60(11), pp. 3008–3021.
34. Shilnikov, L.P., Shilnikov, A.L., Turaev, D.V., and Chua, L.O. (1998). *Methods of Qualitative Theory in Nonlinear Dynamics*, World Scientific, Part I.
35. Shilnikov, L.P., Shilnikov, A.L., Turaev, D.V., and Chua, L.O. (2001). *Methods of Qualitative Theory in Nonlinear Dynamics*, World Scientific, Part II.

36. Zhang, X., Wu, Z., and Chua, L. (2020). Hearts are poised near the edge of chaos, *Int. J. of Bifurcation and Chaos*, 30(09). pp. 20300231–203002378.
37. Noble, D. (1962). A modification of the Hodgkin–Huxley equations applicable to Purkinje fibre action and pacemaker potentials, *J. of Physiology*, 160, pp. 317–352.
38. Jin, P., Wang, G., Liang, Y., H-C Lu., H., and Chua, L. (2021). Neuromorphic dynamics of Chua corsage memristor, *IEEE Trans. On Circuits and Systems I*, 68(11), pp. 4419–4432.
39. Hassard, B. (1978). Bifurcation of periodic solutions of the Hodgkin–Huxley model for the squid giant axon, *J. Theoretical Biology*, 71, pp. 410–420.
40. Keener, J. and Sneyd, J. (2009). *Mathematical Physiology – I: Cellular Physiology*, Springer, 2nd Edition.
41. Chua, L., Sbitnev, V., and Kim, H., Umraiz, M. *Int. J. of Bifurcation and Chaos,* (to be published).
42. Izhikevich, E.M. (2007). *Dynamical systems in neuroscience*, MIT Press.
43. Jin, P., Lian, Y., Wang, G., Chen, L., Iu, H. H., and Chua, L. O. (2022). Shortcuts to edge of chaos domains of memristive circuits and historic measurement of contiguous triple-branch V-I curve of Chua corsage memristor, *International Journal of Bifurcation and Chaos*, 32(15), pp. 22300351–223003543.
44. Chua, L. O., (2023). Homemade $10 Chua Corsage Memristor, *IEEE Electron Devices Magazine*, 1(2), pp. 10–22.

Chapter 7

Nanostructured Systems Based on the Approach of Biological Engineering

Donald Martin

Université Grenoble Alpes, SyNaBi, TIMC-IMAG/CNRS/INSERM, Grenoble, France

don.martin@univ-grenoble-alpes.fr

This chapter explores the design and assembly of nanostructured systems with inspiration from biology. Here, we consider biological inspiration as not the approach typically adopted to construct synthetic materials inspired by nature. In contrast, the approach described in this chapter utilizes biological components in the assembly of smart nanostructured systems. The term "biological engineering" is used to describe a discipline that embodies this approach. Indeed, in 1970, this term was introduced formally with the intention to integrate engineering with biological systems to move beyond single disciplinary areas such as medicine, agriculture, or fermentation engineering. In this chapter, we further refine the discipline of "biological engineering" to be one that utilizes biological proteins, molecules, and lipids in combination with synthetic materials to assemble smart nanostructured systems. We illustrate this discipline with the assembly of nanostructured systems, which are targeted for

Outlooking beyond Nanoelectronics and Nanosystems: Ultra Scaling, Pervasiveness, Sustainable Integration, and Biotic Cross-Inspiration
Edited by Simon Deleonibus

ISBN 978-981-5129-32-8 (Hardcover), 978-1-003-50990-5 (eBook)
www.jennystanford.com

applications such as diagnostic, therapeutic, biofuel cell, or tissue enhancement/replacement in the body.

7.1 Nanostructured Systems

We must first describe what we mean by a nanostructured system. Many devices incorporate components that are nanostructured, whether by default or by design. However, those devices are usually large enough to be observed, manipulated, and capable of providing measurements that can be interfaced. All of those characteristics would not necessarily define such a system as being truly nanostructured.

We can begin by collating the thoughts from the scientific literature on the constitution of a nanostructured system. The WordCloud of Figure 7.1 illustrates the 50 most frequently used terms related to nanostructured systems. These words were found from a search of publications in the Clarivate® Web of Science (WoS) using the terms *ALL = (biological engineering) AND (ALL = (smart systems)) AND ALL = (nanostructur*).*

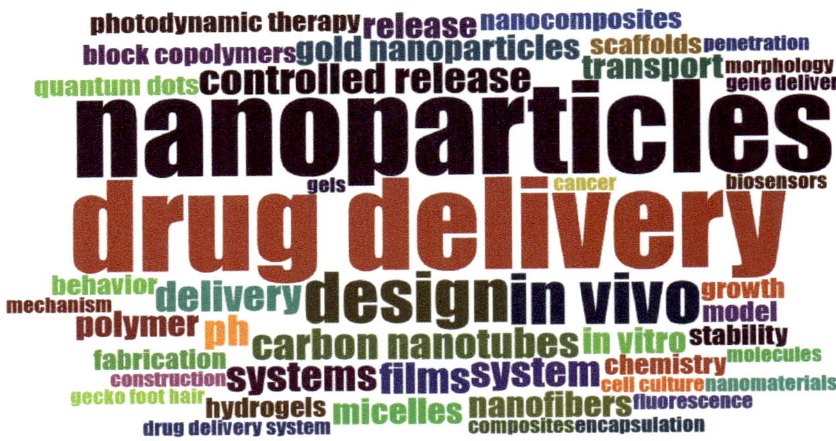

Figure 7.1 WordCloud of the 50 most frequent words from a bibliometric analysis of the literature in the WoS using the search terms *ALL = (biological engineering) AND (ALL = (smart systems)) AND ALL = (nanostructur*)*. The literature search retrieved 121 items. The analysis was performed using the using the "Bibliometrix" library of RStudio v1.2.5033 [1].

This WordCloud does not identify a main unifying theme, most likely because nanostructure refers to a length scale. This does not place it in a specific domain of a particular scientific or engineering discipline. Nonetheless, we can gain further insight into defining smart nanostructured systems that utilize biological engineering by looking in more detail at publications with more than 100 citations. These publications can be considered as being the most influential in the field, and are listed in Table 7.1. Those highly cited articles were published in the 14-year period from 2004 to 2018. We can also consider the current research directions that influence the authors in this field by looking at the articles that have been published in 2022, which are listed in Table 7.2.

To move further toward identifying a main unifying theme to define smart nanostructured systems that utilize biological engineering, we can compare the main WoS categories for those highly cited publications (Table 7.3) with the main WoS categories of the recent 2022 publications (Table 7.4).

Table 7.3 suggests that in the period 2004–2018, a nanostructured system was categorized mainly on a discipline basis, particularly as a system that comprised nanoparticles. The traditional engineering disciplines dominated the categories of highly cited publications from this period.

Table 7.4 suggests that the publications of 2022 are less categorized by the more traditional disciplines in engineering, chemistry, and physics. The 2022 publications are included in more biotechnology and applied chemistry disciplines. Such a distribution broadens the publication of nanostructured systems into applied biological, chemical, and polymer categories.

This bibliometric analysis provides some pointers to what we can consider as a nanostructured system. The literature is moving away from the concept that nanostructured systems are those that comprise ways to utilize nanoparticles, restrained largely to traditional categories in engineering. Indeed, recent literature appears to be shifting the publication of nanostructured systems toward biotechnological application categories.

Table 7.1 The top 14 publications with more than 100 citations from the WoS databases.

Article title	Authors	Reference	Citations
Gold nanocages covered by smart polymers for controlled release with near-infrared light	Yavuz MS, Cheng Y, Chen J, Cobley CM, Zhang Q, Rycenga M, Xie J, Kim C, Song KH, Schwartz AG, Wang LV, Xia Y	*Nat. Mater.*, 8:935-939, 2009	1,179
Bio-inspired design of multiscale structures for function integration	Liu K, Jiang L	*NANO Today*, 6:155-175, 2011	567
A review of atomic force microscopy imaging systems: application to molecular metrology and biological sciences	Jalili N, Laxminarayana K	*Mechatronics*, 14:907-945, 2004	394
Smart Nanoparticles for Drug Delivery Application: Development of Versatile Nanocarrier, Platforms in Biotechnology and Nanomedicine	Lombardo D, Kiselev MA, Caccamo MT	*J. Nanomater.*, 2019:3702518, 2019	385
DNA Nanotechnology Based on i-Motif Structures	Dong Y, Yang Z, Liu D	*Acc. Chem. Res.*, 47:1853-1860, 2014	257
Engineering nanoscale order into a designed protein fiber	Papapostolou D, Smith AM, Atkins EDT, Oliver SJ, Ryadnov MG, Serpell LC, Woolfson DN	*P.N.A.S.*, 104:10853-10858, 2007	200
Silica Coated Upconversion Nanoparticles: A Versatile Platform for the Development of Efficient Theranostics	Liu JN, Bu WB, Shi JL	*Acc. Chem. Res.*, 48:1797-1805, 2015	189
Dynamic Supramolecular Complexes Constructed by Orthogonal Self-Assembly	Hu XY, Xiao T, Lin C, Huang F, Wang L	*Acc. Chem. Res.*, 47:2041-2051, 2014	189
Virus-Directed Design of a Flexible BaTiO3 Nanogenerator	Jeong CK, Kim I, Park KI, Oh H, Paik H, Hwang GT, No K, Nam YS, Lee KJ	*ACS Nano*, 7:11016-11025, 2013	179
Methods of Synthesis, Properties and Biomedical Applications of CuO Nanoparticles	Grigore ME, Biscu ER, Holban M, Gestal MC, Grumezescu AM	*Pharmaceuticals*, 9:75, 2016	172
Hierarchical micro and nano structured, hydrophilic, superhydrophobic and superoleophobic surfaces incorporated in microfluidics, microarrays and lab on chip microsystems	Gogolides E, Ellinas K, Tserepi A	*Microelectr. Eng.*, 132:135-155, 2015	155
RGD-Peptide Functionalized Graphene Biomimetic Live-Cell Sensor for Real-Time Detection of Nitric Oxide Molecules	Guo CX, Ng SR, Khoo SY, Zheng X, Chen P, Li CM	*ACS Nano*, 6:6944-6951, 2012	146
Recent Progress in Photoswitchable Supramolecular Self-Assembling Systems	Yao X, Li T, Wang J, Ma X, Tian H	*Adv. Opt. Mater.*, 4:1322-1349, 2016	133
Controlling Supramolecular Chirality of Two-Component Hydrogels by J- and H-Aggregation of Building Blocks	Liu G, Sheng J, Wu H, Yang C, Yang G, Li Y, Ganguly R, Zhu L, Zhao Y	*J. Amer. Chem. Soc.*, 140:6467-6473, 2018	117

Note: These publications were identified from a literature search of WoS using the search terms *ALL = (biological engineering) AND (ALL = (smart systems)) AND ALL = (nanostructure*)*. The literature search retrieved 121 items.

Table 7.2 The 12 articles published in 2022 that were identified from a literature search of WoS using the search terms *ALL = (biological engineering) AND (ALL = (smart systems)) AND ALL = (nanostructure*)*

Article title	Authors	Reference
Enzyme-Based Mesoporous Nanomotors with Near-Infrared Optical Brakes	Liu MC, Liang Z, Liu M, Zhao T, Ma Y, Zhou Q, Ibrahim YS, Elzatahry AA, Li X, Zhao D	*J. Amer. Chem. Soc.*, 144:3892-3901, 2022
A robust combination of dysprosium vanadate/halloysite nanotubes: the electrochemical system for dimetridazole detection	Kokulnathan T, Chen TW, Chen SM, Ahmed F, Hasan MZ, Bilgrami AL, Kumar S	*Mater. Today Chem.*, 24:100890, 2022
Surface-Enhanced Raman Spectroscopy Substrates for Food Safety and Quality Analysis	Nilghaz A, Mousavi SM, Amiri A, Tian J, Cao R, Wang X	*J. Agric. Food Chem.*, 70,5463-5476, 2022
Stimuli-responsive liposomal nanoformulations in cancer therapy: Pre-clinical & clinical approaches	Ashrafizadeh M, Delfi M, Zarrabi A, Bigham A, Sharifi A, Rabiee N, Paiva-Santos AC, Kumar AP, Tan SC, Hushmandi K, Ren J, Zare EN, Makvandi P	*J. Contr. Rel.*, 351:50-80, 2022
Combining confocal microscopy, dSTORM, and mass spectroscopy to unveil the evolution of the protein corona associated with nanostructured lipid carriers during blood-brain barrier crossing	Battaglini M, Feiner N, Tapeinos C, De Pasquale D, Pucci C, Marino A, Bartolucci M, Petretto A, Albertazzi L, Ciofani G	*Nanoscale*, 14:13292-13307, 2022
waste-free entropy-driven DNA nanomachine for smartly designed photoelectrochemical biosensing of MicroRNA-155	Chen Y, Xu L, Xu Q, Wu Y, Li J, Li H	Biosens. Bioelectr., 215:114569, 2022
Coupled Multiphysics Modelling of Sensors for Chemical, Biomedical, and Environmental Applications with Focus on Smart Materials and Low-Dimensional Nanostructures	Singh S, Melnik R	*Chemosensors*, 10:157, 2022
Water treatment using stimuli-responsive polymers	Abousalman-Rezvani Z, Roghani-Mamaqani H, Riazi H, Abousalman-Rezvani O	*Polym. Chem.*, 13:5940-5964, 2022
Decoding Optical Spectra with Neural Networks to Monitor the Elimination of Carbon Nanoagents from the Body	Sarmanova O, Laptinskiy K, Burikov S, Dolenko S, Trushina D, Dolenko T	*Opt. Mem. Neural Netw.*, 31:256-265, 2022
Microsized Electrochemical Energy Storage Devices and Their Fabrication Techniques For Portable Applications	Bassyouni Z, Allagui A, Abou Ziki JD	*Adv. Mater. Technol.*, 2200459, 2022
Multifunctional maca extract coated CuO nanoparticles with antimicrobial and dopamine sensing activities: A dual electrochemical-Smartphone colorimetric detection system	Ilgar M, Baytemir G, Tasaltin N, Guelluelue S, Yesilyurt IS, Karakus S	*J. Photochem. Photobiol. A*, 431:114075, 2022
Bidirectional Phase Transformation of Supramolecular Networks Using Two Molecular Signals	Cui D, Liu CH, Rosei F, Perepichka DF	*ACS Nano*, 16:1560-1566, 2022

Note: The literature search retrieved 121 items.

Table 7.3 The WoS categories for the top 14 publications with more than 100 citations from the literature described in Table 7.1

Major category (WoS)	Highly cited publications (< 2022)				
Chemistry	Chemistry Multi-disciplinary	Chemistry Physical	Chemistry Medicinal		
Physics	Physics Applied	Optics	Physics Condensed Matter		
Engineering and Materials	Materials Science Multi-disciplinary	Engineering Electrical Electronic	Automation Control Systems	Engineering Mechanical	Robotics
Nanotechnology	Nanoscience Nano-technology				
Biotechnology	Pharmacology Pharmacy				
Polymers					
Other	Multi-disciplinary Sciences				

Table 7.4 The WoS categories for the 12 articles published in 2022 from the literature described in Table 7.2

Major category (WoS)	Published in 2022				
Chemistry	Chemistry Multi-disciplinary	Chemistry Physical	Chemistry Analytical	Electro-chemistry	Chemistry Applied
Physics	Physics Applied	Optics	Biophysics		
Engineering and materials	Materials Science Multi-disciplinary	Instruments Instru-mentation			
Nano-technology	Nanoscience Nano-technology				
Bio-technology	Pharmacology Pharmacy	Biotechnology Applied Microbiology	Agriculture Multi-disciplinary	Food Science Technology	
Polymers	Polymer Science				
Other					

This shift toward biotechnological applications allows us to better define a smart nanostructured system as one that is designed to solve a specific problem, using the skills from many disciplines to control the utilization of objects that have dimensions on the order of nanometers. Such nanostructured systems could include the utilization of nanoparticles. However, it does appear that a smart nanostructured system is better considered as a problem-driven solution, including the field of biotechnology, which relies upon an engineering approach with a strong basis in science.

7.1.1 Nanostructured Systems for Biomedical and Biological Applications

In the application of a smart nanostructured system to solve a biological problem, we can further consider these systems in the context of engineering that is based strongly on science that may not be exclusively physics. We can illustrate this point with examples of implantable systems that are required to exchange materials or energy with the body. We can define these to be nanostructured symbiotic systems [2].

A symbiotic system comprises an implanted medical device that needs to be designed with an intended two-way duplex communication of energy and materials with the body. We can define a new kind of medical device that is capable of duplex communication and defined as a symbiotic device (a "symbio-bot"). Such a concept of a "symbio-bot" extends the existing concepts of what is understood as a biomaterial and biocompatibility, to provide a definition that includes the notion of the intended functional use of the implanted medical device. Such a functional definition includes the important need for an implanted medical device to be a biocompatible system that is intended and designed to have a two-way (duplex) communication of materials with the body. Such duplex communicating systems include biofuel cells or open-loop feedback devices where a molecule from the body is utilized by the duplex communicating systems to produce different materials (e.g., molecule or energy) for use by the body. Such a duplex level of communication is considered to be a necessary characteristic of symbiotic implanted medical devices that are designed to function for long periods of time inside the body to restore and assist the function of the body.

We now reach the point where we can consider that a smart nanostructured system can truly benefit from combining components of biology with synthetic materials. These components of biology include proteins and lipids that provide biological membrane transport capabilities for the interface of the implanted medical device with the body. Such a system could be termed a "symbio-bot". Now we can introduce at least four advantages of combining biological components with synthetic materials in smart nanostructured systems to create a "symbio-bot" such as (i) the capability for self-assembly, (ii) the easy scalability that follows from using self-assembly, (iii) easy assembly into 3D structures, and (iv) up to 1000-fold less energy requirements for switching functions compared to state-of-the-art ENODe systems. We will develop these points further in the following sections.

7.1.2 Role of Biological Engineering in the Assembly of Smart Nanostructured Systems

At this point, it is worth considering a rational approach for the design, assembly, and engineering of "symbio-bots". This rational approach is introduced here as the discipline of "biological engineering". To think about a role in biological engineering, we first need to be clear as to what we mean by biological engineering. Even if the discussions of this chapter are in the context of assembling nanostructured systems, there is a wide range of applicable descriptions used for the discipline of biological engineering. For example, a search of publications in the Clarivate® Web of Science (WoS) using the terms *ALL = (biological engineering) AND (ALL = (bionanotechnology OR nanobiotechnology)) AND ALL = (system*)* found 449 published articles. Those articles utilized a large range of words to describe biological engineering in this context, and the 50 most frequently used words are illustrated in Figure 7.2.

Figure 7.2 WordCloud of the 50 most frequent words from a bibliometric analysis of the literature in the WoS using the search terms *ALL = (biological engineering) AND (ALL = (bionanotechnology OR nanobiotechnology)) AND ALL = (system*)*. The analysis was performed using the using the "Bibliometrix" library of RStudio v1.2.5033 (Aria & Cuccurullo, 2017).

There is a similarity between the WordClouds illustrated in Figures 7.1 and 7.2, most likely due to the inclusion of the search terms "biological engineering" for both. Nonetheless, there are subtle but important differences between those two WordClouds. The frequently used words in the WordCloud of Figure 7.2 provide an insight into a broad range of materials and research outcomes that emanate from research that includes systems that relate to bionanotechnology/nanobiotechnology. It is interesting to note that the terms *bionanotechnology* and *nanobiotechnology* imply for most authors in the literature that these smart systems use nanoparticles. This bias is apparent if we compare the WordCloud of Figure 7.2 to that for a smart system that is nanostructured, as is shown in Figure 7.1. There remains a predominate concept that nanoparticles are the key common feature of a smart system that is nanostructured.

However, neither the WordCloud in Figure 7.1 nor Figure 7.2 provides a sense of what is the essence of biological engineering as a discipline. One may ask why is it necessary to define a discipline of biological engineering when it is apparent that scientists developing nanostructured systems rely on many different disciplines to create research outcomes that utilize nanoparticles. Defining the discipline of biological engineering applied to nanostructured systems is important for the conceptualization and realization for the type of human and material resources that are required to conduct research in the field of nanostructured systems.

As early as the 1930s, biological engineering was used to describe the incorporation of biological knowledge into traditional engineering. In 1970, the term biological engineering was introduced formally, and intended to provide an integrated approach to combine engineering with biological systems to move beyond single disciplinary areas such as medicine, agriculture, or fermentation engineering. During the 1980s and 1990s, biological engineering education, philosophy, and practice continued to grow and mature with engineers from agricultural, food, chemical, and biomedical backgrounds increasingly recognizing the underlying scientific basis of biology as the driving force for their engineering. Compared to the traditional applications-driven engineering disciplines, perhaps the most prominent aspect that defined this maturing discipline of biological engineering was to let the engineering design process be driven by the unique properties of a biological system (e.g., non-linearity of system dynamics, time-variant behavior) [3]. Those roots for biological engineering placed it as a science-based discipline in engineering that was not based on the science of physics, such as mechanical engineering or electrical engineering. Moreover, neither was biological engineering an industrial applications-based discipline, such as mining engineering, power engineering, and agricultural engineering [4]. Having said that, one can further clarify the typology of engineering disciplines based on a different application focus that incorporates traditional engineering disciplines with some aspect of a living system [5]. This typology is summarized in Table 7.5.

Table 7.5 Typology of engineering disciplines that have some relationship to biological engineering [5]

Engineering discipline	Focus of applications
Civil engineering	Waste treatment remediation
Electrical/computing engineering	Hybrid electronic/living system interfaces and sensors
Chemical engineering	Expertise with fermentations
Agricultural engineering	Food processing
Biomedical engineering	Physical and energetic interfaces between physical engineered systems and human living systems

This diversity of examples contains a common theme, which is the focus on developing externally engineered systems/environments that interact with a living system to alter its behavior/properties. A more fundamental definition of biological engineering would focus on the alteration of existing internal biological processes that control physical attributes/behaviors. Novel advances in micro- and nano-technologies with electrical/mechanical systems provide the opportunity to work at cellular and molecular scales [5].

For the development of smart nanostructured systems, we can consider that the discipline of biological engineering will be further differentiated from other engineering disciplines by specializing in the usage of biological elements as components in the design of smart nanostructured systems. Such biological components will include lipids, proteins, enzymes, and other molecules that mimic their natural function in a biological system. Such biologically engineered smart nanostructured systems may include nanoparticles, but it is not the nanoparticle per se that defines the assembly of a smart nanostructured system utilizing biological engineering.

7.2 Naturally Smart Nanostructured Systems in Biology

This section will focus on biological systems that are capable of providing bioinspiration for the assembly of smart nanostructured systems. For this, our approach here is to consider the cell as an individual component of a larger engineering system. In that context, an individual cell is the smallest functional unit that is capable of working as a stand-alone system, but each such functional unit contributes to the overall function of a complete biological system. Essentially, the cells can function individually, but the communication together provides the biological function of a complete tissue in the body. The membrane of each cell provides the interface for each of these functional units within the overall system of cells in the tissue. The membrane proteins have important roles to provide the output effectors for each cell to communicate with its neighboring cells. Concomitantly, the membrane proteins also provide the neighboring cells with inputs that assist the communication between cells to support the coordinated overall function of the tissue.

Clearly, inside each cell, there are complex interactions of genes and molecules with the cell nucleus and other organelles. These complex interactions are essential for the biological life of the cell. Also, essential for biological life is the immune system, comprising the reticuloendothelial system. The immune system protects the tissues of the body from infection and is thus an essential aspect for maintaining the ongoing health of biological tissues.

It is not possible to engineer life, but it is possible to engineer an individual functioning nanostructured "artificial cell" to mimic the membrane transport functions of a living biological cell [6]. That is, the artificial cell is engineered with the capability of a biological cell for membrane transport communication. If the defining feature of a biological system is the capability of the individual cells to communicate together, then a smart nanostructured system can thus be engineered to mimic the overall function of a biological system. In this context, the biological system responses are the important functional outputs to be mimicked by a smart nanostructured system.

In describing naturally occurring biological systems in this chapter, we will utilize the terminology of a "network" in a similar sense to that of a "system". In strict engineering usage of these terms, a "network" usually refers to a system that connects two or more computing devices for transmitting and sharing information. The computing devices are connected using either physical wires, including fiber optics, or wireless transmission of signals. A "system" is usually defined as a group of interacting or interrelated elements that act according to a set of rules to form a unified whole to produce a certain outcome.

The interaction between biological cells does not occur via wires, but by various molecules that are secreted by cells and then sensed by receptors on the cell membrane. In excitable cells, such as nerves or muscles, there is also a capacitive coupling between cells that relies on non-Faradaic interactions of electrical potential. Hence, the communication in a network of cells is due to combined electrochemical forms of transmission. Biological cells are capable of self-assembly into complex 3D structures. This implies that a network of biological cells is also a system, but one in which there may not be an overarching set of predefined rules for the output of the unified whole of cells to produce a certain outcome.

We will explore different systems of biological cells as providing the inspiration for the biological engineering of smart nanostructured systems.

Each cell of the body comprises a complex network of second messenger molecules that are essential for specific functions of specific cells. Nonetheless, the general function of an intracellular signaling network is illustrated in Figure 7.3.

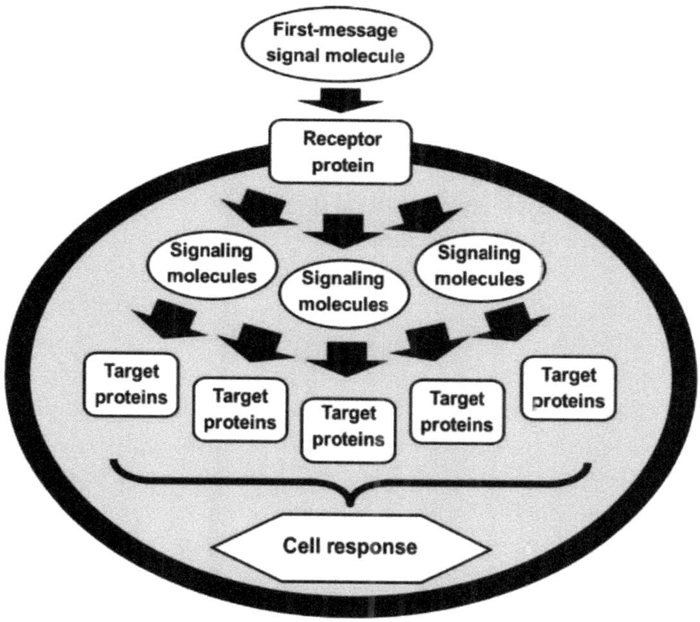

Figure 7.3 A general system of an intracellular second-messenger signaling network. The thick black line represents the cell membrane that encloses the interior of the cell (colored gray). The receptor protein incorporated in the cell membrane binds a ligand molecule from the extracellular fluid. This binding triggers the release of signaling molecules that are usually the α or $\beta\gamma$ subunits of G-proteins associated with the receptor protein. The target proteins inside the cell can include protein kinases, enzymes, or calcium-binding proteins. This type of intracellular signaling network can lead to a multitude of cell responses, which include the activation of ion-transporting proteins, the phosphorylation of proteins, cell contraction, secretion of other molecules, changes in cell metabolism, immune responses, or regulation of gene expression.

The second messenger network inside the cell is triggered by the binding of a first-message signal molecule from the extracellular fluid to a protein receptor incorporated in the cell membrane. The first-message molecule can be either naturally occurring (e.g., neurotransmitters, cytokines) or therapeutic molecules to treat disease. This ligand/receptor binding then triggers a cascade of intracellular biochemical events that contribute to an intracellular networked system. The advantage of such a second-messenger signaling network is that there is a biochemical amplification of the initial ligand/receptor first-message so as to ensure that there is an appropriate response from the cell, which can be sustained.

To provide a detailed description of a functioning second messenger network, we will utilize the example from the kidney. This is an extremely important tissue that utilizes membrane transport to control fluid balance and nutrients/waste in the body. The response effectors are membrane transport proteins. The kidney provides an excellent model for a smart nanostructured biological system that is capable of providing the inspiration for the biological engineering of a smart nanostructured system.

We describe this kidney system in the context of the cell physiology of clear cell carcinomas (CCC). Such CCCs are identified by a clear cytoplasm in neoplastic cells during histological analyses, which is due to histological processing that dissolves the high lipid content of the CCC [7]. The clear-cell morphology of CCC is caused by highly active lipid and glycogen synthesis, and deposition in the cytoplasm as an increase in lipid droplets, which reflects a unique metabolic state of the CCC. The CCCs constitute the most frequent and metastatic form of kidney malignancy and the most therapy-resistant form of ovarian carcinomas [8]. With regard to cell physiology of clear cell renal cell carcinoma (ccRCC), membrane ion channels provide a cell response following the activation of second-messenger pathways that are stimulated by therapeutic medications used for ccRCC.

Despite being a smaller subset of the total number of proteins encoded by the human genome, membrane-bound proteins provide a major target for pharmaceutical drugs. Uhlén et al. provide an analysis of the total human genome that indicates there are 20,344 putative protein-encoding genes [9]. Notwithstanding that there are differences in protein expression between different tissues and that around two-thirds of protein-encoding genes have splice variants for different cellular locations, there are about 5,500 genes that are predicted to encode membrane-associated proteins, with the proportion of specifically identifiable membrane-bound proteins being 14% (2,804 genes). Those membrane-bound proteins include ion channels or molecular transporters, enzymes, and anchors for other proteins [9]. Membrane-bound proteins are essential for several important cell functions that include the transmembrane transport of ions and small molecules, signal transduction, and enzymatic activity [10]. Indeed, such membrane-bound proteins provide the targets for more than 60% of the drugs in clinical use [11].

As a clinical example of the role of membrane ion channels in ccRCC, the ATP-gated membrane ion channel (P2X7) is involved in inflammation, apoptosis, and cell proliferation, and thus plays a crucial role during the oncogenic transformation in various malignancies. In a retrospective clinical study of 273 patients with ccRCC undergoing nephrectomy, it was found that the high intratumoral expression of P2X7 was associated with shortened cancer-specific survival. From that retrospective clinical study, the increased intratumoral expression of P2X7 appeared to be a potential independent adverse prognostic indicator for postoperative cancer-specific survival in ccRCC [12].

P2X receptors are a family of ionotropic ATP-gated ion channels that are equally permeable to Na^+ and K^+, and with significant Ca^{2+} permeability. Of this family, P2X7 channels are unusual in that there are extra amino acids in the intracellular C terminus. P2X7 are key mediators of inflammation and cancer invasion/metastasis. Antagonists to P2X7 are potential anti-metastatic agents. Moreover, the use of P2X7 antagonists has a potential role in other types of kidney failure, since there is an increased expression of P2X7 in renal hypertension, polycystic kidney disease (PKD), and glomerulonephritis [13].

Autosomal dominant polycystic kidney disease (ADPKD) is characterized by multiple renal cysts, hypertension, and a decline in renal function, which leads to renal failure in 50% of all AHPDK patients. Cysts can develop in each nephron segment through a combination of aberrant epithelial cell proliferation and abnormal fluid secretion. Cystic fluid is produced by a cAMP-dependent mechanism similar to that found in secretory epithelia. An important step in cAMP-dependent fluid secretion is the activation of the cystic fibrosis transmembrane conductance regulator (CFTR), which secretes chloride into the cyst lumen followed by sodium and water. CFTR is a cAMP-dependent chloride channel that is best known for its role in cystic fibrosis (CF). A drug, VX-809, used to correct CF, was found to reduce the growth of cysts in ADPKD and to increase the expression of CFTR [14].

The role of cation channels in renal cell carcinoma (RCC) was demonstrated by using the natural drug product Englerin A to inhibit the growth of the renal tumor cell line A-498 by activating transient receptor potential cation channels (TRPC4 and TRPC5) [15].

A large proportion of cancer patients fail to respond to conventional chemotherapy because of the intrinsic resistance of their cancer to pro-apoptotic stimuli and/or the acquisition of a multidrug-resistant (MDR) phenotype during chronic chemotherapy. A significant proportion of kidney cancers overexpress the alpha-1 subunit of the Na^+-K^+-ATPase as compared to corresponding normal tissues, while colon cancers overexpress the alpha-3 subunit. For example, cardiotonic steroids including cardenolides and bufadienolides have anti-cancer effects but with limited cardiotoxicity. An approach to improving the clinical outcomes of chemotherapy could be to target the specific ion channels and subunits of the Na^+-K^+-ATPase that are overexpressed in cancer cells compared to normal cells [16].

The function of the kidneys depends on the maintenance of an epithelial barrier that is capable of regulating paracellular permeability to solutes and water. The tight junction (TJ) provides that epithelial barrier function, which can be measured using techniques such as transepithelial resistance (TER) or electrical impedance spectroscopy (EIS) [17]. The control of permeability is attributed to the insertion of discrete paracellular tight junction channels (PTJC) using evidence from elegant biophysical experiments, although, at the time of that work, it was not known how the TJ proteins could assemble to form PTJC. Such TJ proteins were known to include occludin (3 isoforms), claudin (> 21 orthologs), and junction-associated molecule (JAM), although the stoichiometry of these proteins in a single TJ fibril and their oligomerization in heteromeric and heterotypic assemblies were not known [18].

Using quantitative reverse transcriptase–polymerase chain reaction (rtPCR) and immunohistochemistry, it was shown that claudin-2 mRNA was expressed in the kidney, liver, pancreas, stomach, and small intestine. The highest level of which was detected in the kidney. Also, colorectal cancers (CRCs) express claudin-2 mRNA at high levels [19]. The presence of claudin-2 causes the formation of cation-selective channels that then increase the permeability of the TJ, whereas claudin-1 and claudin-4 decrease the permeability of the TJ [20]. The overexpression of occludin increases cancer cell sensitivity to apoptogenic factors, and consequently, lowers tumorigenicity and invasive capacity [21].

The control of TJ permeability is complex, as shown by the experiments that used epidermal growth factor (EGF) extracted from human urine to decrease TJ permeability. This decreased permeability was due to decreased expression of claudin-1 and claudin-2, and increased expression of claudin-4, with the resulting permeability changes being transient despite the continuous presence of EGF. These complex kinetic responses may be understood by considering that EGF also stimulates the secretion of prostaglandins in kidney cells, and it is known that the prostaglandin E_2 inactivates the extracellular-regulated kinases 1/2 (ERK1/2). Thus, prostaglandin E_2 could provide negative feedback to turn off the signal initiated by EGF, by acting through the cross-talk of the intracellular $G\alpha s$ and ERK1/2 pathways [22, 23].

Subsequently, the scaffolding protein, IQGAP1, was shown to regulate claudin-2 and claudin-4. Claudin-2 forms a cation-selective pore at the TJ and increases the permeability (measured by decreased TER). Claudin-4 reduces the Na^+ permeability of the

TJ (measured by increased TER). Moreover, measurements of claudin 4 expression have a prognostic value in several tumor types and in the association between the loss of TJ and metastatic behavior. Taking all that evidence together suggested that IQGAP1 could be a potential therapeutic target for the treatment of advanced cancers [24].

7.3 Biological Engineering of Smart Nanostructured Systems

7.3.1 Genetic Manipulations to Transform the Function of Cells to Achieve Therapeutic Benefits

As part of our research into the liver-directed gene therapy of Type-I diabetes, we have genetically engineered a human hepatoma cell line (HEPG2ins/g cells) to store and secrete insulin to a glucose stimulus. The engineered cell line (HEPG2ins/g) responds to glucose via intracellular signaling pathways that depend on K_{ATP} ion channels, similar to a normal pancreatic β cell [25]. We used this genetic engineering approach to induce a capability for intracellular pathways to control the function of the induced K_{ATP} ion channels. There are functional K_{ATP} channels expressed in the transfected liver cell-line HEPG2ins/g. These channels are activated only in the parent HEPG2 cell line by pharmacological stimulation. We have demonstrated, by using patch-clamp electrophysiology, confocal laser scanning microscopy with labeling of insulin-containing secretory granules, measurement of intracellular Ca^{2+} levels, and biochemical studies of glucose-stimulated insulin secretion, that the HEPGins/g cells secrete insulin in response to glucose in a manner consistent with the classical triggering pathway present in pancreatic β cells. There is potential value for the HEPG2ins/g cells as a genetically engineered "artificial β cell" in the treatment of Type-I diabetes.

7.3.1.1 Biological control of insulin section in pancreatic β cells

Type-I diabetes or IDDM is caused by the autoimmune destruction of pancreatic β cells [26]. Current treatment of the disease requires daily injections of insulin to control blood glucose levels. Results from "The Diabetes Control and Complications Trial Research Group" [27] show that the onset of diabetic complications, which greatly reduce the quality and longevity of life in IDDM patients, is reduced by tight glucose control.

In a normal pancreatic β cell, a rise in extracellular glucose initiates the inhibition of ATP-sensitive potassium channels (K_{ATP}), which leads to depolarization, the influx of extracellular Ca^{2+} ions, induction of a rise in $[Ca^{2+}]i$ from intracellular stores, exocytosis, and secretion of insulin [28–31]. The K_{ATP} channel has been cloned and found to be a complex of a K^+ channel (Kir6.2) and an ATP-binding cassette protein (SUR1) that functions as a high-affinity receptor for sulphonylureas [32–35].

Although potassium channels have been described in many cell types, apart from pancreatic islets, including skeletal muscle, cardiac, and vascular myocytes; neurons; and renal epithelial cells [36–40], characterization of the role of K^+ channels in hepatocytes has been limited. Henderson et al. [41] described the presence of inwardly rectifying K^+ channels in primary rat hepatocytes that were not affected by voltage or Ca^{2+} stimulation. Capiod and Ogden [42] described the presence of Ca^{2+}-activated and delayed-rectifier K^+ currents, and Lidofsky [43] reported the presence of Ca^{2+}-sensitive and cAMP-dependent K^+ channels in HTC rat hepatoma cells.

7.3.1.2 Genetic engineering of an "artificial β cell"

Glucose control could theoretically be improved by genetically engineering "an artificial β cell" that is capable of synthesizing, storing, and secreting insulin in response to metabolic signals. In pursuit of this goal, hepatocytes have been shown to be a suitable target cell [44–49]. As the starting cell for our genetic engineering, we chose hepatocytes (liver cells) since these cells are known to play a crucial role in intermediary metabolism and the synthesis of proteins in the liver. Most importantly, liver cells express the high-capacity glucose transporter GLUT2 [50] and the glucose phosphorylation enzyme glucokinase [51], which comprise the key elements of the "glucose sensing system," which regulates insulin secretion from pancreatic β cells in response to small external nutrient changes. The insertion of insulin cDNA into a human hepatoma cell line (HEPG2) that lacked native GLUT 2 expression to produce the cell-line HEPG2ins, resulted in the synthesis, storage, and release of insulin to β cell secretagogues, but not to glucose [44]. A second insertion of the glucose transporter GLUT 2 resulted in the near-physiological release of insulin to glucose and other stimuli in the doubly transfected HEPG2ins/g cells [45].

As a smart nanostructured system, the novelty of the HEPG2ins/g cells is that dual expression of the glucose transporter GLUT 2, and the insulin gene has induced functional K_{ATP} channels in the resting cells. Transfection of the parent HEPG2 cell line with both the GLUT2 transporter and insulin cDNA resulted in cells (HEPG2ins/g) that stored insulin in secretory granules and secreted insulin in response to glucose stimulation, at near-physiological levels [45]. That genetic manipulation created a smart nanostructured system that expressed functional K_{ATP} channels in the resting HEPG2ins/g cells to provide a mechanism to trigger insulin secretion in response to stimulation of the cells by glucose [52, 53].

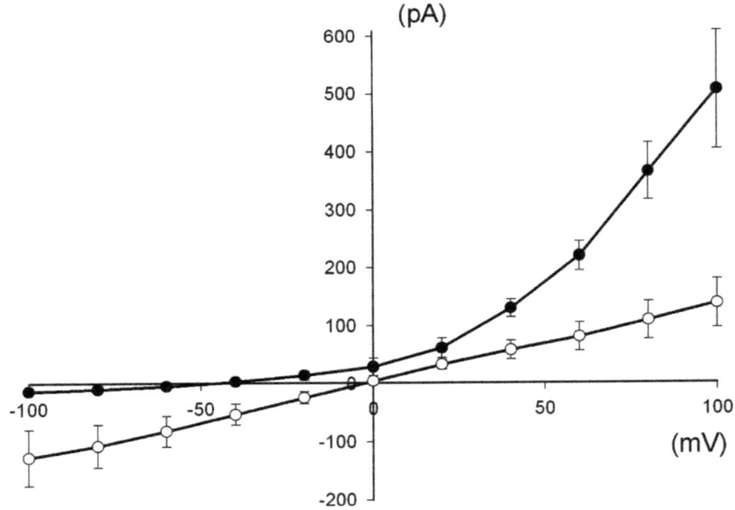

Figure 7.4 Summary of whole-cell *I-V* curves for the samples of HEPG2ins/g (n = 3) and HEPG2 cells (n = 5). Vertical bars indicate SEM. The position of the reversal potential ($E_R \approx -50$ mV) for the HEPG2ins/g cells indicates a potassium-selective current, whereas the current for the HEPG2 cells is non-selective ($E_R \approx 0$ mV). Figure reproduced from [53].

At this point, it is important to highlight the steps in the genetic manipulations to show how this "artificial β cell" was developed. Patch-clamp electrophysiology was used to measure that the doubly transfected HEPG2ins/g cells, in which cDNA for insulin and GLUT2 genes had been inserted, had a more negative membrane potential (−68.9 ± 7.1 mV) than either the untransfected HEPG2 cells (−18.2 ± 5.8 mV) or the HEPG2ins cells that had only the insulin gene inserted (−45.3 ± 4.4 mV). Furthermore, the increased membrane potential was due to the functional activity of K_{ATP} ion channels induced in the HEPG2ins/g cells (Figure 7.4).

The activity of the K_{ATP} channels was inhibited cAMP by ATP (Figure 7.5) and glucose (Figure 7.6).

Figure 7.5 Representative single-channel recordings of K_{ATP} activity in an inside-out membrane patch of HEPG2ins/g cell (*n* = 8 experiments in total). The closed channel level is shown by the horizontal arrows, and open channels are upward deflections. ATP (1 mM) reduced the frequency of open channel activity. Transmembrane potential = +40 mV. The vertical scale bar represents 10 pA, and the horizontal scale bar represents 200 ms. Figure reproduced from [53].

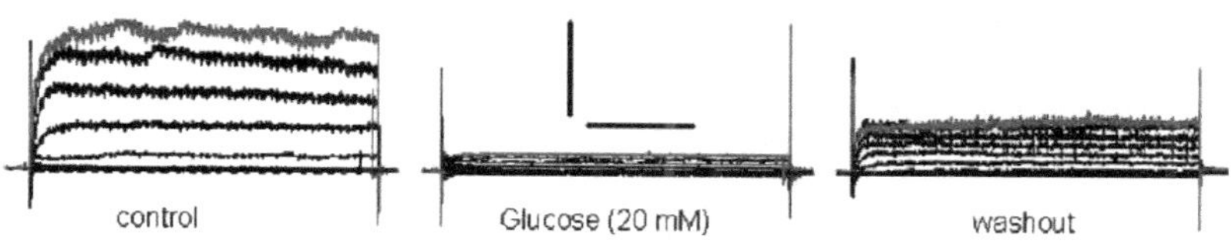

Figure 7.6 Sensitivity of K_{ATP} ion current to high-glucose concentration. Representative whole cell patch clamp currents of the HEPG2ins/g cells (*n* = 4 experiments in total). Glucose (20 mM) was added to the superfusing solution and reversibly inhibited the K_{ATP} currents. The vertical scale bar represents 200 pA, and the horizontal scale bar represents 200 ms. Figure reproduced from [53].

To provide a pharmacological means to control the K_{ATP} channels during experiments, we also demonstrated that the

K_{ATP} channels were activated by diazoxide and inhibited by glibenclamide (Figure 7.7).

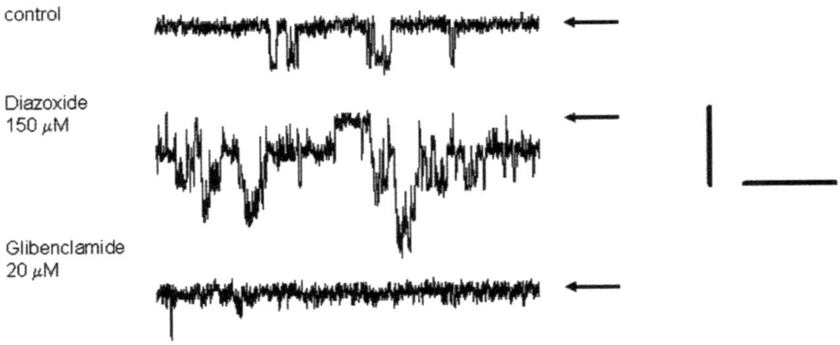

Figure 7.7 Single-channel activity of K_{ATP} in an outside-out patch was increased by the channel opener diazoxide (150 μM) and inhibited by the channel blocker glibenclamide (20 μM). The closed channel level is indicated by the horizontal arrows, and open levels are downward deflections. Transmembrane potential = –40 mV. The vertical scale bar represents 5 pA, and the horizontal scale bar represents 100 ms. Figure reproduced from [53].

The activation by diazoxide would be expected to inhibit the glucose-stimulated cell depolarization, and thus provide an important control for the experiments. Furthermore, we confirmed by Western blotting that the K_{ATP} channels were induced in the HEPG2ins/g cells. This confirmation was achieved with the use of an antibody against the conducting pathway of the K_{ATP} channels, Kir6.2 (Figure 7.8).

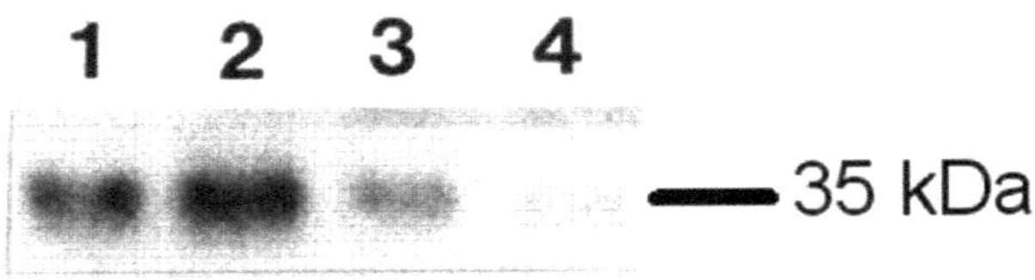

Figure 7.8 Qualitative Western blot analysis for Kir6.2 in HEPG2ins/g (lane 1), MIN 6 (lane 2), HEPG2ins (lane 3), and HEPG2 cells (lane 4). The molecular weight of Kir6.2 is 35 kDa. The MIN 6 cells (positive control) had the strongest staining for Kir6.2. There was a strong presence of Kir6.2 in the HEPG2ins/g cells. There was weaker staining for Kir6.2 in the HEPG2ins cells and almost no staining for Kir6.2 in the HEPG2 cells. This pattern of Kir6.2 expression follows the electrophysiology data, which indicated that the functional presence of the KATP channels was greatest in the HEPG2ins/g cells and absent in the HEPG2 cells. Figure reproduced from [53].

The HEPG2ins/g cell ("artificial β cells") also responded to stimulation by glucose with an increase in the intracellular Ca^{2+} concentration (Figure 7.9).

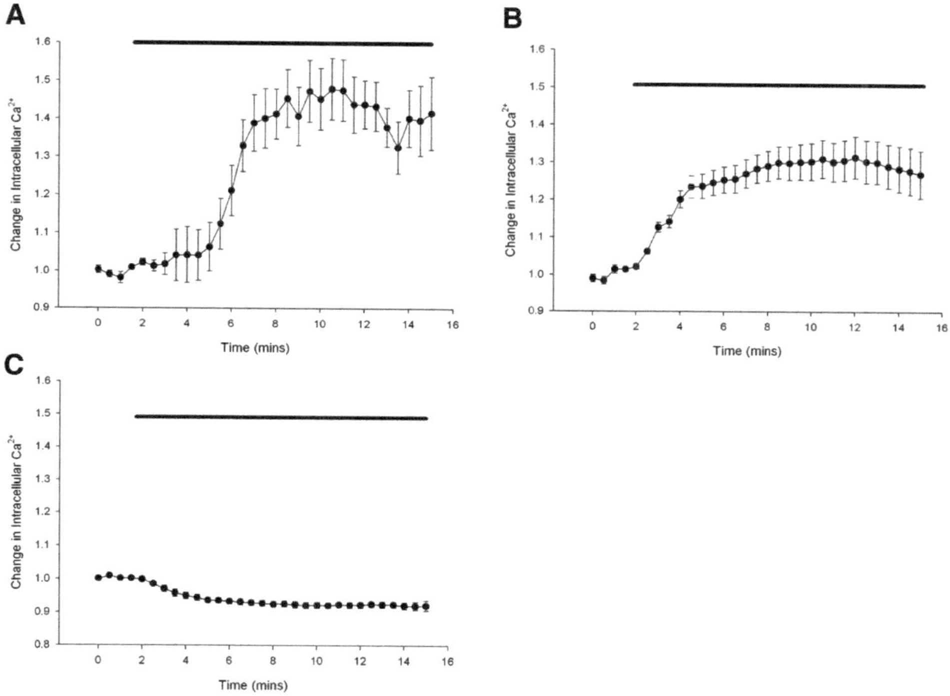

Figure 7.9 Intracellular free Ca^{2+} changes in HEPG2 and HEPG2ins/g cells induced by glucose and the K_{ATP} channel antagonist glibenclamide. The vertical bars on each data point indicate SE where this is larger than the symbol. **(A)** Glucose 20 mM stimulated an increase in the intracellular free Ca^{2+} level in HEPG2ins/g cells (n = 12). **(B)** Glibenclamide 100 μM also stimulated an increase in the intracellular free Ca^{2+} level in HEPG2ins/g cells (n = 9). **(C)** Glucose 20 mM failed to stimulate an increase in the intracellular Ca^{2+} level in the HEPG2 cells (n = 10). The horizontal bars indicate the time of application of glucose or glibenclamide. Figure reproduced from [53].

Ca^{2+} is an important intracellular messenger that is needed to trigger the secretion of insulin. We confirmed that the HEPG2ins/g cells were capable of secreting inclusion in response to glucose, which was controlled by the activity of the induced K_{ATP} channels (Figure 7.10). Note that activating K_{ATP} channels leads to hyperpolarization of the cell, which then does not lea to an increase in intracellular Ca^{2+} concentration and hence there is an inhibition of insulin secretion.

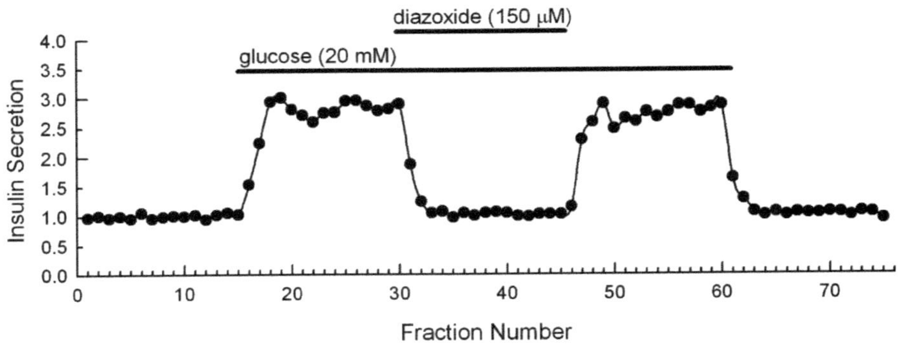

Figure 7.10 Static stimulation of insulin secretion in cultures of HEPG2ins/g cells. Data have been normalized to the average insulin secretion measured in samples between 0 and 15 min (1.00 ± 0.001 pmoles insulin/108 cells), during which cells were perfused with basal medium without any stimuli. The y-axis shows the increase in insulin secretion as a multiple of the normalized basal secretion (=1.0). Data are representative of three independent experiments. Figure reproduced from [53].

Thus, this nanostructured system of the HEPG2ins/g cells follows the classical triggering pathway whereby glucose enters the cell and there is a rise in the ATP-to-ADP ratio leading to closure of KATP channels, which causes membrane depolarization and subsequent activation of insulin secretion by exocytosis of secretory granules [54]. The combined transfection of the GLUT2 and insulin genes induces the activation of functional K_{ATP} channels that do not require pharmacological stimulation to generate currents in this "artificial β cell".

The function of this nanostructured "artificial β cell" is illustrated in Figure 7.11. The expression of functional K_{ATP} channels in the resting HEP G2ins/g cells provides a mechanism for the rapid triggering of insulin secretion. Glucose in the extracellular environment is sensed and transported across the membrane by the facilitative glucose transporter GLUT2 and converted to glucose-6-phosphate by glucokinase; there is a rise in the ATP:ADP ratio leading to the closure of K_{ATP} channels that leads to cell depolarization, influx of extracellular Ca^{2+} to increase the intracellular concentration, increased release of Ca^{2+} from intracellular stores, exocytosis, and secretion of insulin.

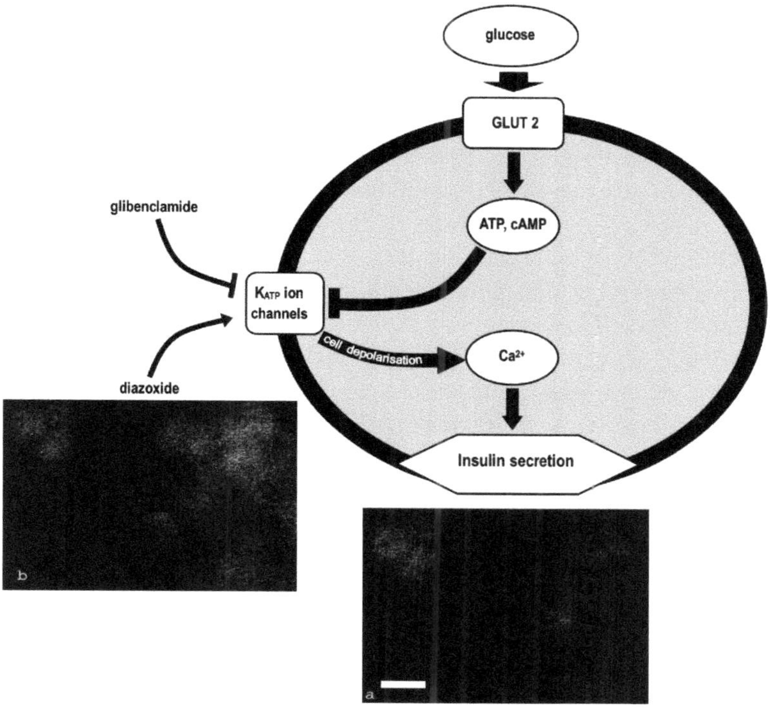

Figure 7.11 Schematic diagram of the biochemical pathway of glucose-stimulated insulin secretion from HEP G2ins/g cells. A rise in extracellular glucose initiates the inhibition of KATP channels, which leads to cell depolarization, exocytosis, and secretion of insulin. In normal pancreatic β cells, depolarization is due to an influx of Ca^{2+} ions. The secretion of insulin is illustrated by the confocal microscopy images (a) and (b) of Zinquin-E fluorescent labeling of secretory granules in HEPG2ins/g cells [53]. Image **(a)** shows Zinquin staining of secretory granules in HEP Gins/g cells superfused with 5 mM glucose in DMEM medium (control). Image **(b)** shows the same HEPGins/g cells as in (a), after changing the super-fusing solution to one containing 150 μM diazoxide in DMEM medium followed by the addition of 20 mM glucose in DMEM medium 15 min later. The image was taken after 45 min of exposure to 20 mM glucose medium. The greater punctate staining in (b) indicates that the KATP channel activator (diazoxide) has inhibited the glucose-stimulated secretion of insulin-containing granules in the HEP Gins/g cells despite prolonged exposure to 20 mM glucose.

As for the situation of a normal pancreatic β cell, it seems likely this increase in Ca^{2+} would also have a wide range of effects including the activation of glycerol phosphate dehydrogenase and other mitochondrial enzymes and activation of protein kinases and phospholipase C. The subunits that comprise the K_{ATP} channel in natural pancreatic β cells are SUR1 and Kir6.2. The subunits comprising the K_{ATP} channel in the parent HEP G2 cells were reported to be SUR2 and Kir6.2, whose mRNA expression was unaltered by glibenclamide. In this "artificial β cell", following overexpression of the insulin and GLUT 2 cDNA, the HEPG2ins/g cells express K_{ATP} channels that are blocked by both tolbutamide and glibenclamide and activated by diazoxide. This pharmacology suggests that the expression of the insulin and GLUT 2 genes in these cells possibly modified the SUR2 subunit of the existing channels to enable the K_{ATP} channels to be functional. This "artificial β cell" supports the possibility of engineering liver cells to mimic the physiology of pancreatic β cells and the eventual use for the treatment of Type-I diabetes.

7.3.2 Ion Transport Proteins Are the Smart Components for Nanostructured Systems

We have seen how a living cell can be used as the basis for genetically engineering a nanostructured system to achieve a specific and different purpose compared to the original function of the living cell. However, for a biological engineering approach, we need to consider how to construct a nanostructured system that does not need to modify an existing living cell. In other words, we need to utilize the principles we have learned from the genetic engineering approach, but apply these to synthetic nanostructured systems using a biological engineering approach. We will illustrate such a biological engineering approach using the example of an artificial cell that was constructed using synthetic and biological components [6].

The synthetic component for that artificial cell was a hollow PSS/PAH polyelectrolyte polymer capsule of 6 μm in diameter (Figure 7.12). The nanostructured biological components comprised a lipid bilayer membrane to enclose the hollow polyelectrolyte capsule, and ion channels incorporated into the lipid bilayer membrane. This complete system provides a structurally stable micro-capsular system and addresses the concern expressed previously about liposomes that incorporate a porin to facilitate membrane permeability [55].

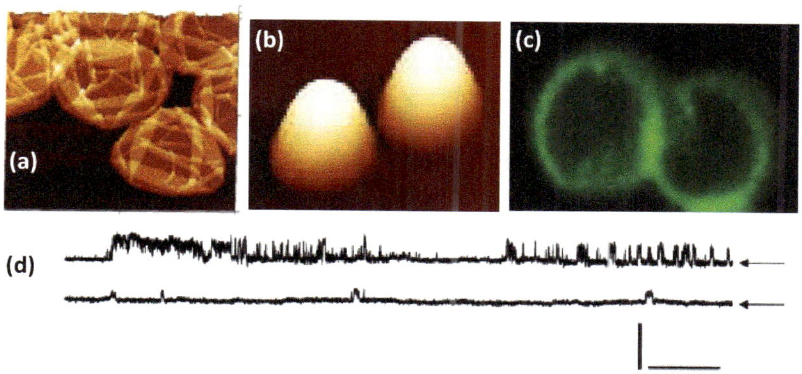

Figure 7.12 Structures to form an artificial cell. (a) 3D rendered AFM images of hollow PAH/PSS polyelectrolyte microcapsules showing a folded and layered structure in an aqueous environment. (b) The microcapsules with a lipid coating do not collapse and maintain a spherical structure in an aqueous environment. (c) Confocal microscopy images of hollow lipid-coated microcapsules loaded with dextran-FITC by adding 10 μL of 10 kDa dextran-FITC to 1 ml of a 1 mg/ml. After washing, the lipid-coated microcapsules retained the fluorescence, which showed the lipid membrane formed an effective insulating barrier to seal the microcapsules. (d) Continuous 10 s patch-clamp recordings of the gating of the bis-gA channels incorporated in the lipid coating of the capsules by MBC (10 μM). MBC is an acronym for the compound N,N_0-dimethyl-N-(2-[2-(methyl-4-[1,1,3,3-tetramethylbutyl]phenoxy)ethoxy]ethyl)benzylammonium chloride. The upper trace is the control condition, where the upward deflections indicate the ion current permeating across the wall of the capsules to the opening of the bis-gA channels incorporated in the lipid coating. The lower trace is recorded 2 min after the addition of the drug to the solution bathing the capsules. The MBC blocked the opening of bis-gA and hence the ion permeation across the capsule wall, as shown by the reduction in the upward deflections. The horizontal arrow indicates the nonconducting condition of the bis-gA channels where there is zero current recorded. The horizontal scale bar represents 1 s and the vertical scale bar represents 5 pA. The four panels were compiled from figures in [6].

The ~20 nm thick wall of the hollow polyelectrolyte capsule allows the transfer of transition metal complexes with overall large positive or negative charges as well as neutral species while providing enough structural strength to maintain a spherical shape when coated with a lipid bilayer. The lipid coating provides a seal, containing both large and small solute species within the microcapsule. The incorporation of the engineered ion channel, gramicidin (bis-gA), provides a platform that can be used to effectively control the transport of molecules from the microcapsule. This engineered micro/nanosystem has potential applications in physiologically relevant delivery or sensing applications. Indeed, we have reported an algorithm and method that would enable such a system to be used efficiently in a biosensing application [56]. The opening of bis-gA suggests that the bis-gA ion channels incorporated rapidly into the lipid bilayer around the microcapsules and provided functional and gateable pathways to control ion transport across the microcapsules.

In this type of "artificial cell" as a smart nanostructured system, the intriguing aspects of variations in the kinetics of the gating of incorporated membrane-spanning bis-gA ion channels provide the possibility to use this "artificial cell" to gain further fundamental insights into the mechanisms underlying instability in the lipid coating of polyelectrolyte supports. Measuring the kinetics of channel open/close activity could be a valuable addition to techniques such as colloidal force spectroscopy [57] in enhancing an understanding of the mechanisms of interaction between lipid bilayers and polyelectrolyte supports. Notwithstanding, the "artificial cell" system provides a platform for the development of a sealed microcontainer for drug delivery or smart sensing that can navigate in the physiological environment and be controlled with physiologically relevant stimuli. Our evidence and formulation of a Hidden Markov Chain Model for automatic detection of changes in the ion channel gating reinforce the potential physiological applications of this "artificial cell" system that incorporates ion channels [56]. The controlled transport in this "artificial cell" system could be tuned for selective release biomimetically by controlling the gating of incorporated ion channels. This system provides a platform for the creation of "smart" biomimetic delivery vessels for the effective and selective therapeutic delivery and targeting of drugs.

Membrane transport proteins also provide the ion gradients that are the basis for cells to develop differences in electrical potentials across their membranes. Such gradients provide the energy for many functions, including a triggering mechanism for the inflow of Ca^{2+} ions through voltage-gated ion channels into muscle cells to aid cell contraction, and importantly for the generation of action potentials in excitable cell membranes. A good example of a body tissue that contains excitable membranes is the retina of the eye. The state-of-the-art electronic neuromorphic visual systems aim to mimic the human visual system and to reproduce the complete functionality of a synapse. Those devices have some limitations in achieving the typical synaptic process that underlies learning in biological systems (e.g., activity-dependent synaptic plasticity), which may be resolved by biological engineering with ion channels in an "artificial cell" system.

In order to mimic synaptic plasticity, an electronic device is assumed to be able to change its resistance (synaptic strength, or weight) upon proper electrical stimuli (synaptic activity) and show several stable resistive states throughout its dynamic range (analog behavior). The electronic component called a memristor provides that functionality [58, 59]. Indeed, memristors provide the CMOS-based electronic building blocks for artificial devices capable of pattern learning and recognition [60–63]. In 2018, the use of multiple memristive devices made a significant advance in creating the basis for building the next generation of intelligent computing systems [64]. Despite the considerable advances in electronic design to develop the memristor, those electronic devices still consume significantly greater energy to perform neuromorphic computing functions and remain much larger in size than biological neurons and synapses.

Most electronic memristors are volatile in the sense that, after the power is turned off, the device reverts to a fixed memristance (or memductance) state. However, an ideal memristor is non-volatile and remembers its most recent memristance (or memductance) state when the power is turned off. This is due to the properties of internal state variables in the memristor, which are physical parameters such as temperature, pressure, impurity concentration, chemical moiety, etc., that do not depend on any external variables such as the voltage or current associated with other devices. The internal state-variables render the memristor with a strictly monotonically increasing transconductance constitutive relation, and hence a non-volatile memory because the memristor remains in the last memristance (or memductance) value from the continuously tunable slope along its transconductance curve is due to the ability of the memristor to store its transconductance curve. It follows that the ideal memristor is capable of the simplest form of learning, namely, habituation [58]. The internal molecular configuration of membrane ion-conducting proteins defines their ion-conducting properties. Thus, their transconductance is defined by an internal state-dependent Ohm's law, which is a basic requirement for their function as a non-volatile memristor.

Ion-transporting proteins are known to be memristors [58] and have the distinct advantages of transconductance driven by electrochemical ion gradients, not applied electrical energy, and being naturally nanostructured as compared to electronic memristors. Moreover, ion channels mediate changes in ion fluxes across the membrane to provide the fundamental basis for membrane depolarizations and hence the generation of action potentials. Since the transconductance is driven by electrochemical ion gradients, not by applied electrical energy as for electronic memristors, the advantage of using ion channels as memristors is that they require 1000-fold less energy for switching compared to state-of-the-art ENODe systems. This is a biological engineering approach to overcoming the roadblock of internal heating that limits both the miniaturization and the performance of state-of-the-art electronic devices. Hence, the memristor function of membrane ion-transport proteins is the sufficient basis for controlling membrane information transfer.

7.3.3 Biological Engineering of Systems to Provide Power

As an example, for the use of ion channels as memristive components, over several years, we have published the electrophysiological results of ion channel function in secretory and excitable cells, which led to an artificial cell with incorporated K^+-permeable ion channels to control membrane transport [6] that could be detected using a Hidden Markov Chain stochastic model [56]. We used that know-how and biomimetic technology to develop a diffusion device that used the Na^+-dependent membrane transporter NhaA, which we patented as the basis for a Na^+ ion-transporting method of generating electrical energy that utilizes a biomimetic membrane and incorporated ion transport proteins [65].

The research team SyNaBi at the laboratory TIMC in the Université Grenoble Alpes has designed a supported lipid bilayer (SLB) biomimetic membrane system that comprised polyaniline (PANI) to support a lipid bilayer membrane that incorporated NhaA to give the system the capability of controllable electrogenic ion transport (Figure 7.13). The high turnover rate of NhaA (~10^5 per min) provides this PANI-SLB-NhaA system with the function of a high-speed rechargeable biocapacitor, which also provides a low-energy-consuming fast switch for biological engineering applications [66].

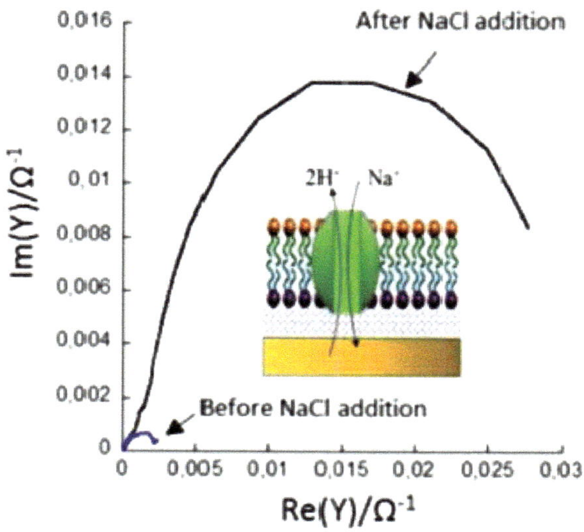

Figure 7.13 Electrochemical impedance spectroscopy response (Cole–Cole) of SLB-active NhaA-PANI modified gold electrode before and after the addition of NaCl 10 mM. Reproduced from [66].

The research team SyNaBi has also developed an implanted biofuel cell (IBFC) that utilizes the redox functions of enzymes to produce electrical energy, particularly when implanted inside the body [67]. We have experience in the development of bioelectrodes that function for long periods when implanted inside the body [68]. Our development of these implanted biofuel cells has provided us with a great deal of know-how in understanding and designing such active autonomous systems for implantation in the body [69].

An implanted biofuel cell (IBFC) is a device that produces power only from the chemicals that are naturally present inside the body, such as glucose and oxygen. It generates electrical power from electrochemical reactions occurring on bioelectrodes. As these sources of energy are always present in the organism, the IBFC can potentially work for an infinite time, as long as the substrates can freely diffuse and the bioelectrodes are not destroyed by the organism. We believe that in the near future, this new generation of green power sources will power implantable electronic devices in order to compensate for the failure of function in humans for the rest of their lives, from brain stimulators to artificial urinary sphincters. The principle of the IBFC is the oxidation of glucose at the anode by glucose oxidase to supply electrons to the electrical circuit. After passing through the electrical circuit, the electrons are collected at the cathode. At this electrode, the enzyme laccase is used to reduce dioxygen to water. The in vivo proof of concept in a mammal was achieved by Cinquin et al. in 2010 [70]. Several technical problems have been overcome for successfully implanting a full enzymatic biofuel cell. The IBFC has been carefully designed to be miniaturized and made able to operate in the abdominal cavity of a rat. As previously said, since the fuels are not embedded in the IBFC, a very high power/volume ratio is achieved. The bigger the electrodes are, the stronger the power is. This principle helps to miniaturize the electrodes but not the complete device, since the electrodes have been wrapped in several layers of materials in order to provide a biocompatible system. Nonetheless, this biofuel cell supplies enough energy (40 μW) to power electronic devices like a pacemaker with electrodes of 240 μl volume [71]. We recently implanted an enzymatic biofuel cell capable of operating for 2 months in a rabbit [67]. To our knowledge, this is the first long-term implantable device with enzymes implanted with success in animals. Although we have overcome technical problems like miniaturization, sterilization, [72] and monitoring, the issue of sterile inflammatory response and later encapsulation or degradation [68] remain roadblocks to its long-term success as an implanted device. The IBFC is the first implantable device to communicate with the body via molecules.

7.3.4 Can a Biofuel Cell be Defined as a Symbiotic Medical Device "Symbio-Bot"?

A communicating implant such as an IBFC is required to be a biocompatible system. This leads us to the notion of establishing a nonaggression pact with the body. The IBFC is not only in direct relation with organism's fluids for a long time but it has to communicate with the body to operate. This is the first kind of implant that exchanges molecules with the behavior of a biochemical redox factory that can output energy or molecules. In a sense, the IBFC functions like an organ, a cell, or an energy factory, similar to mitochondria but much less complex and productive.

The two-way duplex communication is not more an observation but a requirement to provide fuel to the IBFC and enable it to produce electricity, which will indirectly restore the function of the body. Operating in this intimate way, it can be considered as a symbiotic robot. Thus, it has to integrate the body in such a way as to establish stable duplex communication. A long-term integration of this type of device in the body will only succeed if the body accepts at least a "porous encapsulation," preferably transporting the appropriate molecules with a sufficient flow rate.

The development of "symbio-bots," will surely benefit from progress made in the field of medical devices but even more importantly in the field of regenerative medicine. It also paves the way for studies in human and social science. The acceptance of such devices by patients opens a field of interest for philosophy in the context of self-image, for example, with the question of how a "symbio-bot" will be accepted by the patient.

7.4 Overcoming Constraints to Allow Nanostructured Systems to Function in the Body

There are many applications for smart nanostructured systems based on biological engineering, both for stand-alone devices and systems that are designed to function as implanted medical devices. Examples where such systems have advantages for stand-alone devices include ultra-miniaturized light-sensing systems for drone or robotic vision, and renewable energy sources. The overarching engineering advantage is the extremely favorable power-to-weight ratio that can be achieved using biological components in a nanostructured system designed for sophisticated biosensing (e.g., vision) or delivery of power. For those two applications, a biologically engineered nanostructured system provides the distinct advantages of (i) easy scalability, (ii) easy assembly into 3D structures, and (iii) up to 1000-fold less energy requirements for switching compared to state-of-the-art ENODe systems.

Those are the same advantages for biologically engineered smart nanostructured systems designed to function as implanted medical devices. For an implanted application, an additional and highly significant advantage of utilizing biological components is the ease of providing a more natural interface to living tissues. This is because a smart nanostructured system based on biological engineering utilizes biological components that include lipids and proteins. We return to the concept of the "symbio-bot" introduced in the previous section. The "symbio-bot" embodies the advantages of utilizing a biological engineering approach to develop an implanted medical device.

The major challenges to overcome for the function of an implanted "symbio-bot" are the biocompatibility and the interface of the implanted device with the body [69]. In this section, we discuss a roadmap for the design of an implantable "symbio-bot" based on biological engineering. This type of roadmap would be useful, for example, to increase the longevity of bioelectrodes that are designed for implantation in the body [68].

We can consider the flowchart illustrated in Figure 7.14 as a roadmap of the decision-making themes needed for the development of nanostructured systems that are designed to function in the body. The themes are summarized in the box below, with the colors of the themes used to highlight the relevant corresponding parts of the decision-making flowchart shown in Figure 7.14.

Blue zone	**In the body** *(diagnostic and imaging systems)*
Green zone	**Fluid and tissue samples from the body** *(diagnostic and imaging systems)*
Purple zone	**Appropriate medical therapies** *(therapeutic systems and devices)*
Yellow zone	**The patient in society** *(human, ethical, and societal aspects of diagnostic, imaging, and therapeutic devices)*

Note that a "diagnostic medical device" for diagnostics or imaging can also be created following the stage of regulatory approval and appropriate sterilization that is prior to the input of a doctor for a medical diagnosis. The "diagnostic medical device" at this stage would be one that assists the doctor in achieving an appropriate medical diagnosis. For clarity, a separate symbol to illustrate the "diagnostic medical device" at this stage has been omitted from the flowchart.

Note that this flowchart does not emphasize "technology or materials push", but instead describes "medical device pull" that is based on clinical needs and questions. Nonetheless, such a "medical device pull" will need the skills of materials and engineering disciplines to design the "symbio-bot" that is required to solve clinical needs. The major questions to be answered at the various stages of this roadmap to "pull" the medical device are:

(1) What are clinical problems to be solved?

(2) Are diagnostic devices inside the body or using fluid or tissue samples from the body (blue and green zones)?

(3) Are diagnostic devices easily sterilized, and passed regulatory affairs?

(4) Are diagnostic devices helped clinicians with medical diagnosis?

(5) Are diagnostic devices needed, or is pharmaceutical therapy preferred (purple zone)?

(6) Are diagnostic devices understood and tolerated by patients (yellow zone)?

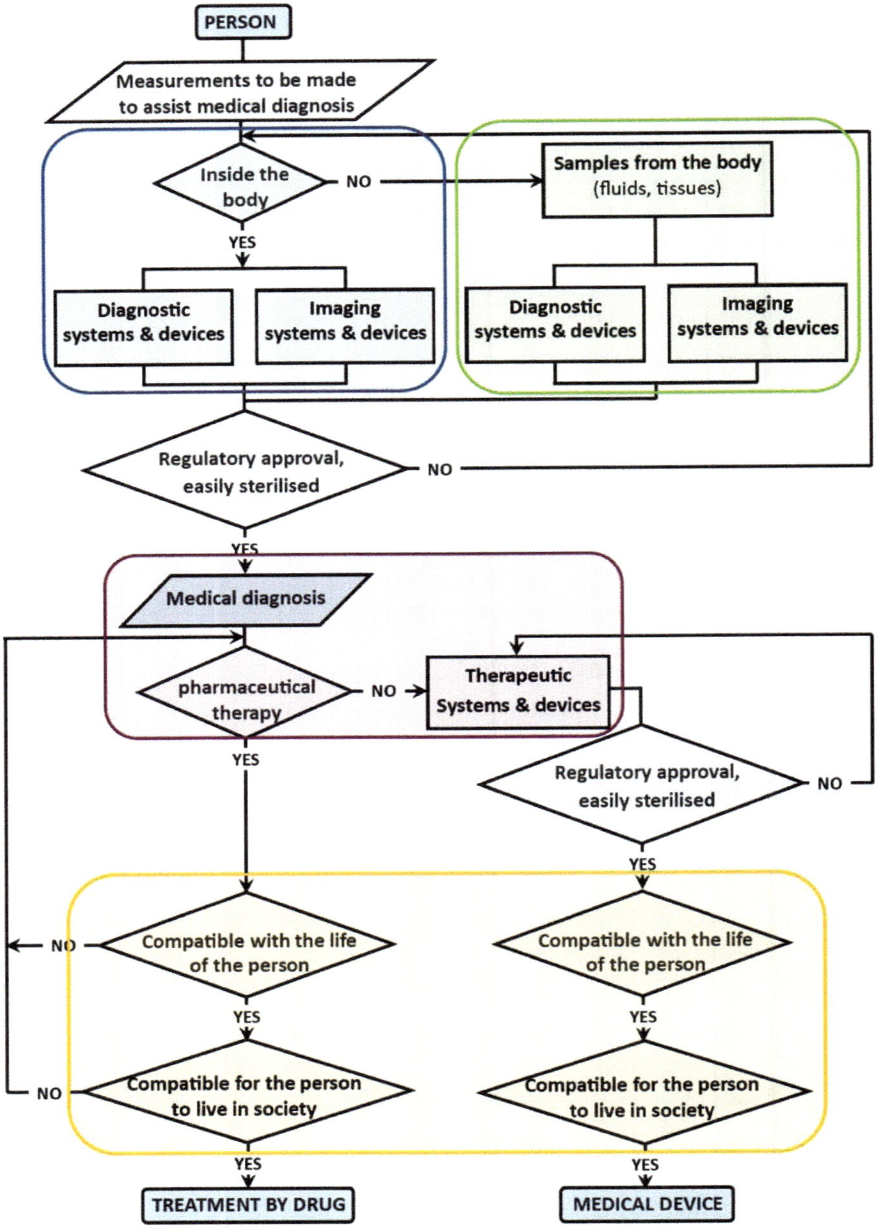

Figure 7.14 Flowchart of decision-making to develop implantable medical devices.

The progression along this roadmap requires at least the interdisciplinary expertise of clinicians, engineers, biologists, biochemists, biophysicists, materials scientists, and regulatory affairs professionals. To follow this roadmap from patient to device or pharmaceutical treatment requires a multidisciplinary approach, where the primary driver is the needs of the patient.

The vision outlined in the roadmap of Figure 7.14 for the development of an implanted "symbio-bot" is predicated on two questions, which are the basis for transforming the paradigm for medical devices and health. The first question is at the level of the patient, who needs to understand, consider, and accept a specific medical symbiotic device that is intended to be used inside their body. The second is at the level of the medical device, which needs appropriate technology so that it understands, considers, interacts with, and is accepted by the body tissues of the patient.

7.4.1 Why Does This "Symbio-Bot" Vision Propose a Transformation of the Paradigm for Medical Devices and Health?

In the current research environment of nanobiotechnologies, the proposed paradigm shift for a "symbio-bot" provides the opportunity to exploit the advances achieved already with symbiotic medical devices but within a framework of human factors. The "symbio-bot" will have biomimetic membrane interfaces to provide a two-way communication of molecules with the tissues of the body. That is, the "symbio-bot" is provided with an unprecedented technical level of authority to communicate with the tissues of the body. The human aspect of delegating such technical authority requires that the patients understand and permit the autonomous function of such a symbiotic medical device with their bodies. One of the best-known examples of such acceptance is the implantation of pacemakers, which require a very high level of confidence in the technology and a largely positive benefit/risk balance.

There are three overarching issues to define this paradigm shift for a "symbio-bot":

ISSUE (1) The need to enhance the psycho-social representations/perceptions of symbiotic medical devices ("symbio-bot") and to develop tools to enhance targeted patients' practical understanding of these novel "symbio-bot".

ISSUE (2) The need to enhance the interface between the "symbio-bot" and the body to enable stable biocompatible function during long-term implantation. This needs to be achieved by defining a detailed molecular-level understanding of the nanostructure and function of a "symbio-bot", and utilizing "smart" matrices with integrated biosensors for the materials used to assemble a "symbio-bot".

ISSUE (3) The need to develop the discipline of biological engineering with biological proteins, synthetic materials, and biomimetic membranes to create a "symbio-bot" for long-term implantation.

7.4.2 What Benefit Would This Shift in Paradigm Have for the Development of Symbiotic Medical Devices?

This multidisciplinary approach of "symbio-bot" necessitates a paradigm shift away from the traditional partitioning of scientific disciplines into linear subsets for the development of medical devices. In the classic linear development progression for a medical device, there are five phases that move from planning, design, prototype, design for manufacturing, and manufacturing. This linear progression does not easily provide the necessary and constant feedback with end users (i.e., medical doctors and patients), which results in inappropriate medical devices that require re-design and adjustments if these are indeed possible. The multidisciplinary approach of "symbio-bot" easily provides a circular production process to ensure the end users interact with the different scientific disciplines in all production phases.

Advances in nanobiotechnology provide a rich resource for the development of nanomedicine. We can position "symbio-bot" by investigating the conceptual structure of the state-of-the-art nanobiotechnologies applied to the field of nanomedicine. We searched the published literature that is available in the Clarivate® Web of Science within the time period of 1900–2021. To focus on specific aspects of the various nanobiotechnologies, the published literature was searched using combinations of the "symbio-bot" keywords: "biomedicine", "nanomedicine", "biomimetic", "bioinspiration", "disease", "therapy", "human", "microbiota", "biotechnology", "nanobiotechnology", "nanosensor", and "membrane". The results from these literature searches were analyzed using the "Bibliometrix" library of RStudio v1.2.5033 to determine the conceptual structure of the nanomedicine literature in both a clinical context and a technological context [1].

The state-of-the-art conceptual structure of clinical nanomedicine is clustered around three central themes of research, which are fundamental biology, clinical biology, and therapeutic applications. Within those clustering themes, the dominant keyword nodes are drug delivery, nanoparticles, and gut microbiota.

The state-of-the-art conceptual structure of technological nanomedicine is clustered around four central themes of research, which are fundamental biology, clinical biology, nanoparticles, and sensors (performance, fabrication). Within those clustering themes, the dominant keyword nodes are in vitro, differentiation, cells, transplantation, nanoparticles, drug delivery, sensors, and sensitivity.

Of significance to a "symbio-bot", we can identify the emerging conceptual structure for nanomedicine by analyzing the non-central keywords and themes. These non-central keywords have indices of centrality of $W < 5$, and provide a rational basis to identify an emerging conceptual structure for nanomedicine. This is illustrated in Figure 7.15, which for clarity includes only the overlapping keywords that appear in multiple themes. Indeed, these overlapping keywords provide pointers to the evolution of the conceptual structure, particularly if we consider that biomedical membrane technology is synonymous with theme proteins and membranes. The emergence of biological engineering as a central theme would then supersede proteins and membranes for the growth of the field of nanomedicine.

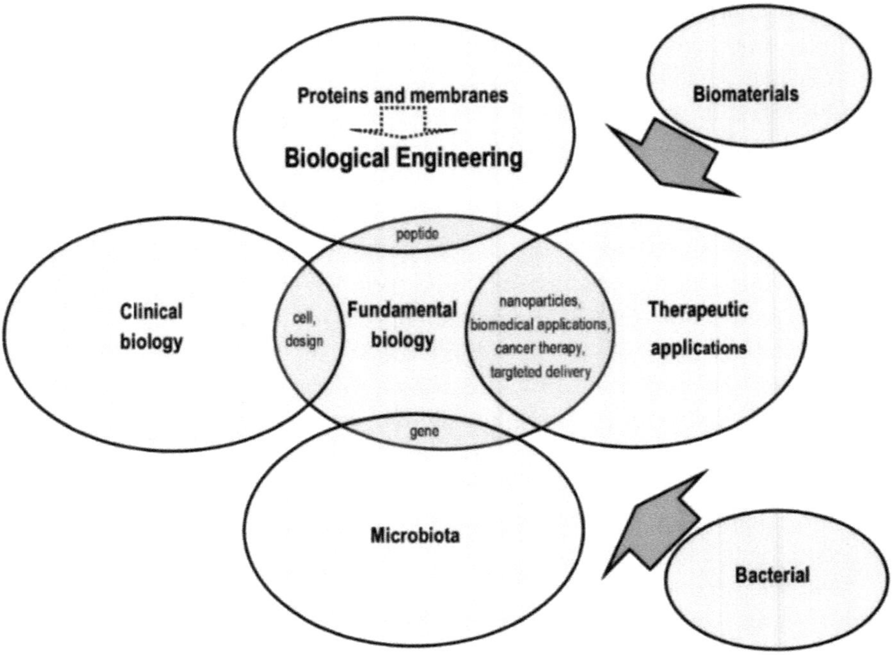

Figure 7.15 Non-central keywords and associated themes in the conceptual structure of the nanomedicine literature. These keywords were identified by indices of centrality of $W < 5$. Keywords with $W = 0$ are not included in these lists. The major thematic axis of biological engineering in nanomedicine is most likely to emerge and supersede the theme of proteins and membranes, as indicated by the dashed arrow in the uppermost ellipse. The themes of biomaterials and bacteria are most likely to contribute to the emergence of other themes that will consolidate the overall conceptual structure of the nanomedicine field.

The definition of biological engineering includes technologies that are oriented toward bioinspired and biomimetic technologies. One could consider that such technologies have a hierarchy that includes the three levels of (i) targeting the genome with DNA, (ii) targeting the membrane with proteins, and (iii) targeting the microbiota with bioinspired devices. The topic of microbiota is an exploding field where there is scope for introducing such new technologies for diagnosis and therapy. It is clear that the emergence of biological engineering as a major theme, or indeed as a discipline, relies upon the combination of expertise from the traditional disciplines of chemistry, physics, biology, engineering, and mathematics. The significant feature of the growth of nanomedicine and the emergence of biological engineering is that these require a multidisciplinary approach, and not be driven only as a collection of expertise from the traditional disciplines.

The Bibliometrix analysis indicated that the positioning of the three overarching issues defined in 7.4.1 is of central relevance to the growth of nanomedicine. Managing the emergence of biological engineering to optimize its multidisciplinary contribution to outcomes in nanomedicine requires a non-traditional approach. Otherwise, the growth of nanomedicine will remain constrained by the dichotomy of whether the starting point for advances in nanobiotechnology and nanomedicine is in the realm of fundamental scientists in traditional disciplines to first seek knowledge in the basic functions of molecules that comprise biomimetic membranes, or in the realm of clinicians and engineers who first seek to observe and characterize defined medical problems and needs. A biological engineering approach, which includes biomimetic membrane technology, commences with the premise that the value of nanomedicine is to improve the health of individuals. A related premise is that the value of biomimetic membrane technology provides a symbiotic interface to improve the function of systems and devices that interact with tissues and cells of the body [2]. Indeed, such a biological engineering approach allows contributions from both a fundamental and an applied approach within the same project.

The added value of following a roadmap, such as in Figure 7.14, is to introduce the paradigm shift that consolidates human factors with health technologies, which advantageously provides a research program that will lead to more acceptable symbiotic medical devices, from both a physiological perspective and a patient perspective.

7.4.3 What Benefit Would This Shift in Paradigm Have for Socio-Ethics Issues?

Current approaches for health innovations are mostly based on a techno-determinist paradigm. Even if some user-centered approaches exist, introducing human motivations, perceptions, representations, and practices into innovative uses in health is not sufficient to develop new devices' full potential and prevent their unintended consequences. A focus on transformations of body representations will be done in the context of modern societies dominated by the cybernetic model. About the specificities of these new symbiotic systems of control and communication, the paradigm of the post-human body as a "machine" will be completed and debated. This new paradigm of representations of the body and its capabilities needs to be modeled including maps of the transformations of mental frontiers between natural and artificial, normal and deviant, power and weakness, etc.

The main challenge of co-designing with targeted patients is to identify and characterize all these transformations and to address the associated ethical and social issues. Although end-user-centered approaches have included "human-in-the-loop", the key ethical and social issues related to the use of "symbio-bots" in health settings are not been clearly considered. These ethical and social issues include the autonomy of the device, delegation of authority and responsibility, social connectedness, health literacy, patient engagement, and integration in daily self-care practices.

7.4.4 What Is the Position of "Symbio-Bots" Within the Industrial, Healthcare, and Medical Device Framework in Europe?

Healthcare sustains a major revolution to evolve from reactive management (dealing with the management of an established pathology validated by an ad hoc medical diagnosis) to preventive management. According to the WHO, prevention can be defined as all measures that are aimed at avoiding or reducing the number and severity of diseases, accidents, and disabilities through reliable diagnostics, monitoring, or treatment. Prevention can be divided into three areas dealing with (i) early detection and treatment of a disease, (ii) the actions to reduce the progression and complications of a proven disease or health problem, and (iii) the actions related to medical and rehabilitative care.

External factors such as the growing and aging population along with a greater prevalence of chronic diseases and the emergence of medical deserts require to rethink the medical habits. These societal evolutions, as pointed out by the European Union [73], require healthcare solutions that are coordinated, convenient (able to adapt to tissue conformation and movement without disconnections and/or tissue injury), ensure accurate and real-time measurements, enable delocalized patient monitoring/treatment, and accessible at an affordable price [74]. Among them, wearable and implanted medical devices are of primary interest to address these requirements.

The multidisciplinary approach to the development of "symbio-bots" provides the shift in paradigm to efficiently deliver healthcare in the context of these re-definitions of healthcare.

At an industrial level, the conjunction of the aforementioned desired evolutions and undergone constraints has strongly reinforced the need for new medical devices leading to a continuously increasing healthcare market (+2.9% in 2013–2017, +5.4% forecasted for 2018–2022). By itself, the wearable healthcare market is expected to be worth US$ 46.6 billion by 2025. If numerous delocalized (self)diagnostic tools have already entered the healthcare market with blockbuster systems such as glucose biosensors for diabetes management or I-Stat multiparametric analyzer for in-field emergency physiological diagnostics, wearable and implanted connected medical devices dedicated to real-time patient monitoring and/or treatment are less developed. More particularly, the emerging sector of smart, hybrid, and conformable medical devices constitutes great market opportunities and differentiators. It can boost the French med-tech industry, which is suffering from competition with low-cost production countries (India, China) but also with countries that have succeeded in establishing strategic, technological, and economic transformations (USA, Switzerland, Germany, Italy). In particular, a very dynamic pool of French start-ups and SMEs from both the health/med-tech and electronics sectors could play a decisive role in the development of these future MDs. The medical devices sector is characterized by a very strong dominance of SMEs (92% of companies in France), reflecting a rich and highly diversified industrial network. The 1,500 companies, including nearly 200 start-ups, registered on national territory in 2019, have a turnover of 30 billion euros and represent

90,000 jobs. The SNITEM analysis conducted in 2019 confirms the dynamism of the sector with a growth of +200 companies over the last 2 years, and an export turnover (9 billion euros) growing strongly by 10% per year [75]. The sector is highly innovative with 3,750 patents filed per year and two-thirds of companies engaged in R&D programs. However, this sector suffers from the lack of dedicated equipment and specific know-how to support rapid prototyping at an affordable cost.

The significant opportunity is to frame a new health paradigm, which will provide exceptional material to challenge all the scientific disciplines, including those from the humanities and social sciences. This is particularly important in constructing the philosophical and ethical framework for this e-Health, developing expertise in user experience design, including co-design with patients, and in corporal/emotional communication/design. The explicit response for the position of "symbio-bot" is the process of understanding and appropriating this new paradigm for patients, to then demonstrate the relevance of cross-disciplinarity at theoretical and methodological complementarity.

The challenge is to demonstrate whether technologies such as "symbio-bots" transform habits and expectations, traditionally built around the natural limits of the human being and body that are intuitively obvious to people. Such limits are also shifted so that e-health and connected health take on meaning with regard to a realization of the model of the vision of medicine as predictive, preventive, personalized, and participatory ("P4 medicine").

Of important relevance to this vision of "symbio-bots" to transform the paradigm for medical devices and health, in the context of the healthcare sector, Deloitte [74] has commented that:

> *"Is the long-awaited seismic shift in healthcare finally here? A collision of forces—a global pandemic of historic proportions; exponential advances in medical science; an explosion of digital technologies, data access, and analytics; informed and empowered consumers; and a movement from disease care to prevention and well-being—proving to be the catalyst for the clinical, financial, and operational transformation that healthcare has long promised to the world."*

The Deloitte report [74] defines the major issues confronting the healthcare sector as:

- **Health Equity**: It is the ability to fulfill our human potential in all aspects of health and well-being. It is an opportunity to achieve an overall state of well-being encompassing clinical, mental, social, emotional physical, and spiritual health, and it is influenced by not just healthcare, but also social, economic, and environmental factors.
- **Environmental, Social, and Governance (ESG)**: The public health community has dubbed climate change the 21st century's greatest threat to public health.
- **Mental Health and Well-Being Health**: Recent health, climate, and political crises have both highlighted and exacerbated the world's mental health challenges, which are moving up the priority list in government, provider, health plan, and employer health equity discussions.
- **Digital Transformation and Healthcare Delivery Model Convergence**: Travel bans are making it difficult to recruit foreign staff, so governments have been enlisting retired doctors and nurses, and medical students to assist with patient care. One solution lies in digital transformation and healthcare delivery model (HCDM) convergence, which is a trend that has accelerated during the COVID-19 pandemic.
- **Future of Medical Science**: Medical science is currently being transformed by scientific discoveries that will dramatically advance the way we diagnose and treat different diseases. Breakthroughs in digital medicines, nanomedicine, genomics, microbiometrics/metabolomics, and others are occurring at an unprecedented and exponential pace, building on transformed clinical trial structures and timing, disruptive market entrants, and increased consumer participation.
- **Public Health Reimagined**: The pandemic has ignited growing recognition of the need to invest in population health. We are only as strong as our most vulnerable populations, and early health promotion, detection, and intervention are essential to prevent, reduce, or delay the onset of chronic disease.

The roadmap illustrated in Figure 7.15 provides the vision and multidisciplinary research program that addresses directly the issue of the "future of medical science". We can construct a research program to answer the question of transforming the paradigm for medical devices and health by enhancing the interface between a patient and a symbiotic medical device at two levels. The first "patient-level" directly implicates human and societal needs and wants in the research and innovation for symbiotic medical devices. The second "device-level" directly implicates the need to improve its interface with the tissues of the body.

The roadmap of Figure 7.15 includes human factors as a concurrent central theme alongside the technical research that is required to develop medical devices. The roadmap overcomes the perennial problem of existing medical device developments, which is the reluctance of many groups of people who do not want to engage with things of a scientific basis, including medicines and treatments. The COVID-19 vaccine hesitancy is a very recent example. The roadmap paradigm of concurrent co-design with human and technical factors is a significant step forward to overcome the reluctance of healthcare providers, insurers, and patients to engage with new medical interventions and devices, even after they have been cleared by regulatory authorities. Consequently, the roadmap also addresses the societal issues of "health equity" and "mental health and well-being health".

An important consequence of the "symbio-bot" transformation of the medical device and health paradigm is that it naturally includes the ideals of responsible research innovation (RRI) in its research program [76].

The most common definition of RRI was provided by von Schomberg [77] as:

> *"RRI is a transparent, interactive process by which societal actors and innovators become mutually responsive to each other with a view to the (ethical) acceptability, sustainability and societal desirability of the innovation process and its marketable products (in order to allow a proper embedding of scientific and technological advances in our society)."*

In the context of the roadmap to illustrate the development of "symbio-bots" responding directly to societal issues, it is important to consider that this roadmap is not simply "technology-push" but has a multidisciplinary research program that embeds the concerns of "societal needs and wants" directly into the co-design processes for novel symbiotic medical devices. Thus, research and innovation tasks that lead to the development of "symbio-bots" go beyond simply being driven by the technology readiness level (TRL), but concurrently include the needs for the societal readiness level (SRL).

7.5 Concluding Remarks

This chapter has explored the approach to the design and assembly of nanostructured systems with inspiration from biology. The approach described in this chapter utilizes biological components in the assembly of smart nanostructured systems. The term "biological engineering" is used to describe a discipline that embodies this approach. Indeed, in 1970, this term was introduced formally with the intention to integrate engineering with biological systems to move beyond single disciplinary areas such as medicine, agriculture, or fermentation engineering. We have further refined the discipline of "biological engineering" to be one that utilizes biological proteins, molecules, and lipids in combination with synthetic materials to assemble smart nanostructured systems. We have illustrated this discipline with the assembly of nanostructured systems that are targeted for applications such as diagnostic, therapeutic, biofuel cell, or tissue enhancement/replacement in the body.

References

1. Aria M, Cuccurullo C (2017). Bibliometrix: An R-tool for comprehensive science mapping analysis. *Journal of Informetrics*, 11, 929–975.

2. Alcaraz JP, Cinquin, P, Martin DK (2018). Tackling the concept of symbiotic implantable medical devices with nanotechnologies. *Biotechnology Journal*, 13, 1800102.

3. Hall SG, Lima M (2001). Problem-solving approaches and philosophies in biological engineering: Challenges form technical, social, and ethical arenas. *Transaction of the ASAE*, 44, 1037–1041.
4. Johnson AT (2006). The making of a new discipline. *Int. J. Engng Ed.*, 22, 3–8.
5. Tao BY, Allen DK, Okos MR (2006). The evolution of biological engineering. *Int. J. Engng Ed.*, 22, 45–52.
6. Battle AR, Valenzuela SM, Mechler A, Nichols RJ, Praporski S, di Maio IL, Islam H, Girard-Egrot AP, Cornell BA, Prashar J, Caruso F, Martin LL, Martin DK (2009). Novel engineered ion channel provides controllable ion permeability for polyelectrolyte microcapsules coated with a lipid membrane. *Advanced Functional Materials*, 19, 201–208.
7. Rini BI, Campbell, SC, Escudier B (2009). Renal cell carcinoma. *Lancet*, 373, 1119–1132.
8. Zou Y, Palte MJ, Deik AA, Li H, Eaton JK, Wang W, et al (2019). A GPX4-dependent cancer cell state underlies the clear-cell morphology and confers sensitivity to ferroptosis. *Nat Commun*, 10, 1617.
9. Uhlén M, Fagerberg L, Hallström BM, Lindskog C, Oksvold P, Mardinoglu A, et al (2015). Tissue-based map of the human proteome. *Science*, 347, 1260419.
10. Bolla JR, Agasid MT, Mehmood S, Robinson CV (2019). Membrane protein-lipid interactions probed using mass spectrometry. *Annu Rev Biochem*, 88, 85–111.
11. Rask-Andersen M, Almén MS, Schiöth HB (2011). Trends in the exploitation of novel drug targets. *Nat Rev Drug Disc*, 10, 579–590.
12. Liu Z, Liu Y, Xu L, An H, Chang Y, Yang Y, et al (2015). P2X7 receptor predicts postoperative cancer-specific survival of patients with clear-cell renal cell carcinoma. *Cancer Sci*, 106, 1224–1231.
13. Burnstock G, Knight GE (2018). The potential of P2X7 receptors as a therapeutic target, including inflammation and tumour progression. *Purinerg Sign*, 14, 1–8.
14. Yanda MK, Liu Q, Cebotaru L (2018). A potential strategy for reducing cysts in autosomal dominant polycystic kidney disease with a CFTR corrector. *J Biol Chem*, 293, 11513–11526.
15. Carson C, Raman P, Tullai J, Xu L, Henault M, Thomas E, et al (2015). Englerin A agonizes the TRPC4/C5 cation channel to inhibit tumor cell line proliferation. *PLosONE*, 10, e0127498.

16. Mijatovic T, Dufraasne F, Kiss R (2012). Cardiotonic steroids-mediating targeting of the Na(+)/K(+)-ATPase to combat chemoresistant cancers. *Curr Med Chem*, 19, 627–646.

17. Srinivasan B, Kolli AR, Esch MB, Abaci HE, Shuler ML, Hickman JJ (2015). TEER measurement techniques for in vitro barrier model systems. *J Lab Autom*, 20, 107–216.

18. Tang VW, Goodenough DA (2003). Paracellular ion channel at the tight junction. *Biophys J*, 84, 1660–1673.

19. Aung PP, Mitani Y, Sanada Y, Nakayama H, Matsusaki K, Yasui W (2006). Differential expression of claudin-2 in normal tissues and gastrointestinal carcinomas. *Virchows Arch*, 448, 423–434.

20. Singh AB, Harris RC (2004). Epidermal growth factor receptor activation differentially regulates claudin expression and enhances transepithelial resistance in Madin-Darby Canine Kidney Cells. *J Biol Chem*, 279, 3543–3552.

21. Cereijido M, Contreras RG, Flores-Benitez D, Flores-Maldonado C, Larre I, Ruiz A, et al (2007). New diseases derived or associated with the tight junction. *Arch Med Res*, 38, 465–478.

22. Flores-Benitez D, Ruiz-Cabrera A, Flores-Maldonado C, Shoshani L, Cereijido M, Contreras RG (2007). Control of tight junctional sealing: Role of epidermal growth factor. *Amer J Physiol Renal Physiol*, 292, F828–F836.

23. Flores-Benitez D, Rincon-Heredia R, Razgado LF, Larre I, Cereijido M, Contreras RG (2009). Control of tight junctional sealing: Roles of epidermal growth factor and prostaglandin E_2. *Amer J Physiol Cell Physiol*, 297, C611–C620.

24. Tanos BE, Andres E. Perez Bay AE, Salvarezza1 S, Vivanco I, Mellinghoff I, Osman M, et al (2015). IQGAP1 controls tight junction formation through differential regulation of claudin recruitment. *J Cell Sci*, 128, 853–862.

25. Liu GJ, Simpson AM, Swan AM, Tuch BE, Martin DK (2003). ATP-sensitive potassium channels induced in liver cells after transfection with insulin receptor and GLUT2 transporter regulate glucose-stimulated insulin secretion. *FASEB Journal*, 17, 1682–1684.

26. Eisenbarth, GS (1986). Type I diabetes mellitus: A chronic autoimmune disease. *New Engl. J. Med.* 14, 1360–1368.

27. The Diabetes Control and Complications Trial Research Group (1993). The effect of intensive treatment of diabetes on the development and progression of long-term complications in insulin-dependent diabetes mellitus. *New Engl. J. Med.*, 329, 977–986.

28. Cook DL, Hales CN (1984). Intracellular ATP directly blocks K^+ channels in pancreatic β cells. *Nature,* 311, 271–273.

29. Ashcroft FM, Harrison DE, Ashcroft SJH (1984). Glucose induces closure of single potassium channels in isolated rat pancreatic β cells. *Nature,* 312, 446–468.

30. Pralong WK, Bartley C, Wollheim CB (1990). Single islet β cell stimulation by nutrients: Relationship between pyridine nucleotides, cytosolic Ca^{2+} and secretion. *EMBO J.,* 9, 53–60.

31. Sakurada M, Kanasuka Saitoh T, Makino H, Yamamura K, Miyazaki J-I, Kikuchi M, Yoshida S (1993). Relation between glucose-stimulated insulin secretion and intracellular calcium accumulation studies with a super fusion system of a glucoseresponsive pancreatic β cell line MIN6. *Endocrinology,* 132, 2659–2665.

32. Aguilar-Bryan L, Nichols CG, Wechsler SW, Clement JP, Boyd AE, Gonzalez G, Herrera-Sosa H, Nguy K, Bryan J, Nelson DA (1995). Cloning of the β cell high affinity sulfonylurea receptor: A regulator of insulin secretion. *Science,* 268, 423–426.

33. Inagaki N, Gonoi T, Clement JP IV, Namba N, Inazawa J, Gonzalez G, Aguilar-Bryan L, Seino S, Bryan J (1995). Reconstitution of I_{KATP}: An inward rectifier subunit plus the sulfonylurea receptor. *Science,* 270, 1166–1170.

34. Sakura H, Ammala C, Smith PA, Gribble FM, Ashcroft FM (1995). Cloning and functional expression of the cDNA encoding a novel ATP-sensitive potassium channel expressed in pancreatic β cells, brain, heart and skeletal muscle. *FEBS Lett.,* 377, 338–344.

35. Babenko AP, Aguilar-Bryan L, Bryan J (1998). A view of sur/KIR6.X K_{ATP} channels. *Annu. Rev. Physiol.,* 60, 308–315.

36. Spruce AE, Standen NB, Stanfield PR (1985). Voltage-dependent ATP-sensitive potassium channels of skeletal muscle membrane. *Nature,* 316, 736–738.

37. Noma A (1983). ATP-regulated K+ channels in cardiac muscle. *Nature,* 305, 147–148.

38. Wellman GC, Barrett-Jolley R, Koppel H, Everitt D, Quayle JM (1999). Inhibition of vascular K(ATP) channels by U-37883A: A comparison with cardiac and skeletal muscle. *Br. J. Pharmacol.,* 128, 909–916.

39. Lee K, Dixon AK, Freeman TC, Richardson PJ (1998). Identification of an ATP-sensitive potassium channel current in rat striatal cholinergic interneurones. *J. Physiol.,* 510, 441–453.

40. Russ U, Rauch U, Quast U (1999). Pharmacological evidence for a KATP channel in renin-secreting cells from rat kidney. *J. Physiol.*, 517, 781–790.

41. Henderson RM, Graf J, Boyer JL (1989). Inward-rectifying potassium channels in rat hepatocytes. *Am. J. Physiol.*, 171, G1028–G1035.

42. Capiod T, Ogden DC (1989). The properties of calcium-activated potassium ion channels in guinea-pig isolated hepatocytes. *J. Physiol.*, 409, 285–295.

43. Lidofsky SD (1995). Convergent and parallel activation of low-conductance potassium channels by calcium and cAMP-dependent protein kinase. *Proc. Natl. Acad. Sci. USA.*, 92, 7115–7119.

44. Simpson AM, Tuch BE, Swan MA, Tu J, Marshall, GM (1995). Functional expression of the human insulin gene in a human hepatoma cell line (HEPG2). *Gene Ther.*, 2, 223–231.

45. Simpson AM, Marshall GM, Tuch BE, Maxwell L, Szymanska B, Tu J, Beynon S, Swan MA, Camacho M (1997). Gene therapy of diabetes: Glucose stimulated insulin secretion in a human hepatoma cell line (HEPG2ins/g). *Gene Ther.*, 4, 1202–1215.

46. Simpson AM, Szymanska B, Tuch BE, Marshall GM (1999). Secretion and storage of insulin from a human hepatoma cell line: Huh7-ins. *Transplant. Proc.*, 31, 812.

47. Ferber S, Halkin A, Cohen H, Ber I, Einav Y, Goldberg I, Barshack I, Seijffers R, Kopolovic J, Kaiser N, Karasik A (2000). Pancreatic and duodenal homeobox gene I induces expression of insulin genes in liver and ameliorates streptozotocin-induced hyperglycaemia. *Nat. Med.*, 6, 568–572.

48. Soria B, Roche E, Berna G, Leon-Quinto T, Reig JA, Martin F (2000). Insulinsecreting cells derived from embryonic stem cells normalize glycaemia in streptozotocininduced diabetic mice. *Diabetes,* 49, 157–162.

49. Tuch BE, Szymanska B, Yao M, Tabiin MT, Gross DJ, Holman S, Swan MA, Humphrey RKB, Marshall GM, Simpson AM (2003). Function of a genetically modified human liver cell line that stores, processes and secretes insulin. *Gene Ther.*, 10, 490–503.

50. Permutt MA, Koranyi L, Keller K, Lacy PE, Scharp DM, Mueckler M (1989). Cloning and functional expression of a human pancreatic islet glucose-transporter cDNA. *Proc. Natl. Acad. Sci. USA,* 86, 8688–8692.

51. Weinhouse S (1976). Regulation of glucokinase in liver. In *Current Topics in Cellular Regulation.* (B. L. Horecker and E. R. Stadtman, Eds.) Academic Press: New York, San Francisco, London, pp. 1–50.

52. Liu GJ, Simpson AM, Swan MA, Tao C, Tuch BE, Crawford RM, Jovanovic A, Martin DK (2003). ATP-sensitive potassium channels induced in liver cells after transfection with insulin cDNA and the GLUT2 transporter regulate glucose-stimulated insulin secretion. *FASEB Journal,* 17, 1682–1684.

53. Liu GJ, Simpson AM, Swan MA, Tao C, Tuch BE, Crawford RM, Jovanovic A, Martin DK (2003). ATP-sensitive potassium channels induced in liver cells after transfection with insulin cDNA and the GLUT2 transporter regulate glucose-stimulated insulin secretion. *FASEB Journal Express,* July 18:10.1096/fj.02-0051fje.

54. Henquin J-C (2000). Triggering and amplifying pathways of regulation of insulin secretion by glucose. *Diabetes,* 49, 1751–1760.

55. Lindemann M, Winterhalter M (2006). Membrane channels as a tool to control nanoreactors. *IEE Proc. Nanobiotechnol.,* 153, 107–111.

56. Krishnamurthy V, Luk KY, Cornell B, Prashar J, di Maio IL, Islam H, Battle A, Valenzuela S, Martin DK (2007). Gramicidin ion channel based nano-biosensors: Construction, stochastic dynamical models and statistical detection algorithms. *IEEE Sensors Journal,* 7, 1281–1288.

57. Köhler G, Moya SE, Leporatti S, Bitterlich C, Donath E (2007). Stability and fusion of lipid bilayers on polyelectrolyte multilayer supports studies by colloidal force spectroscopy. *Eur. Biophys. J.,* 36, 337–347.

58. Chua LO (1971). Memristor – The missing circuit element. *IEEE Transactions on Circuit Theory,* CT-18, 507–519.

59. Strukov DB, Snider GS, Stewart DR, Williams RS (2008). The missing memristor found. *Nature,* 453, 80–83.

60. Ziegler M, Riggert C, Hansen M, Bartsch T, Kohlstedt H (2015). Memristive Hebbian plasticity model: Device requirements for the emulation of Hebbian plasticity based on memristive devices. *IEEE Trans Biomed Circuits Syst.,* 9(2), 197–206.

61. Li Y, Zhong Y, Zhang J, Xu L, Wang Q, Sun H, Tong H, Cheng X, Miao X (2014). Activity-dependent synaptic plasticity of a chalcogenide electronic synapse for neuromorphic systems. *Sci Rep.,* 4, 4906.

62. Saïghi S, Mayr CG, Serrano-Gotarredona T, Schmidt H, Lecerf G, Tomas J, Grollier J, Boyn S, Vincent AF, Querlioz D, La Barbera S,

Alibart F, Vuillaume D, Bichler O, Gamrat C, Linares-Barranco B (2015). Plasticity in memristive devices for spiking neural networks. *Front Neurosci.*, 9, 51.

63. Boybat I, Le Gallo M, Nandakumar SR, Moraitis T, Parnell T, Tuma T, Rajendran B, Leblebici Y, Sebastian A, Eleftheriou E (2018). Neuromorphic computing with multi-memristive synapses. *Nat Commun.*, 9(1), 25.
64. Boybat I, Le Gallo M, Nandakumar SR, Moraitis T, Parnell T, Tuma T, Rajendran B, Leblebici Y, Sebastian A, Eleftheriou E (2018). Neuromorphic computing with multi-memristive synapses. *Nat Commun.*, 9(1), 25.
65. Cinquin P, Martin DK (2017). Biomimetic artificial membrane device. *US 9,577,280.*
66. Ben Tahar A, Zebda A, Alcaraz JP, Gayet L, Boualam A, Cinquin P, Martin DK (2019). A PANI supported lipid bilayer that contains NhaA transporter proteins provides a basis for a biomimetic biocapacitor. *Chemical Communications*, 55, 13152–13155.
67. El Ichi-Ribault S, Alcaraz JP, Boucher F, Boutaud B, Dalmolin R, Boutonnat J, Zebda A, Cinquin P, Martin DK (2018). Remote wireless control of an enzymatic biofuel cell implanted in a rabbit for 2 months. *Electrochimica Acta*, 269, 360–366.
68. El Ichi S, Zebda A, Alcaraz JP, Laaroussi, A, Boucher F, Boutonnat J, Reverdy-Bruas N, Chaussy D, Belgacem MN, Cinquin P, Martin DK (2015). Bioelectrodes modified with chitosan for long-term energy supply from the body. *Energy Environmental Science*, 8, 1017–1026.
69. Zebda A, Alcaraz JP, Vadgama P, Shleev S, Minter S, Boucher F, Cinquin P, Martin DK (2018). Challenges for successful implantation of biofuel cells. *Bioelectrochemistry*, 124, 57–72.
70. Cinquin P, Gondran C, Giroud F, Mazabrard S, Pellissier A, Boucher F, Alcaraz JP, Gorgy K, Lenouvel F, Mathe S, Porcu P, Cosnier S (2010). A glucose biofuel cell implanted in rats. *PloS One*, 5, e10476.
71. Zebda A, Cosnier S, Alcaraz JP, Holzinger M, Le Goff A, Gondran C, Boucher F, Giroud F, Gorgy K, Lamraoui H, Cinquin P (2013). Single glucose biofuel cells implanted in rats power electronic devices. *Scientific Rep.*, 3, 1516.
72. Alcaraz JP, El Ichi-Ribault S, Cortella L, Guimier-Pingault C, Zebda A, Cinquin P, Martin DK (2016). La biopile enzymatique à glucose/oxygène: Quelques nuances de Grays. *Médecine/Sciences*, 32, 771–773.

73. http://ec.europa.eu/growth/sectors/medical-devices_en.
74. Deloitte (2022). Global Health Care Outlook: Are we finally seeing the long-promised transformation? https://www2.deloitte.com/global/en/pages/life-sciences-and-healthcare/articles/global-health-care-sector-outlook.html.
75. https://www.snitem.fr/.
76. Schuijff M, Dijkstra AM (2020). Practices of responsible research and innovation: A review. *Science and Engineering Ethics*, 26, 533–574.
77. European Commission, Directorate-General for Research and Innovation, *Towards responsible research and innovation in the information and communication technologies and security technologies fields*, Schomberg, R. (editor), Publications Office, 2011. https://data.europa.eu/doi/10.2777/58723.

Index